Pipeline Engineering Monograph Series

Pipeline System Automation and Control

Mike S. Yoon
C. Bruce Warren
Steve Adam

ASME
SETTING THE STANDARD

Library of Congress Cataloging-in-Publication Data

Yoon, Mike.
Pipeline system automation and control / by Mike S. Yoon, C. Bruce Warren, and Steve Adam.
p. cm.
Includes bibliographical references.
ISBN 978-0-7918-0263-2
1. Pipelines--Automatic control. 2. Supervisory control systems. I. Warren, C. Bruce. II. Adam, Steve, 1970-III. Title.

TJ930.Y66 2007
621.8'672--dc22

2007027259

Table of Contents

Preface to Pipeline Engineering Monograph Series

The editorial board of the ASME Pipeline Engineering Monograph series seeks to cover various facets of pipeline engineering. This monograph series puts emphasis on practical applications and current practices in the pipeline industry. Each book is intended to enhance the learning process for pipeline engineering students and to provide authoritative references of current pipeline engineering practices for practicing engineers.

Pipeline engineering information is neither readily available from a single source nor covered comprehensively in a single volume. Additionally, many pipeline engineers have acquired their knowledge through on-the-job training together with short courses or seminars. On-the-job training may not be comprehensive and courses or seminars tend to be oriented toward specific functional areas and tasks.

The editorial board has tried to compile a comprehensive collection of relevant pipeline engineering information in this series. The books in this monograph series were written to fill the gap between the basic engineering principles learned from the academic world and the solutions that may be applied to practical pipeline engineering problems. The purpose of these books is to show how pipeline engineering concepts and techniques can be applied to solve the problems with which engineers are confronted and to provide them with the knowledge they need in order to make informed decisions.

The editorial board has sought to present the material so that practicing engineers and graduate level pipeline engineering students may easily understand it. Although the monograph contains introductory material from a pipeline engineering viewpoint, it is reasonably comprehensive and requires a basic understanding of undergraduate engineering subjects. For example, students or engineers need to have basic knowledge of material corrosion mechanisms in order to understand pipe corrosion.

Each book or chapter starts with engineering fundamentals to establish a clear understanding of the engineering principles and theories. These are followed by a discussion of the latest practices in the pipeline industry, and if necessary, new emerging technologies even if they are not as yet widely practiced. Controversial techniques may be identified, but not construed as a recommendation. Examples are included where appropriate to aid the reader in gaining a working knowledge of the material. For a more in-depth treatment of advanced topics, technical papers are included. The monographs in this series may be published in various forms; some in complete text form, some as a collection of key papers published in journals or conference proceedings, or some as a combinations of both.

The editorial board plans to publish the following pipeline engineering topics:

- Pipe Material

- Pipeline Corrosion
- Pipeline Integrity
- Pipeline Inspection
- Pipeline Risk Management
- Pipeline System Automation and Control
- Pipeline System Design
- Geo-technical Engineering
- Pipeline Project Management
- Pipeline Codes and Standards

Other topics may be added to the series at the recommendation of the users and at the discretion of the editorial board.

The books in this monograph series will be of considerable help to pipeline engineering students and practicing engineers. The editorial board hopes that pipeline engineers can gain expert knowledge and save an immeasurable amount of time through use of these books.

Acknowledgments

We, on the editorial board, wish to express our sincere gratitude to the authors, editors and reviewers for their great contributions. They managed each volume, wrote technical sections, offered many ideas, and contributed valuable suggestions. Financial support from the Pipeline Systems Division (PSD) of ASME enabled us to create this monograph series, providing the crucial remainder to the time and expenses already incurred by the editors and authors themselves. We are indebted to the organizing and technical committees of the International Pipeline Conferences (IPC), which have provided an excellent forum to share pipeline engineering expertise throughout the international pipeline community. We were fortunate to have the skillful assistance of the publication department of ASME not only to publish this series but also to undertake this non-trivial task.

Editorial Board

Preface

Pipeline System Automation and Control discusses the methods for monitoring and controlling a pipeline system safely and efficiently. Pipeline technologies are advancing rapidly, particularly in the area of automation and control, and pipeline operation engineers and managers have to be familiar with the latest automation technologies to decide whether they are suitable for the requirements of the pipeline system they are overseeing. They should have sufficient knowledge to enable them to make informed decisions on the technical aspects of the proposed system, the selection of contractors and/or suppliers, and the operation of the installed system. This book reviews the various automation technologies and discusses the salient features involved in the design, implementation and operation of pipeline automation with emphasis on centralized automation system. The goal of this book is to provide pipeline automation engineers with a comprehensive understanding, rather than expert knowledge, of pipeline automation, so that they can seek expert advice or consult professional literature.

The key role of pipeline companies is to transport the products from various product sources to designated markets safely and in the most economical manner possible. During the past few decades, pipeline systems have grown in size and complexity, driven by business requirements consolidating pipelines in fewer entities and by more interconnections between pipeline systems. At the same time, environmental concerns and safety issues require more sophisticated monitoring and control system.

As a consequence, the pipeline operation and commercial transactions have become more complicated, with products being exchanged from one pipeline system to another, either physically or virtually on paper. Also, shippers and producers demand accurate information expediently, particularly information on custody transfer and transportation data. In short, the business cycle is becoming shorter, the number of users are increasing, different users require different information, users are spread-out geographically, and accurate information has to be exchanged at a much faster rate.

In the past, a SCADA system was used to monitor and control compressor/ pump and meter stations. The system users were typically the pipeline dispatchers, system engineers, local operators, and maintenance staff. They were located at one or more dispatching centers and local operation centers, requiring limited sets of information. Due to the development of communication and computer technologies, potential users of the automation system have increased significantly, covering both internal and external customers. Now, the internal customers include not only traditional users such as the pipeline dispatchers and special interest groups such as management, accounting and marketing, but also external users such as shippers and producers. To make the matters more complex, the information requirements of these groups are different from those of the dispatching group.

In order to meet these requirements, centralized pipeline monitoring and system automation is necessary. Such an integrated system allows the pipeline company to manage transportation services effectively and to improve its operating efficiency and profitability. At the core of the centralized system is a SCADA system.

A centralized SCADA system renders numerous benefits. It enables the pipeline operators to perform operating tasks remotely by providing accurate and real-time information, assists them to monitor product movements accurately, and allows for safe operation of the pipeline system including pump or compressor stations. In addition, the SCADA system can facilitate efficient operation and satisfies the pipeline customers by providing reliable and timely information. In short, the SCADA system can help optimize the pipeline system operation. Through the SCADA system, the pipeline operators can monitor and control the entire pipeline system remotely from a central location. It provides the timely information necessary for the operators to perform their operational duties and allow them to take corrective action to achieve the operating objectives.

Chapter 1 discusses the functionality, architecture, communication systems and system capability of SCADA. A SCADA system is the key element to satisfy the integrated and centralized automation requirements. A typical centralized control system consists of various sub-systems, which are monitoring and controlling local stations. Most modern day SCADA systems incorporate the latest instrumentation, computer, and communication technologies, in order to provide the dispatchers and local operators with the capability to make timely responses to constantly changing environments and shipping requirements. It is connected to remote local stations via a communication network. A local control system such as a PLC controls the main systems such as a compressor/pump and/or meter station. These control and monitoring systems are instrumented with appropriate measurement devices. The field instrumentation provides various measurements including pressure, temperature, flow rate or volume, and densities or gas compositions. Remote terminal units (RTU) collect the measured values and send them to the host SCADA system through various communication networks. Reliable communication systems are essential for proper operation of a SCADA system. The communication systems handle both data and voice traffic between the central control center and remote sites. This system can consist of one or a combination of communication media such as telephone networks, fiber optic cables, satellite communication, and radio.

Without measuring devices, no automation and control system can work. The key measuring devices are flow meters, pressure and temperature transducers, and a densitometer or chromatograph. Chapter 2 briefly discusses the basics of instrumentation required for automation.

Chapter 3 discusses the local control of compressor/pump stations and meter station. Fundamentally, the differences in control systems between a gas and a liquid pipeline are minimal. The measurement and control equipment are similar, and the control and information requirements are similar. The main differences lie in how the fluid is moved and stored. A compressor station is required for a gas pipeline, whereas a pump station is used for a liquid pipeline. Compressor and pump stations are the most complex facilities in pipeline systems. They include compressor/pump units and drivers, auxiliary equipment, various valves, and the power system. In order to operate such complex machineries, monitoring and controlling equipments are a necessity. Meter stations play an important role in pipeline operations, because custody transfer and billings are based on volume measurements. Each meter station contains a meter run with measuring devices, including auxiliary equipment and valves, from which the flow computer or RTU determines the corrected volumes. Most stations are in remote locations, so they are often unmanned in order to operate the pipeline system as economically as possible.

Normally, the SCADA measurement system provides real-time measured data and their historical trends. In order for pipeline companies to charge for their transportation services, a computerized volume accounting system is required for custody transfer and billing purposes. The accounting system provides improved measurement accuracy with audit trails, immediate availability of custody transfer volumes, and enhanced security and flexible reporting capability. Since the measurement and accounting of gas and liquid are processed differently, they are addressed in separate chapters; gas accounting in Chapter 4 and liquid accounting including batch operation in Chapter 5.

Petroleum liquid pipelines are designed and operated in batch modes to transport multiple products in a single pipeline system. Transportation of multiple products along a batch pipeline system requires batch scheduling and unique operations. In addition to batch volume accounting, Chapter 5 discusses automation issues related to batch operation such nomination, scheduling and batch tracking.

The field instrumentation, station control system and host SCADA are the basic components of an automated pipeline system. Chapters 1 to 3 of this book deal with the basic automation system components to meet the minimum but critical requirements for safe pipeline operations. In order to improve operating efficiency, however, several advanced applications are utilized. Chapter 6 introduces these applications, which include a pipeline modeling system with the capability of detecting abnormal operating conditions and tracking various quantities, automated energy optimization with unit selection capability, computerized batch tracking, and station performance monitoring functions. Also, this chapter briefly addresses operator training functional requirements and system architecture.

The real-time monitoring capability has been extended to detect abnormal conditions such as leaks. API 1130 discusses the computational pipeline monitoring (CPM) methodologies and addresses implementation and operation issues. Since all the CPM methodologies are based on the data received from a host SCADA system, they are included in this book as part of pipeline automation and discussed in Chapter 7. This chapter describes the nature and consequences of pipeline leaks and various leak detection techniques with an emphasis on the CPM methodologies. It details the CPM techniques and their limitations, their design and implementation including performance evaluation, system testing, and operational considerations and emergency responses. In addition to the CPM methodologies, Appendix 3 discusses other leak detection methods based on inspection tools such as magnetic flux or sensing devices such as acoustic sensor. It introduces emerging technologies using an artificial intelligence or fiber optic cables.

Pipeline companies have started to employ geographical information systems (GIS) for their pipeline system engineering, construction, and operations. The U.S. National Pipeline Mapping System Initiative requires pipeline companies to submit the location and selected attributes of their system in a GIS compatible format. (National Pipeline Mapping System, Pipeline Safety and Hazardous Materials Administration (PSHMA), DOT, Washington D.C., U.S.A.) It uses a database that is referenced to physical locations and provides the analysis and query capability with detailed visual displays of the data. GIS capability has helped pipeline companies engineer pipelines, enhance safety, improve operations, and address emergency situations efficiently. Chapter 8 covers the fundamental concepts of GIS and spatial data management, change management, GIS tools for design and operations, web-based services, and regulatory considerations. The chapter closes with a discussion on the growing use of centralized databases for pipeline facility data management.

Acknowledgment

The authors are greatly indebted to the Executives of the Pipeline Systems Division (PSD) of ASME International for their encouragement to write this book as part of the Pipeline Engineering Monograph series. We thank the organizing and technical committees of the International Pipeline Conferences (IPC), which have provided an excellent forum to share pipeline engineering expertise throughout the international pipeline community. We were fortunate to have the skillful assistance of the publication department of ASME International not only to publish this series but also to undertake this non-trivial task.

Many people and companies were very helpful in shaping the content and style of this book. We thank the following people for their suggestions and reviews as well as providing valuable information and displays; Kevin Hempel, Doug Robertson, Brett Christie, and Ross Mactaggart of CriticalControl Energy Services Inc; Guenter Wagner, Heribert Sheerer and Martin Altoff of LIWACOM Informationstechnik; Warren Shockey of Enbridge Inc.; Jack Blair, formerly of TransCanada Pipelines Ltd.; Ian Clarke of Quintessential Computing Services Inc.; Jim Enarson, an independent consultant; and Shelly Mercer of Colt WorleyParsons for the book cover design.

We are greatly indebted to Robin Warren for her assistance in the final editing of the manuscript. The authors acknowledge Larry Stack, Jason Konoff, Bill Morrow and others at Telvent for their contribution to this publication.

Mike Yoon is deeply indebted to Alykhan Mamdani and Don Shaw of CriticalControl Energy Services Inc. for providing him with office space, displays and secretarial services and to Guenter Wagner of LIWACOM for providing him with office space and displays.

The GIS chapter authored by Steve Adam is co-authored by Barbara Ball and David Parker. The authors would like to acknowledge their colleagues at Colt Geomatics for their contributions to this chapter. The section on Spatial Data Management was built on contributions by Cathy Ha, Michael Jin, Scott MacKenzie, Maria Barabas, and Kevin Jacquard. Assistance on the GIS Tools section was provided by Aaron Ho, Dan Hoang, Craig Sheridan, Jocelyn Eby, Scott Neurauter, Yan Wong, Sabrina Szeto and Robin Robbins. The glossary was meticulously compiled by Yan Wong.

Finally, we want to dedicate this book to our wives and families for supporting our efforts and putting up with us through yet another project.

About the Authors

Mike S. Yoon, Ph.D., has served several pipeline companies as an engineering specialist, manager, consultant and/or teacher. Over the past 32 years, he worked in various pipeline system design and project management as well as in management of automation system suppliers. He published several papers including a report on leak detection technologies. He served as Co-Chairman of the International Pipeline Conference (IPC), Chairman of Pipeline Systems Division (PSD) of ASME, and currently serves as Editor-in-Chief of the ASME Pipeline Engineering Monograph series.

C. Bruce Warren, B.S. and P. Eng., graduated from University of Saskatchewan. Over the past 11 years, he served as private technology management consultant providing strategic planning and project management services. Previously, he worked in various design, commissioning and maintenance of control systems for electrical generating and pipeline facilities as well as in management roles for pipeline application software supplier for more than 20 years.

Steve Adam has spent most of his career applying geomatics and GIS technologies to hydrocarbon and environmental engineering. He has published scientific papers, articles, and book chapters on topics ranging from using satellite imagery for UN peacekeeping to financial analysis of leveraging GIS on pipeline projects. His current focus is to innovate engineering processes using GIS technology. Steve holds a Ph.D. in environmental engineering and is the Manager of Geomatics Engineering with Colt WorleyParsons (Canada).

1 SCADA Systems

This chapter discusses the history, architecture, and application of supervisory control systems in general and how they apply to gas and liquid pipelines in particular. The components and structure of SCADA (Supervisory Control and Data Acquisition) are similar for both gas and liquid pipelines; the major differences are in the associated operating and business applications. Chapters 4 through 8 discuss applications in detail.

SCADA is an acronym for Supervisory Control and Data Acquisition. A SCADA system is a computer-based data acquisition system (often referred to as a SCADA host) designed to gather operating data from an array of geographically remote field locations, and to transmit this data via communication links to one or more control center location(s) for display, control and reporting. Operators at one or more control centers monitor this data. They may then issue commands of a supervisory nature to the remote locations in response to the incoming data. Additionally, software programs implemented within the SCADA host can provide for specific responses to changes in field conditions, by reporting such changes or automatically sending commands to remote field locations. SCADA systems are used for controlling diverse networks such as electrical generating, transmission and distribution systems, gas and oil production distribution and pipeline transmission and water distribution systems. It must be noted that a SCADA system is designed to assist pipeline operators in the operation of the pipeline system using real-time and historical information, but not to provide a closed-loop control function.

1.1 Introduction

The operational nerve center of today's pipelines is the pipeline control center. It is from this central location that a geographically diverse pipeline is monitored and operated. It is also the center for gathering information in real time that is used for real-time operation, for making business decisions and for operational planning.

In order to accommodate a rapidly changing business condition or environment, corporate-wide information access has become critical to the efficient operation and management of a pipeline system. Not only is it important to provide accurate information to operation and management staff, but timely access to this information is of vital importance. Companies that are able to acquire, process and analyze information more efficiently than their competitors have a distinct market advantage.

Looking at the information requirements of a pipeline company and considering both operational and business/economic aspects, the key requirements can be broadly grouped into the following categories:

- **Measurement information** – Measurement information is used for the safe and efficient operation of the pipeline system. It includes pipeline data acquired from field telemetry equipment such as volumes, flows, pressures, temperatures, product quality, measurement, equipment status. It would also include any calculated data originating from the SCADA host. A SCADA system gathers this elemental data from a variety of field instrumentation, typically connected to an RTU (Remote Terminal Unit) or PLC (Programmable Logic Controller).
- **Simulation information** – Simulation information incorporates measurement data and simulated data to diagnose current pipeline states and predict future behaviour of the pipeline. The simulation information can be used for system optimization, line pack and capacity management, storage management, product scheduling, and training-related applications on the pipeline system. This data would originate from a modeling application that utilizes SCADA measurement information.
- **Business information** – Business information combines measurement data and possibly simulated data along with business and economic data. The information is used in business applications related to custody transfer, preventative maintenance, cost tracking, contracting, marketing, inventory, scheduling and accounting. This is where SCADA and simulation data is aggregated with other business data to feed into business processes.
- **Decision support information** – Decision support information is a summary of the key measurement, simulation, and business data required for executive level decision support. Extracting this key data is generally the function of a Management Information System (MIS). Such a system has the ability to gather and aggregate data from numerous corporate and operational databases to supply key performance data.

Pipeline supervisory control systems typically regulate pipeline pressure and flow, start and stop pumps (or compressors) at stations, and monitor the status of pumps/compressors, and valves. Local equipment control systems monitor and control the detailed process for the compressor/pump and its associated driver.

The division of control between a central location and the local compressor or pump station varies widely. A large complex pipeline system may be divided into multiple control sections defined in terms of size of the pipeline network, complexity of the network, or number of shippers. This division allows the operators, assigned to each section, to efficiently monitor and safely control the pipeline system. A control center houses most of the equipment used by the operators on a daily basis. The equipment required includes the SCADA system computers and terminals, printers, communication devices, and network equipment used to implement LANs (Local Area Networks) and/or WANs (Wide Area Networks).

Air conditioning units to control room temperature and auxiliary power unit for a backup source of power are also required. In addition, pipeline system maps and schematics may be displayed, and operator manuals and other information required for performing dispatching functions can be made available.

Since the control center provides real-time information, it may also include an emergency situation-room adjacent to the control room. This room may be dedicated to addressing dispatching issues and particularly to resolving emergency or upset conditions. Several stakeholders, including technical support and management, may be assembled to address emergencies.

A backup control center may be required in order to operate the pipeline system continuously in the event that the main control center is severely disrupted. The backup center is equipped with not only the same equipment and devices as the main control center, but also receives the same real-time data and keeps the same historical data to maintain the continuity of operation and the integrity of the control system. This backup is normally in a physically separate location from the main control room.

A properly designed, installed, and operating SCADA system is a keystone in the operation and management of a pipeline in today's competitive deregulated pipeline market (1).

1.2 History

SCADA systems were first developed for use in the electrical industry for control of high voltage transmission systems. Electrical systems have special requirements for response, speed, and reliability that have driven the development of SCADA system capabilities.

The first field control systems in the pipeline industry were based upon pneumatics and confined to a particular plant facility with no remote control or centralized control. The first step towards centralized automation was the introduction of remote telemetry. This allowed a central location to monitor key pipeline parameters on a remote meter. There was no, or limited, remote control. Operators at such centers had an overview of the complete pipeline operation. They would contact local operators by telephone or radio to make any adjustments or to start or stop equipment such as pumps or compressors.

Initially, controlling pipelines was a labour intensive process. Local pump (or compressor) stations were monitored and controlled by local staff on duty 24 hours a day, seven days a week. System requirements were monitored by frequent measurements and conditions relayed by telephone or radio to a dispatch control group. The group would then determine the need for any local control actions or setting changes, which in turn were relayed back to the local stations for implementation. Data logging was a manual process of recording readings onto paper log sheets.

With the advent of the electrical measurement of process data, important data could be sent to a central dispatch or monitoring center via rudimentary data transmission systems using leased telephone lines or private radio links. Control changes, however, were still required to be sent by voice and implemented manually.

The next step in the evolution of control systems was the development of simple local logic controllers that used electromechanical relays to implement the logic. This allowed interlocks to be used to ensure proper sequencing of equipment and prevent the operation of equipment if a key component was not operational, in an incorrect state or locally locked out (in the case of equipment maintenance, for example). It also made it possible to issue commands from the dispatching center and receive equipment status and key analog data at the dispatching center. This was the first SCADA system.

Because of the limited bandwidth of the radio systems and constraints on the capacity of equipment, these systems were limited in the number of measurements and control and alarm points they could control. These early systems typically consisted of proprietary hardware and radio/communication systems from a single vendor. The job of the control engineer was to assemble all of the bits and pieces and integrate them. The SCADA vendor in the early days sold a system that was very basic in nature; it could receive a limited number of analogue points and send a limited number of digital control actions (start/stop/open/close, etc). The "integrator" needed to design the local systems to interface to this simple SCADA system. Likewise, at the dispatch/control center, the integrator needed to construct dedicated panel boards for displaying the status of the system.

With the advent of the integrated circuit, these systems became "solid state" (i.e. no longer used electromechanical relays) and the capability of the system increased. They were still purpose-built, with no capacity for data storage, etc. Data logging was done manually, albeit at a central location.

The development of the mini-computers in the 1970's especially Digital Equipment Corporation's "PDP" series (bought out by Compaq in 1998 and subsequently by Hewlett-Packard in 2002) provided a huge kick-start to many of the automation systems seen today. Machine automation, plant automation, remote control, and monitoring of pipelines, electrical transmissions systems, etc were now made technically and economically feasible. These systems were now able to provide storage (albeit limited and very expensive) as well as display status and analogue readings on a CRT screen rather than on dedicated panel instruments. The cornerstone components of a modern SCADA system were now in place: local control and data gathering, centralized master unit, central storage on disk and display on computer screens.

The personal computer, first available as a practical device in the 1980's, may prove to have been the single biggest advance in the development of pipeline automation technology. It was quickly adopted for use by a growing number of high tech companies to meet the production and service needs of the marketplace. In parallel with the development of the personal computer, the 1980's saw the introduction of local and wide area networks and thus the potential for more advanced communication. Companies now had an efficient method of sharing information between various locations.

Systems that were once considered prohibitively expensive for many business operations had now become affordable. The advancements within the computer industry during the past five decades have laid the foundation for where we are today. During the 1990's information technology firmly established itself in almost all areas of the oil and gas industry. We have seen significant advancements in automation systems in the pipeline industry as evidenced by electronic measurement systems, controller devices, logic controllers such as RTUs and Programmable Logic Controllers (PLCs), and SCADA system hosts.

Located at one or more strategic control centers, SCADA provides operations and management personnel with full access to current and historical data through computer terminals that feature a full set of graphic displays, reports, and trends. Together they consolidate and summarize information on the measurement and calculation as well as remote control capability of facilities and equipment. Modern systems can be configured in various ways from small-scale single host computer setups to large-scale distributed and redundant computer setups. Remote sites may also contain smaller operator stations that offer local monitoring and control capabilities.

Along with the computer and communication network technologies, we have been witnessing great advancements since the late 1990's and early 2000's in internet technology and its applications to the pipeline industry. Even though the pipeline industry has not yet fully utilized the potential of internet technology, closer integration between the field and office information systems has been accelerating and internet-enabled applications are proliferating.

1.3 System Architecture

1.3.1 General

A SCADA host or "master" is a collection of computer equipment and software located at the control center and used to centrally monitor and control the activity of the SCADA network, receive and store data from field devices and send commands to the field.

The architecture of SCADA systems can vary from a relatively simple configuration of a computer and modems to a complicated network of equipment. Most modern SCADA systems use an open architecture. This not only accommodates different hardware and operating systems, but also allows for easy integration and enhancement of third-party software.

In whatever form it takes however, SCADA architecture will incorporate the following key hardware and software capabilities:

1. Ability to interface with field devices and facilities for control and/or monitoring, usually through a remote terminal unit (RTU)
2. Provision of a communication network capable of two-way communication between the RTU and the control center. This network might also provide communication between the control center and a backup control center.
3. Ability to process all incoming data and enable outgoing commands through a collection of equipment and software called the SCADA host
4. Provide support to pipeline operations through application software such as leak detection, inventory applications and training (Refer to the other chapters in this book for details pertaining to these applications)
5. Ability to interface to corporate systems
6. Provision of some business applications such as meter ticketing, nomination management, etc.

Reliability and availability requirements particular to individual installations will determine the configuration of redundant SCADA servers, redundant database servers, network redundancy, and routing considerations. It is important to remember that reliability and availability are not the same thing. Reliability provides an indication of how frequently a system or device will fail. Availability is the amount of time a system is fully functional divided by the sum of the time a system is fully functional plus the time to repair failures. Thus two systems with the same failure rate (i.e. reliability) may have very different times to effect a repair and therefore, very different availability performance figures.

Modern computing environment encourages a client/server architecture, because it allows client functions to be flexible while enabling server functions to be made robust. Typically, the human-machine interface works as a client and SCADA and application computers as servers in a client/server architecture. The SCADA servers access all RTUs, PLCs and other field devices through a communication server by connecting the communication devices to the host SCADA computers. The real-time and historical databases reside in the SCADA server computers.

There are three basic tiers in a SCADA system as shown in Figure 1, namely: field, control room, and corporate. The field to SCADA connection is some form of a telecommunications network, and the connection between SCADA host and the corporate or enterprise environment is made with a Wide Area Network (WAN). A backup control system located at an offsite may be connected via a WAN to the main control system.

The network is normally an internal private network. However, there are now SCADA systems that utilize secure connections to the Internet that replaces the private network (2). Web-based SCADA systems are ideal for remote unattended applications, assuming that an RTU or flow computer is available. In other words, they are suitable to pipeline systems or remote locations where centralized computing or control requirements are not intense and the primary function is remote data gathering. For example, it can be economically installed on gathering pipeline systems in which control changes are infrequent, remote locations where it is expensive to install a communication line, or on the locations with small amount of volumes where a traditional SCADA installation cannot be economically justified.

A web-based SCADA system offers several benefits. The main advantages are:

- It provides an economical solution with wireless technology using the Internet infrastructure.
- It allows data access from anywhere without extra investment in communication and software.

Internet connectivity, typically in the form of a web portal or a limited Virtual Private Network (VPN) protected by a firewall may also be provided to allow for selected access from outside the corporate network for other remotely located staff, for customer access or to gather data from other systems.

1.3.2 Host Hardware Architecture

A typical SCADA hardware architecture is shown in Figure 1. The host computer equipment generally consists of:

- One or more SCADA servers
- Network component(s) (routers/hubs)
- Storage Server(s) for storage of historical data
- Application server(s)
- Communication server

Older SCADA systems had dedicated custom-engineered operator consoles and an engineering/system manager's console. In today's networked environment, these have been replaced with standard workstations that are then configured to be an operator workstation or other system console with a graphical user interface. In a distributed process environment, host functionality can be split among multiple computers in single or multiple locations.

Whereas in the past SCADA computers were purpose-built or modified versions of commercial systems, systems today use high-end commercial servers and desktops, as are used in most IT environments.

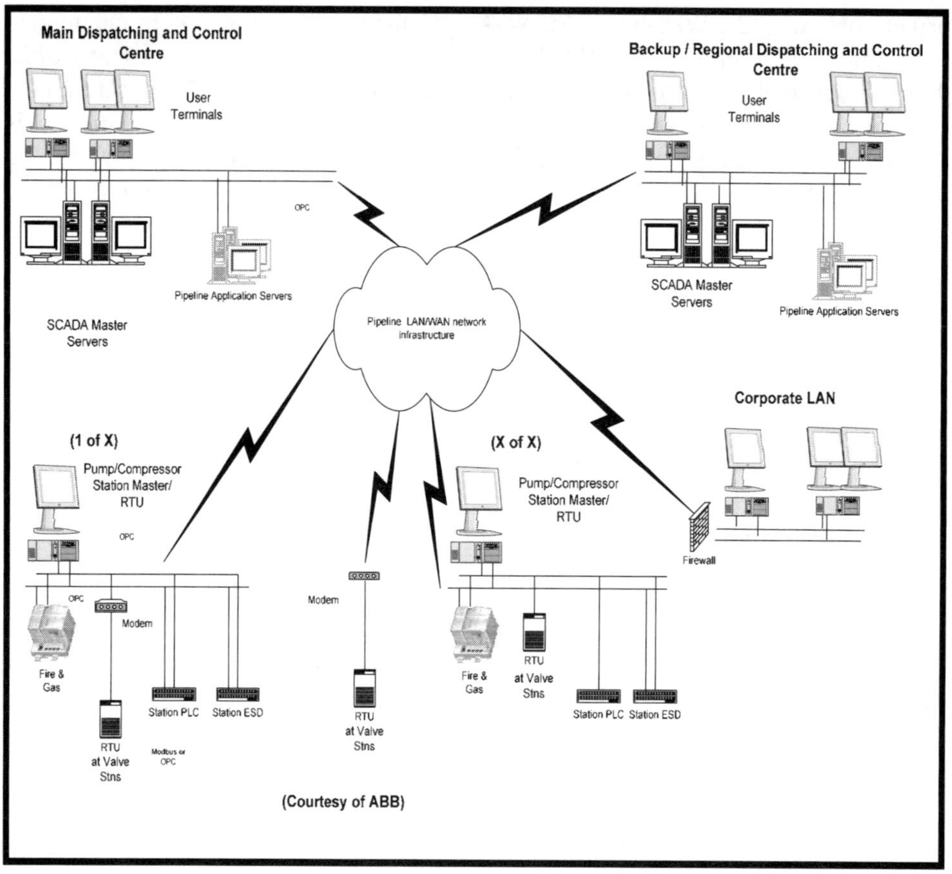

Figure 1 Typical SCADA Architecture

The SCADA host equipment is made up of hardware not suited for an industrial environment. The computer itself is often installed in a separate computer room and the operations console in a control room with properly controlled temperature and humidity, and suitably conditioned power.

The requirements for redundancy and reliability will dictate the final configuration and variations to this basic architecture. The distributed nature of a networked SCADA host allows for the distribution of functionality between servers. This provides some increase in reliability by being able to move non-critical applications to dedicated servers. This allows routine maintenance to occur with minimal impact on core SCADA applications.

The critical SCADA functionality can reside on a primary and a backup server in two general modes. These are "hot standby" or "cold standby":

- "Hot standby" means that two servers are continuously operating in parallel and the operating system will automatically switch to the functioning server in the event of a failure. This will appear seamless to the operator with no loss of data or operating capability.

- "Cold Standby" means that in the event of a failure of the primary server, the idle server takes over operation. This will result in a delay before the system is back to full functionality. As well, there may be some data loss. For these reasons, the general approach in a modern networked SCADA host is to provide hot standby servers to critical functions.

Other considerations regarding the infrastructure to support the SCADA host system include:

- Power supplies: Some means of backup power supply needs to be provided to ensure that the central control room can be fully functional in the event of a domestic power outage.
- Heating and ventilation: The design of these systems needs to incorporate redundancy in the event of failures of primary equipment and domestic power and include provisions for its physical security.
- Physical security: Access to the control room, SCADA equipment room(s) and associated equipment needs to limit and monitor admission only to authorized personnel.
- System Maintenance: This requires a secure method of allowing for vendor upgrades to system software and vendor access for troubleshooting purposes without compromising the reliability of the system.

1.4.1 Host Software Architecture

SCADA host software architecture is different for every product. However, they all have the following key components as indicated in Figure 2:

- Operating system such as Unix, Windows or Linux
- Relational database for historical data management, interfacing with corporate databases
- real time database (RTDB) for processing real time data quickly
- Real time event manager, which is the core of the SCADA server
- HMI manager for user interfaces

The following software components are important for system development, configuration, and maintenance:

- Utilities for configuring and loading the SCADA database and for analyzing communications and troubleshooting
- Development software, such as Microsoft VisualBasic, to facilitate easy user applications programming

- Applications including third party software

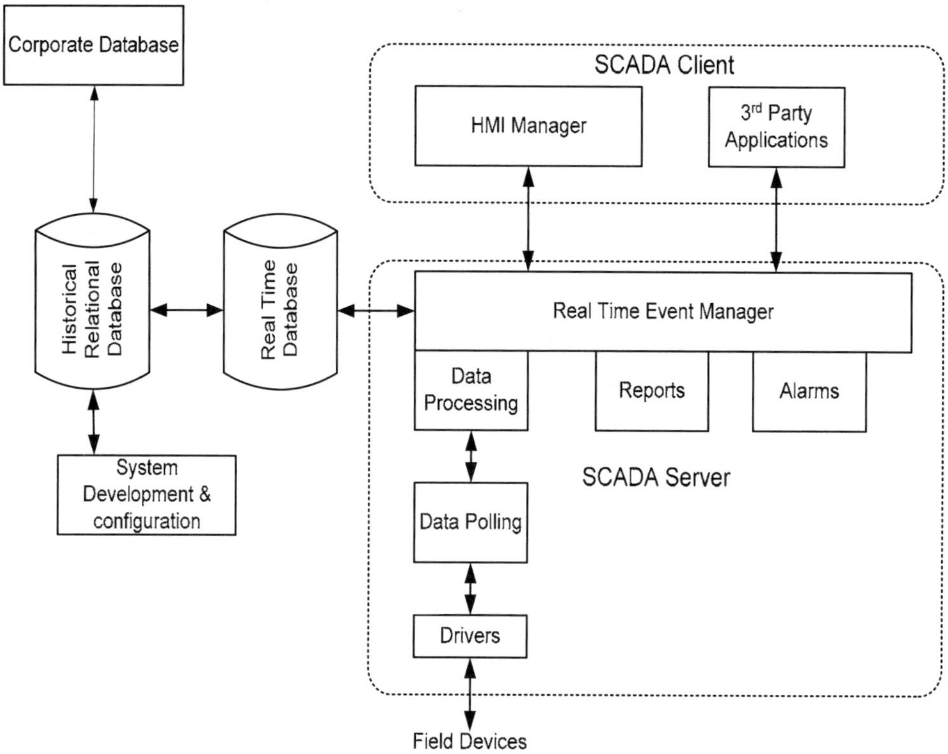

Figure 2 High-Level SCADA Host Software Architecture

The RTDB will be either an embedded third party product or a proprietary database developed by the SCADA vendor. As discussed later in this chapter, it has different requirements and performance objectives than a standard relational database.

The SCADA server will manage the polling of data, processing of that data and the passing of it to the RTDB. The server will make data available to the presentation layer consisting of the HMI Manager and interfaces to other applications, as well as process control and data requests.

The administration and configuration process will have all of the tools required initially to set up the database, RTUs and displays as well as system administration tools for ongoing maintenance of the system. The capability to write custom internal applications and scripts will be handled by this software.

1.4 Communications

1.4.1 Modems

A modem (3) is generally defined as an electronic device that encodes digital data on to an analog carrier signal (a process referred to as *modulation*), and also decodes modulated signals (*demodulation*). This enables computers' digital data to be carried over analog networks, such as cable television lines and the conventional telephone network (sometimes referred to as the plain old telephone system – POTS, or the PSTN – Public Switched Telephone Network – PSTN).

An *optical modem* is a device that converts a computer's electronic signals into optical signals for transmission over optical fiber and also converts optical signals from an optical fiber cable back into electronic signals for use by a computer.

Optical modems employ a very different type of technology than modems used with copper wire media and thus the use of the word *modem* in their name might not be the most appropriate terminology. Consequently, the terms *optical transceiver* and *optical media adapter* are also sometimes used. Optical modems provide much higher data transmission rates than are possible with other types of modems because of the extremely high capacity of the optical fiber cable through which the data is transmitted.

In general, modems are used for the "last mile" connection between a RTU and the SCADA network or where it is not feasible to have a high speed network, connection directly to the RTU.

1.4.2 Protocols

In the context of data communication, a network protocol is a formal set of rules, conventions, and data structure that governs how computers and other network devices exchange information over a network. In other words, a protocol is a standard procedure and format that two data communication devices must understand, accept, and use to be able to exchange data with each other.

The Open Systems Interconnection (OSI) model is a reference model developed by ISO (International Organization for Standardization) in 1984, as a conceptual framework of standards for communication in a network across different equipment and applications by different vendors. It is now considered the primary architectural model for inter-computing and inter-networking communications. The OSI model defines the communications process into seven conceptual layers, which abstract the tasks involved with moving information between networked computers into seven smaller, more manageable task groups.

The following three levels (4) are described with particular emphasis on the requirements placed on the protocols employed by SCADA applications for pipelines. Note that these levels are a subset of the seven-layer OSI model.

1. Physical Layer

   The physical layer determines the nature of the communications interface at both the Host and at the RTU. The following aspects of the physical layer are determined by the specifics of the application:

   - The network architecture, e.g. multi-drop circuits
   - The electrical interface, e.g. EIA standard RS232-C
   - The mode, e.g. serial asynchronous transmission
   - The character set or code for data and character transmission, e.g. ASCII
   - The size of each byte, e.g. 8 bits/byte

2. Data Link Layer

   The link layer describes the total data packet that is transmitted between the Host and the RTU. In addition to the contents of the actual message, the link layer must include line control, framing, and addressing information.

3. Application Layer

   The details of the application layer are determined by the functions the RTU is required to perform and the data types and formats that must be transferred between the Host and RTUs. Generally, the application should be able to support the following:

   - Multiple data types such as strings, binary, integer and floating point in various word lengths
   - Block transfer of a predefined block of data from the RTU
   - Random data read of data from multiple random locations within the RTUs database
   - Sequential data read of a table of contiguous data elements from the RTU with one request
   - Data write commands where commands and data are transmitted from the Host to the RTU
   - Data write with "check before operate" commands where the data is transmitted from the host to the RTU and buffered until a valid "operate" command is received
   - Time synchronization provides a means of synchronizing the clocks in all of the RTUs on the network simultaneously.
   - Data freeze commands that cause the RTU to capture and freeze field data for later retrieval by the Host. The frozen data is time stamped with the time it was captured.

A wide variety of network protocols exist, which are defined by worldwide standards organizations and technology vendors over years of technology evolution and development. One of the most popular network protocol suites is TCP/IP, which is the heart of Internet working communications and uses a similar but different model to that of OSI (TCP/IP predates the OSI model). A detailed discussion of networking protocols is beyond the scope of this book.

For a retrofit or upgrade project, it is important to ensure that the SCADA system can support all of the protocols that exist in the legacy equipment that will be connected to the SCADA system. In some cases where there may be proprietary protocols, converters may need to be implemented.

On a new SCADA system, there is no need to be concerned about existing equipment and protocols. However, it is important to ensure that the SCADA system utilizes industry standard protocols and not just proprietary ones. This will make expansion and addition of new equipment easier. It will also provide more flexibility in being able to choose equipment from a wide range of vendors and not be tied to a specific vendor's equipment.

1.4.3 Networks

1.4.3.1 History

When SCADA systems were first developed, the concept of computing in general centered on a "mainframe" or mini-computer system – a single monolithic system that performed all computing functions associated with a given process (5). Networks were generally nonexistent, and each centralized system stood alone. As a result, SCADA systems were standalone systems with virtually no connection to other systems.

The wide-area networks (WANs) that were implemented to communicate with remote terminal units (RTUs) were designed with a single purpose in mind - that of communicating with RTUs in the field and nothing else. In addition, the WAN protocols in use today were largely unknown at the time. The protocols in use were proprietary and were generally very "lean", supporting virtually no functionality beyond that required to scan and control points within the remote device.

Connectivity to the SCADA master station itself was very limited; without network connectivity, connections to the master were typically done at the bus level via an adaptor or controller (often proprietary) plugged into the CPU backplane. Limited connectivity to external systems was available through low-speed serial connections utilizing communication standards such as RS-232.

Failover and redundancy in these first-generation systems was accomplished by the use of two identically equipped mainframe systems connected at the bus level. One system was configured as the primary system, while the second was configured as a standby system. This arrangement meant that little or no processing was done on the standby system.

The next generation began to take advantage of improvements in computing technology and local area networking (LAN) technology to distribute the processing across multiple processors. Processors were connected to a LAN and shared information in real-time. These processors were typically minicomputers.

They would serve as communication controllers to handle RTU data, operator interfaces and as database servers. This distribution of functionality across multiple processors provided a system with greater processing capability than would have been possible with a single processor.

1.4.3.2 Current technology (Circa 2006):

A SCADA system will usually incorporate a local area network (LAN) and one or more wide-area networks (WANs). This resembles the second generation described above, but has the added dimension of opening the system architecture through the use of open standards and protocols.

The major improvement in third-generation SCADA systems comes from the use of WAN protocols such as the TCP/IP protocol suite. Not only does this facilitate the use of standard third party equipment but more importantly it allows for the possibility to distribute SCADA functionality across a WAN and not just a LAN.

In some WAN distributed systems, pipeline controls are not assigned to a single central location. Instead, control operations can be switched or shared between numerous control centers. Responsibilities can be divided vertically according to a control hierarchy or horizontally according to geographic criteria. In both cases co-ordination and integration of control commands issued from various centers are maintained. In the event of the loss of one or more control centers, the operation can be switched to another center (6).

1.4.4 Transmission Media

The SCADA network model as shown in Figure 1 requires some form of communication media to implement the WAN connection between the SCADA host and remote locations. This section describes the most popular media and associated issues to be considered.

Ultimately the choice of which media to use to implement a connection to a remote site will be based on cost, availability of a particular medium and technical factors such as reliability, bandwidth, geography, etc. A second choice to be made is whether the communication should be leased from a 3rd party or owned and operated by the pipeline company. This decision needs to be consistent with the corporate IT and operating guidelines.

Another consideration is the bandwidth capability of the medium and technology.

Key Considerations for Communication Networks				
Method	Power	Coverage	Speed	Cost
Telephone Line	Low	Poor	Slow	Low
Fibre Optic	Low	Dedicated	Fast	High
GSM/GPRS	Low	Moderate	Moderate	Low
Radio	Medium	Dedicated	Fast	Moderate
VSAT	Medium	Very Good	Moderate	Moderate

Table 1 - Comparison of Communication Media

1.4.4.1 Metallic Lines

As the name implies, this is a hardwired physical connection between the SCADA host and the remote location. This is a good practical choice in SCADA applications where the distances between the SCADA host and the remote locations are not significant and there may be a limited choice of other media.

It is not usual for a pipeline company to provide this type of connection. An equivalent is usually leasing "lines" from a telephone company. Depending on location and distance, leasing a "line" from a telephone company will likely not be a simple wire pair from location A to location B although it will behave as such. The connection will utilize the internal network of the telephone company and may be any combination of wire, fibre optic cable, and radio.

Another alternative is to utilize mobile telephone networks (GSM and GPRS) which provide good coverage in populated areas. GSM acts like a leased line modem connection with limited bandwidth. GPRS was developed to provide direct internet connection from a remote terminal at speeds up to 56,000 bps. A limiting factor may exist where telephone companies are reluctant to provide static internet IP addresses as are required by SCADA systems (7).

1.4.4.2 Radio

Application of radio transmission on a pipeline SCADA usually takes two forms. The simple case is where a radio link is used as the last communication link between the SCADA and a remote site. The main communication backbone of the SCADA system is some other media other than simple radio. A long distance pipeline that may be geographically located in remote areas as well as near occupied areas may well incorporate a mix of radio links and fixed links (leased lines, fibre optic, etc.)

VHF and UHF radios can be supplied as part of an RTU assembly for a relatively small incremental cost. Unlicensed radios have a range of about 10 km or less and licensed radio system a range of 15 to 20 km. An extensive radio based network would require a number of radio repeaters.

A more complex radio system results when the complete WAN is implemented in a microwave system utilizing a network of point-point stations. These will have towers to mount the parabolic dish antennae and an equipment building to house the radio equipment. High frequency radios operate on a line-of-sight basis. The distance to be covered and the intervening topography will determine the number of sites required in the link.

A number of protection schemes are available to provide increased reliability. These include frequency diversity, space diversity, and monitored hot standby (MHSB).

Both space diversity and frequency diversity provide protection against path fading due to multi-path propagation in addition to providing protection against equipment failure. Such techniques are typically only required in bands below 10 GHz, specifically for long paths over flat terrain or over areas subject to atmospheric inversion layers.

Space diversity requires the use of additional antennas, which must be separated vertically in line with engineering calculations. Frequency diversity can be achieved with one antenna per terminal configured with a dual-pole feed. Frequency diversity has the disadvantage of requiring two frequency channels per link, and the frequency inefficiency of this technique is therefore a major consideration in many parts of the world.

MHSB protection can be used at frequencies below 10 GHz if the path conditions are suitable. It is also the normal protection scheme at the higher frequencies where multi-path fading is of negligible concern. MHSB systems are available using one single-feed antenna per terminal, utilizing only one frequency channel per link. MHSB thus seems an efficient protection scheme in relation to equipment and frequency usage.

If the pipeline configuration is such that the radio equipment can be situated at locations where the pipeline already has buildings and electrical power (pump stations, meter stations, etc) then the economics of a microwave radio system may be favourable. A benefit of a microwave system is that they have multi-channel capability and can for instance provide both data and voice communications.

Most other media will utilize equipment familiar to network technicians who can maintain both the LAN and the WAN. However, the extensive use of radio will incur the additional operating cost of radio technicians to maintain this specialized equipment.

1.4.4.3 Fibre Optic Cable

A fibre optic cable uses coherent laser light sent along a "cable." The cable is essentially a hollow reflective tube and the light is reflected along the tube to emerge at the other end. The cables are not lossless and repeater equipment is required at spacing of up to 100 km.

The growth of the need for bandwidth capability for the internet and private networks has spurred advances in fibre optic equipment. A single fibre optic cable can provide bandwidths of the same order of magnitude or greater as that of microwave radio. In the case of fibre optics, the bandwidth limitation is not a function of the medium, but of the terminal equipment. Bandwidth can be improved by upgrading terminal equipment as the technology improves without the need to upgrade the cabling itself. Because a fibre optic cable uses light and not electricity to transmit data it has the benefit of being unaffected by electromagnetic interference.

On new pipeline projects, some pipeline companies have installed fibre optic cable in the same right of way as the pipeline. This can be a cost effective way of providing a transmission medium to implement the SCADA WAN.

1.4.4.4 Satellite

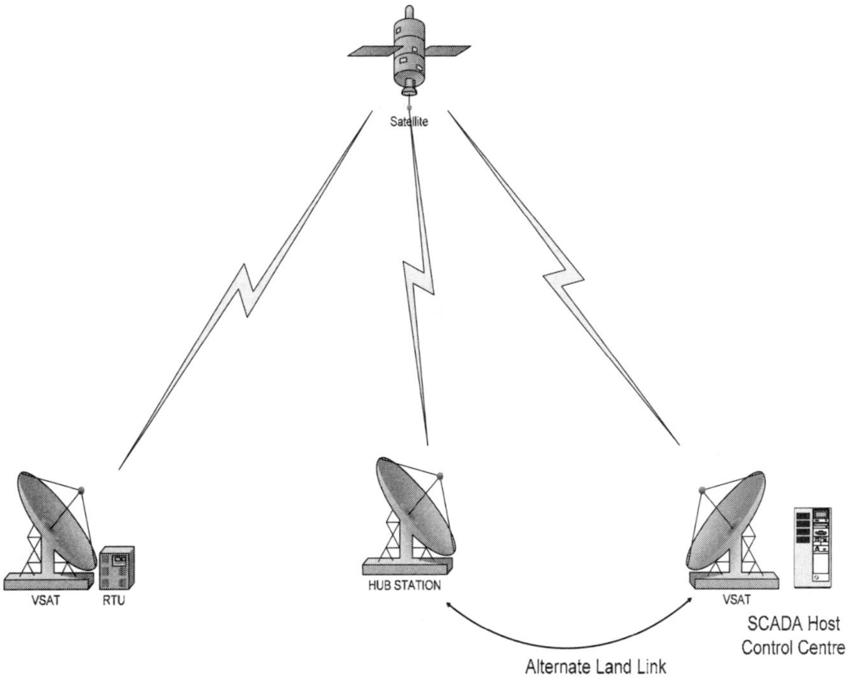

Figure 3 Typical VSAT System

The ability to lease communication channels using relatively small dish antennas, or very small aperture terminals (VSATs) can provide a cost effective communication solution for pipelines under certain conditions. This solution is usually considered when the RTU is in a very remote location where the ability to utilize other media is not practical or very expensive.

The line of sight limitation of radio systems can mean that a remote RTU may require multiple hops to connect to the SCADA system. There is no such limit for a VSAT system.

VSAT equipment is required at the RTU site and at the SCADA master location. Space is leased from a provider who monitors the link and provides the central hub (see Figure 3). Depending on the geography, it may be feasible to have a land link between the hub and the SCADA master location.

The capital cost of a VSAT system is typically more than alternative techniques but when operating costs are factored in, VSAT can be a cost effective solution. However, poor weather conditions can adversely affect the reliability of communications.

1.4.5 Polling

Polling is the term used to describe the process of the SCADA host communicating with a number of RTUs connected on a network and exchanging data with each RTU. The arrangement between the SCADA host and the remote RTU is sometimes referred to as 'master-slave' implying that the SCADA host is in charge of each communication session with a RTU. Three basic types of polling regimes are described in this section.

1.4.5.1 Polled Only

In this arrangement, the SCADA host will sequentially initiate communication with each RTU in sequence on a fixed schedule. There will be a fixed number of attempts to establish communication with an RTU before reporting that communications with the RTU are faulty. One can imagine that for a system with a large number of points to be updated at the SCADA host, this may take some time (depending on the bandwidth of the communication media) and therefore there will be some time lag between the sample time for the first data point and the last.

One variation of this scheme that eliminates this issue is the ability of the master to issue a freeze command to all RTUs. The RTUs then store their data samples and the master begins polling and retrieves the data. This results in a database update at the master where all data was taken more or less at the same time. One way of mitigating this is to have all the RTUs take and store data samples at the same time. This is possible by means of a synchronized system clock.

The major disadvantage of this scheme is that the status and value of all data base points are transmitted every polling cycle, which can represent a significant amount of bandwidth. For example, in the case where VSAT is being used, the user pays for data being transmitted, which may result in significant cost.

1.4.5.2 Polled Report by Exception (RBE)

In this scheme, a local history of each data point is saved and the RTU will only send back those points that have changed since the last poll. In the case of an analogue value, these will have a dead band that the value must exceed before a new value is sent back to the SCADA host.

This reduces the amount of data traffic on the network. The user must be careful in choosing dead bands for analogue values for example to ensure that information is not lost. In some applications, it may be necessary to always refresh some analogue values, for example if they are being used in a real-time model.

1.4.5.3 Unsolicited RBE

In this case, the host does not poll on a regular basis, but each RTU "pushes" data back to the host when it has updated data to send. This can reduce data traffic even more than the polled RBE. However, it has the disadvantage of the host not knowing if data points have not changed or failed. A variation can be to have a system that incorporates a guaranteed polling time. For example, all RTUs may be scanned at least once every 15 minutes.

Each of the above schemes is applied as dictated by the communications protocol and physical limitations of the remote equipment or network. For example, if the remote equipment is powered by batteries that are charged by solar panels, a Polled Only scheme would deplete the power of the remote units more quickly than a RBE scheme. Similarly, if a network has bandwidth limitations, then a RBE would be a good fit.

1.5 Data Management

Typical data required for the safe and efficient operation of pipeline systems from various locations include the following:

- Meter station – values and quality of flow, pressure, and temperature. In addition, a chromatograph may be required for gas pipelines and a densitometer for liquid pipelines, particularly batch pipelines. The status of various valves is also required.
- Pump station – values and quality of suction, casing and discharge pressures. Sometimes, temperature value and its quality are made available. Status of various valves is also required. For variable speed pumps, unit speeds are made available for the operator to review the performance of the unit. Data to allow monitoring of unit operating point is also useful for determining operating efficiency.
- Compressor station – values and quality of suction and discharge pressures and temperatures. Either measured or calculated flow rate may be needed for recycling operation. Status of various valves is also

required. Data to allow monitoring of unit operating point is also useful for determining operating efficiency.

- Control or pressure reducing valve station – values and quality of suction and discharge pressures.
- Pipeline – values and quality of pressures along the pipeline. Sometimes, values and quality of temperature are available. These may be installed at automated block valve sites to take advantage of the need for an RTU for valve control. The incremental cost of pressure and temperature measurements in this situation is minimal.

Alarm messages are generated to signal the potential or real interruption of normal operation at any monitored location on the pipeline.

1.5.1 Real Time Data Base

All SCADA systems have a real time database (RTDB). The RTDB must be able to process large amounts of real time quickly. A typical corporate relational database cannot meet the requirements and demands of a RTDB. Conventional database systems are typically not used in real-time applications due to their poor performance and lack of predictability. Current database systems do not schedule their transactions to meet response requirements and they commonly lock database tables to assure only the consistency of the database. Lock and time-driven schedules are incompatible, resulting in response failures when low priority transactions block higher priority transactions (8).

In the past, most SCADA vendors had their own proprietary database that was optimized for operation in their proprietary operating system. There are now off-the-shelf third party RTDBs available so that it is no longer necessary to have a proprietary SCADA database.

Whichever type of database system is used by the SCADA host it must be able to meet the requirements of a real-time environment and easily interface to standard external databases for the purposes of making key data available to other business processes. Generally, this requires that the SCADA database be SQL compatible to at least a basic degree. Another method used is to utilize some form of a data repository or data historian to store SCADA data for access by other applications (see Section 1.5.5). This reduces the transactions in the real-time database and improves response performance.

Creating the SCADA database consists of populating the database with each of the individual data sources in the SCADA network. Each point will require a number of information fields to be entered to complete a record in the database.

This effort is a time consuming task and must be done accurately. Typically, the SCADA host provides a high-level software utility for interactive creation and modification of the system database. This is probably the most arduous task for a user, as the database must be entered with a great deal of care. The user must have a rigorous method for keeping the database accurate and up-to-date. Some method of checking the input should be a feature of the system so that input errors are minimized.

Features to import the database from a spreadsheet or other flat files may have been designed into the system. A key feature of a SCADA system is the ability to download RTU configuration information from the database thus eliminating the need to re-enter data at each RTU. This eliminates another source of possible error.

Database changes (e.g. addition, deletion, or modification of points) can generally be performed on-line and should not require recompiling the system software.

1.5.2 Data Types

There are four basic data types in a SCADA system namely "discrete", "analogue", "internal", and "parameter".

1.5.2.1 Discrete

This term reflects the fact that these points can only be in one of two (or more) predefined states. Discrete points are generally binary in nature, i.e., they only have two possible states. This can represent open/closed, on/off, normal/alarm, etc. They are referred to as digital, status or binary points and can be either inputs (from field location) or outputs (from SCADA master). Some systems will implement three or four state points, such as a valve status, to indicate that the valve is "open", "in transit", or "closed". Other systems support many more states, such as in the case of pump-off controllers (used in oil production) where the number of discrete states can exceed 50.

Field discrete points are monitored by a SCADA host and used to update display screens, generate alarms, etc. Some points will be simple alarm points that are normal in one state and alarm in another state. Other points will generate an alarm when they change status from one state to another other than by operator command. For example a pump that was running and then stops due to some local problem (loss of lube oil, etc) would generate an alarm based on an unplanned change of state.

1.5.2.2 Analogue

"Analogue" or "Analog" refers to points that have a numeric value rather than two or more discrete states. Analogue inputs are field data points with a value that represents a process value at any given remote location such as pipeline pressure, oil temperature or pressure set point on a control valve. Analogue output points can also be sent as commands from the SCADA host, such as set points for controllers.

1.5.2.3 Internal

A third type of data point is determined internally by the SCADA host as opposed to being sent by a RTU. The internal data type is also called derived data. This can range from a simple calculation to change the engineering units of a field value to more complicated calculations such as the corrected volume measurement in a tank based on tank level, temperature, and product density. A variation of a calculated point is one where the SCADA operator enters in a value manually. For example, this may be used to monitor a value that is not connected to a RTU but is used for reporting, such as a tank level in a customer's tank farm.

Discrete points can be internally generated based on Boolean logic using other points as input. An example may be a logic evaluation of the station block valves to determine if a pump station is on-line or in by-pass mode.

1.5.2.4 Parameter

Parameters or factors are generally used to calculate derived values. Examples include orifice plate sizes, AGA calculation parameters, and performance curves.

1.5.3 Data Processing

All data points will be stored with a time stamp indicating when they were sampled by the RTU. A "quality" flag may also be stored indicating the quality of the value. Some examples of quality indicators are:

- "Good" means that the data is fresh (has been scanned recently) and is within range.
- "Stale" is an indication that the point has not been refreshed for some configurable period.
- "Inhibited" allows the operator to temporarily inhibit points to prevent nuisance alarms for any reason.
- "Overridden" indicates that an operator has overwritten the field value with a manually entered value.
- "Deactivated" means this point will not be updated and control actions will not be allowed. Any calculated point relying on this point will be labelled "suspect".
- "Suspect" or "bad" means that the point's value cannot be relied upon.

Analogue values are processed by the SCADA host and stored in the RTDB, usually along with the original or raw value received from the RTU. Typical processing of analogue points could include:

- Conversion to engineering units: The SCADA system should allow for a mixture of engineering units. It is not uncommon for pipelines to use a mix of English and Metric units.

- Alarm checking against pre-set values for each reading: Alarms will typically be LOW, LOW-LOW, HIGH, and HIGH-HIGH each with configurable limits and dead band settings. Alternate wording for multilevel alarms are Low Warning, Low Alarm, High Warning and High Alarm.
- Rate of change alarm: This will alarm if a field value is changing more rapidly than expected, which may be an indication of a field transducer error.
- Instrument failure alarm: This will alarm if a field value is above or below a preconfigured threshold – It is also an indicator of field transducer or instrumentation failure.
- Averaging: The database will support keeping one or more running averages of values and perhaps store them as separate database points.
- Totalizing: The database will support a running total of a value such as volume going into a tank.

1.5.4 Data Security and Integrity

SCADA data security and integrity features must be consistent with the corporate IT standards and should be outlined during the development of the SCADA requirements. The following list identifies topic areas that need to be addressed with some general methods in use today:

Copying Records	SCADA manuals should include detailed procedures for generating accurate and complete copies of records in both human readable and electronic form.
Limited Access	System should allow for each user's account to limit the access and function the user can execute.
Audit Trails	All SCADA historical records should use secure, computer-generated, time-stamped audit trails to independently record the date and time of operator entries and actions that create, modify, or delete electronic records. Record changes should not obscure previously recorded information.
Training	System administrator training requirements and operator training requirements should be developed. Records of all training and qualifications of personnel with access to the system should be kept current.
System Documentation Controls	The distribution, access, and use of this documentation should be controlled and subject to revision and

	change control procedures that maintain an audit trail that documents system development and modification of systems documentation.
Open System Controls	Systems with any component(s) that are not installed in the SCADA host secure area are considered "Open Systems." Open systems should have procedural controls to ensure the authenticity and integrity of electronic records from the point of their creation to the point of their receipt.

1.5.5 Historical Data Base

An historical database provides for internal analysis and reference as well as meeting the requirements of regulating agencies to review pipeline system operation. For example, operation engineers use the historical data for operational analysis for performance enhancement. The regulator may require emergency scan data to track events leading to and following an emergency condition and eventually to determine the cause/effect relationship.

SCADA historical data includes time-stamped analogue values and other control related analogue values. It can also include digital points and host generated points including alarm and event logs. Operator task logs are also typically included.

Online user interfaces to SCADA historical data generally include all of the following:

- Time-sequenced trending of analog values,
- Query-driven display of alarm, event, and status points, and
- Pre-configured reports.

Access to online (i.e., not yet archived) historical data should be optimized for efficient retrieval. For example, some systems will automatically average data depending on the time horizon of trend displays. A one-year trend of pressure may show a daily average rather than readings for every scan cycle.

The specific user requirements will determine the historical period of data available on-line and is only limited by the amount of disk storage installed. Typical periods would be 1-3 years of online historical data.

No specific access speed specification is applicable due to the diverse nature of potential queries. Instead, the following interface guidelines are recommended:

- For data retrieval that could take more than ten (10) seconds, an on-screen "in progress" indication should be provided.
- For data retrieval that could take more than twenty (20) seconds, an ability to cancel the query should be provided.

- For data retrieval that could take more than thirty (30) seconds, a rough progress indicator (e.g., percent complete bar graph) should be provided.

A popular method of handling historical data that also serves to reduce processing load on the SCADA master is to incorporate a data historian as a data repository of current and historical SCADA data. Queries involving historical data would be handled by the data historian (which could be a standard relational database) and offload the real time data base manager. It would also provides a level of security in so much as it would eliminate the need and ability of outside applications to interact with the RTDB.

In such a system, the RTDB may retain some short term historical data to facilitate operator displays such as short term trends. Again, the time periods for short term historical, long term historical and archiving requirements need to be established at the project definition stage and they must be consistent with corporate IT policy and the pipeline business process requirements. There will likely be regulatory requirements that need to be met and will define the time periods and archive methods associated with historical operating data.

1.5.6 Data Archiving

Since a large amount of data can be accumulated, the historical data needs to be archived periodically. Archive data refers to data that has been stored on archival media (CD, digital tape, etc) and is stored in a separate location from the SCADA host system as required by corporate policy. The period of time after which data should be archived is determined by corporate policy. The data archive should include all analogue and digital data, alarms, events and operator actions.

Existing site or corporate archiving facilities, technologies, and procedures should be exploited if possible. Archive system design should consider the potential to migrate the historical data to ensure that access can be preserved for any future upgrade or replacement of the SCADA system.

SCADA system manuals must include detailed procedures for both sending and retrieving historical data from the archive. Retrieved historical data must include any and all data that was, or may have been, considered for verifying manufacturing and/or product quality. Retrieved data context, format, and/or access must be identical to, or at least comparable to, original data context, formats, and/or access. The SCADA system must be able to retrieve archived data without interrupting ongoing process operations.

1.5.7 Event Analysis

To facilitate analyzing system upsets and events, the SCADA system can have a feature known as "playback." This functions much like a rewind on a VCR and allows a user to replay historical data through an off-line operator terminal in order to more easily analyze and determine the root cause of an upset. It can also be used to do "post-mortems" with operators to provide feedback on actions that were taken and to determine if remedial action taken was done correctly and in a timely fashion.

The SCADA data base manager needs to store and time tag all operator actions (alarm acknowledgment, commands, etc) as well as all incoming and outgoing data to get the most benefit from this feature.

A further enhancement is the ability to export such a file to an off-line simulator (See Chapter 6) and to use it to build training scenarios from actual pipeline events. This allows other operators to benefit from such events and to pass along this body of knowledge.

1.6 Human Machine Interface (HMI) and Reporting

Key features of displays and reports are discussed in this section. Typical data included in displays and reports are as follows:

- Telemetered data, including analogue, digital, and derived values and quality
- Parameter data, such as orifice plate size
- Schematic information, including station yard piping, facility locations on the pipeline system, and other pertinent information

The displays need to be designed to meet the needs of individual operators, because they are the prime users of SCADA displays. Displays need to:

- provide a fixed area on the screen for alarm and emergency annunciation
- refresh the displays dynamically and within a short time (at most a few seconds) after a command is issued
- allow the operators to be able to navigate the displays easily and quickly
- maintain a consistent "look and feel" and use intuitive design industry-accepted display design methodologies and standards.

1.6.1 Human Machine Interface (HMI)

All SCADA vendors will have a comprehensive HMI system, which will include tools for creating and modifying displays and reports. In fact, the capabilities of most systems can be bewildering and intimidating. Since a typical SCADA host will have a large RTDB, the challenge is to design an HMI that presents relevant information to the operator in an easy to understand set of displays.

It is important to develop some guiding principles for each system before the displays are created. These guidelines should include some variation of the following:

1. Have a hierarchical approach: Top-level displays will show key summary information but also have the ability to "zoom" in quickly for more detail. Typically, the top level display is a pipeline system overview as shown in Figure 4 or a pipeline system schematic. The system overview display allows the operator not only to view the current pipeline states including set points and alarms of the system but also to access a particular station for viewing control points and/or modifying their values.

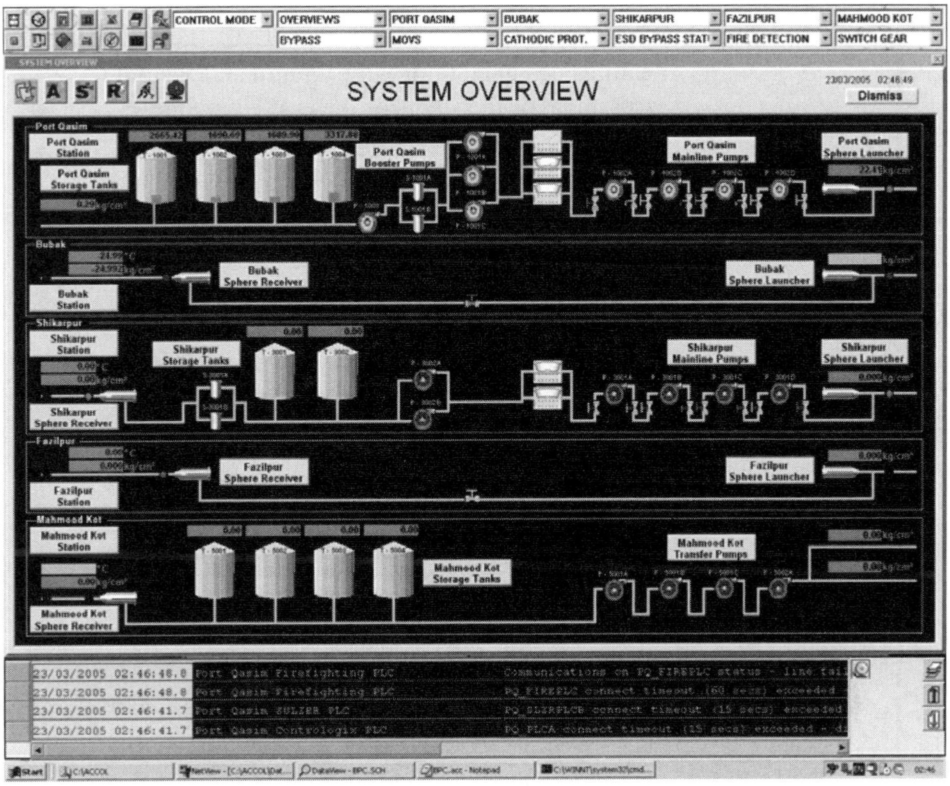

Figure 4 Display of Pipeline System Overview (Courtesy of Telvent)

It not only displays all pump/compressor stations and current alarm messages but also flow, pressure and temperature including set points. In addition, this display may show the link to pump/compressor, meter, or valve station control panels through which the operator can send a control command.

2. Ensure a consistent "look and feel" of displays to minimize training and the chance of operator error. These will include the use of colour (for

example "red" means valve is closed or pump is on, etc) and a consistent and logical approach to the use of buttons, menus and toolbars.

3. Keep screens as uncluttered as possible while still supplying the required information. The possibility of confusion is minimized and of information being lost or "buried" on the screen reduced.

One approach is to utilize a hierarchy of display types with some guidelines such as:

Task	Description
Pipeline Monitoring: Overview	Features designed to provide a rapid and accurate assessment of the status of the entire pipeline. Overview displays typically display key station status, pressures and flows at key locations in a graphical format showing a single-line diagram of the complete pipeline.
Pipeline Monitoring: Unit	Features designed to provide structured access to the more detailed overview summary of a particular station selected from the overview display. Individual equipment status and analogue values in the station will be shown.
Pipeline Monitoring: Detail	Detailed process monitoring is commonly provided through onscreen pop-up windows to provide detailed information on each element in the station display.
Pipeline Monitoring: Analytical	Features designed to display historical and/or statistical information to users. These typically include an historical trend display. These can be a combination of pre-configured displays as well as options for the user to select values to be trended.
Pipeline Control: Detailed	Features that allow users to manipulate pipeline control elements (e.g., by starting or stopping equipment, opening/closing valves or changing set-points, etc.). This control is commonly provided as part of the pipeline monitoring detail interface features.
Alarm Management	Features designed to notify users of monitored alarms, allow acknowledgement of those alarms, and provide a record of alarm-related events, and display summaries and histories of alarms.

The design process should include a review of any existing system used by operators as well as a review of their requirements by performing a task analysis and workflow review.

Prototyping and creation of display mock-ups for review can be an effective method of reviewing the proposed HMI with operators before significant effort is spent in creating the production displays. The goal should be to create an HMI that meets the operator's needs and is intuitive to use with a minimum of training.

Screen navigation should follow the current expected features found in most window-type navigation software to reduce operator-learning time and to make the system as intuitive as possible. For example, selection of a device may be a left mouse click, whereas a right mouse click would display information or operating parameters associated with the device.

The displays are either in tabular or graphical format. In some cases, it may be useful to have both tabular and graphical formats for displaying data. The selection of format depends on how the data is used. For example, a pressure profile in tabular format is useful for verifying line pack calculations, while it is more useful to display pressure drop along the pipeline in graphical format.

Most modern SCADA systems use several display mechanisms, some of which are briefly described below:

- Bar – used to provide a horizontal or vertical bar graph in which a color bar expands or shrinks proportionally to show the current data value on the scale based on defined minimum and maximum scale limits. The value and scale can be shown on bar graphs.
- Slider bar – used to adjust a displayed value by moving the cursor. Manual overrides of analog points typically involve the use of slider bars.
- Plot and trend display types are used to display graphs. *Plot* is used to compare sets of data values to each other, while *trend* is used to examine the changes in data values over time.
- Text – either characters or numbers are accepted and displayed as input.
- Image – graphical images, augmented with real time information governing the image's current color, shape or presentation can be used to relate discrete information in an intuitive manner.

There are other display types such as pushbutton for selecting a button to perform a specific function, meter/gauge for showing a meter/gauge device with values, and region for marking a location on a display.

Some SCADA display systems support display format control. The format control functions include popup and pan/zoom. For example, the functions such as set-point control and communication control can be supported by pop-ups. A large display area can be easily navigated by means of a panning/zooming feature of the display system.

1.6.1.1 Examples of Tabular Displays

1. **Flow rate summary** - Flow rates are measured and reported at every meter and meter station. An example of the flow rate summary is shown below. Rate control panel can be accessed through this summary display.

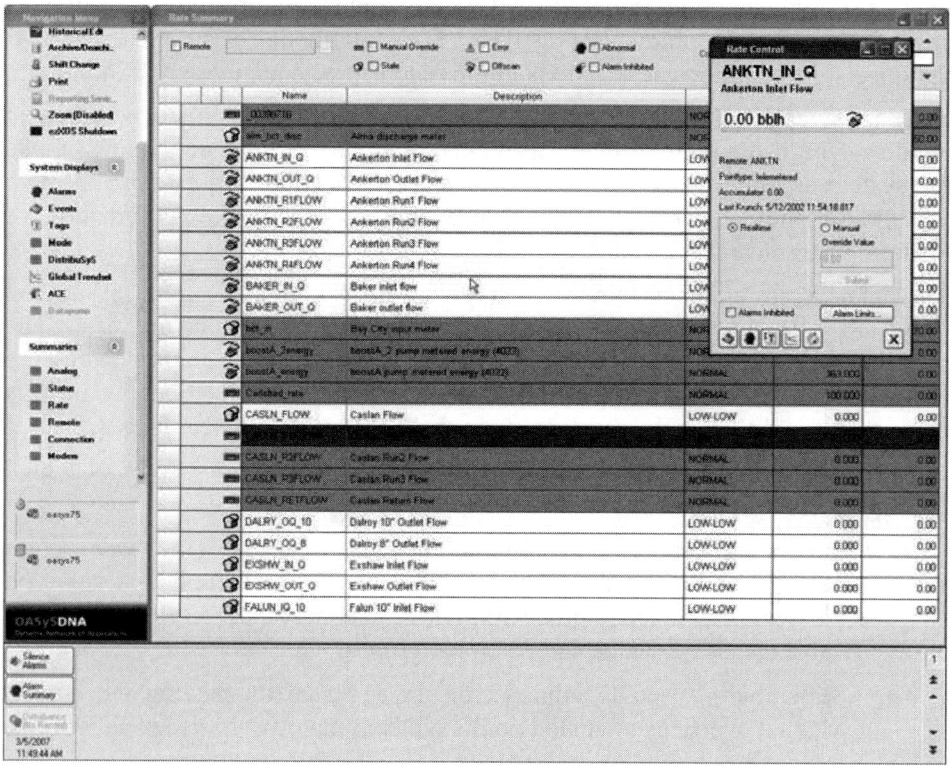

Figure 5 Flow Rate Summary (Courtesy of Telvent)

2. **Event summary** - All events such as database modifications are recorded in the database and reported in the event summary.
3. **Alarm summary** – Normally, both current alarm and alarm history summaries are made available. Shown below is an example of an alarm history summary. An example display of a current alarm summary is shown in Figure 6.

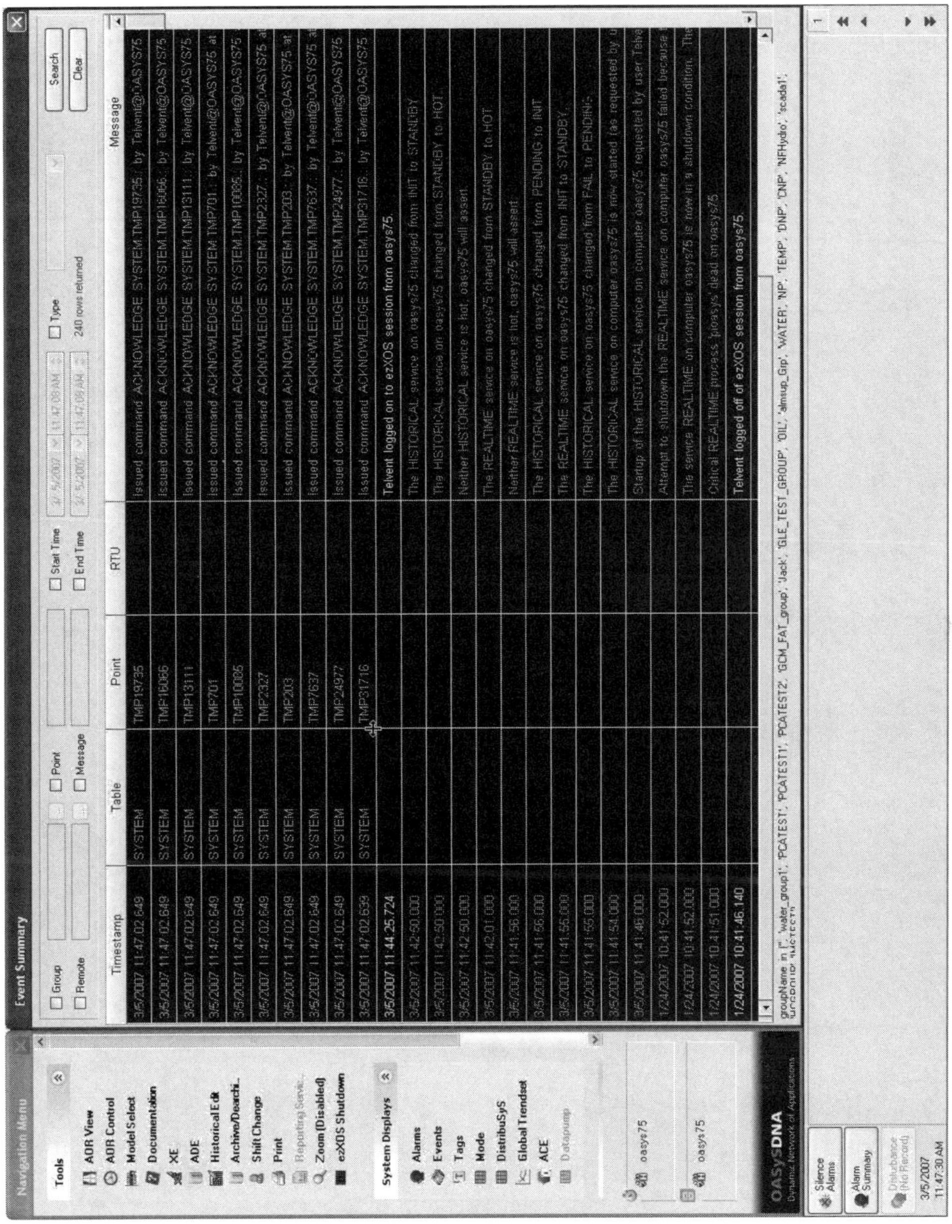

Figure 6 Alarm History Summary (Courtesy of Telvent)

4. **Analog summary** - Analog values are measured and reported at all measurement points. The analog summary is shown below. It shows both controllable and uncontrollable values and allows the operator to change set points and override analog values from this display.

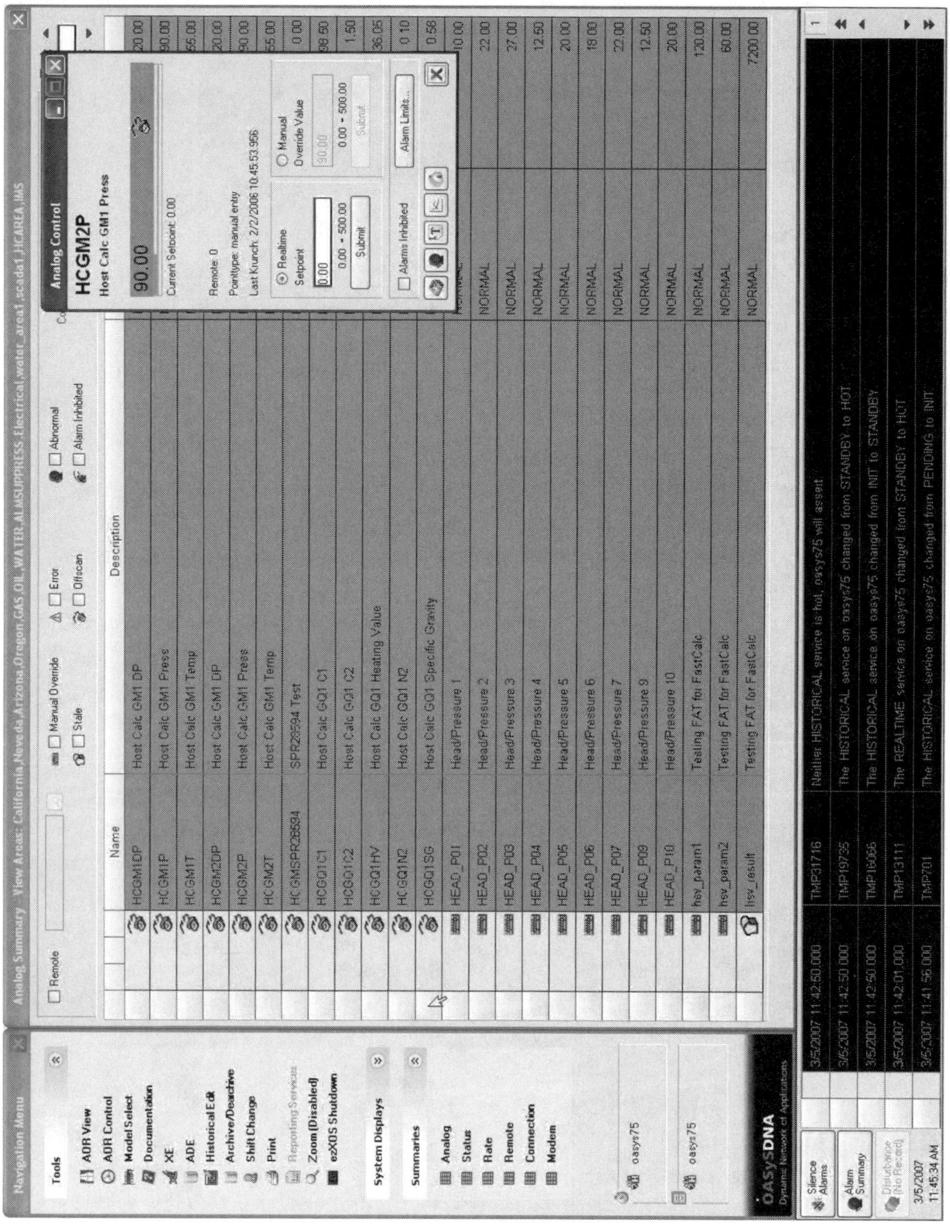

Figure 7 Analog Summary (Courtesy of Telvent)

5. **Status Summary** - Current operating statuses are reported at all measurement and equipment points. The status summary is shown in Figure 8. The status can be changed or overridden through this display.

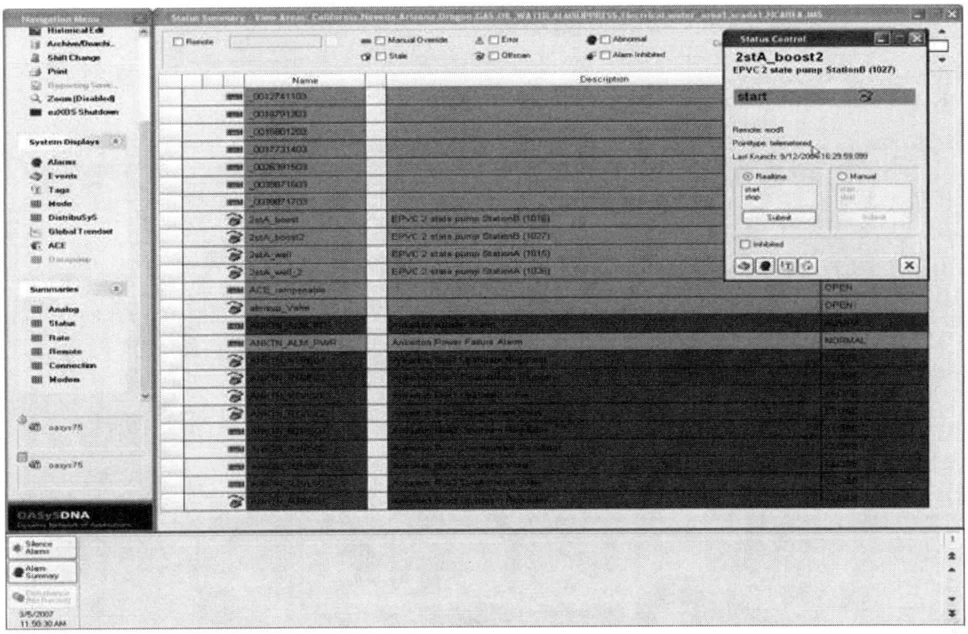

Figure 8 Status Summary (Courtesy of Telvent)

6. Remote Communication Summary

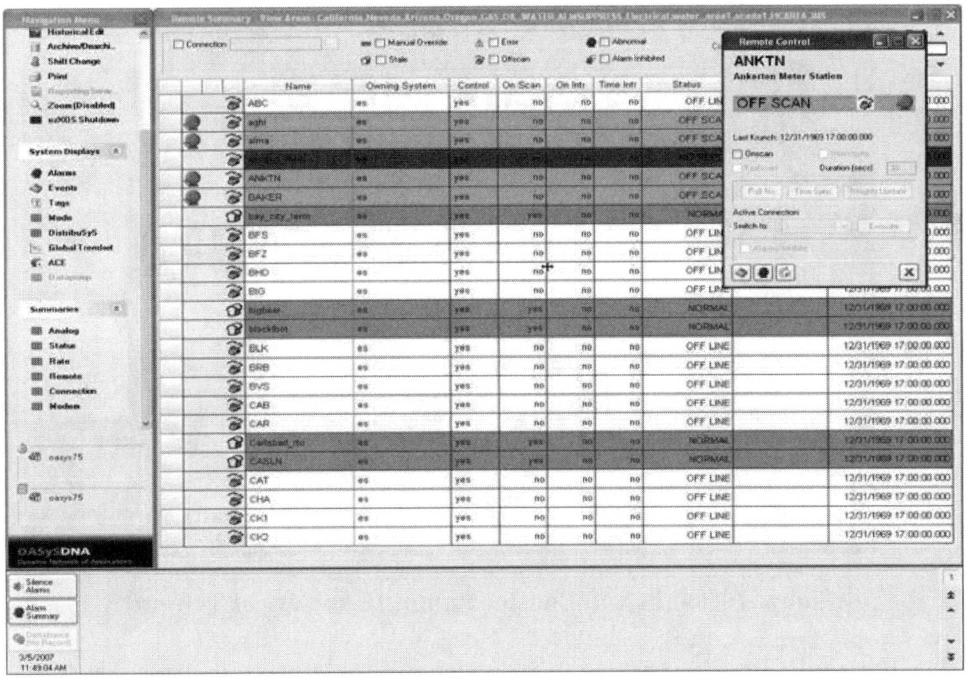

Figure 9 Remote Communication Summary (Courtesy of Telvent)

1.6.1.2 Examples of Graphical Displays

1. **Tanks and Booster Pumps**

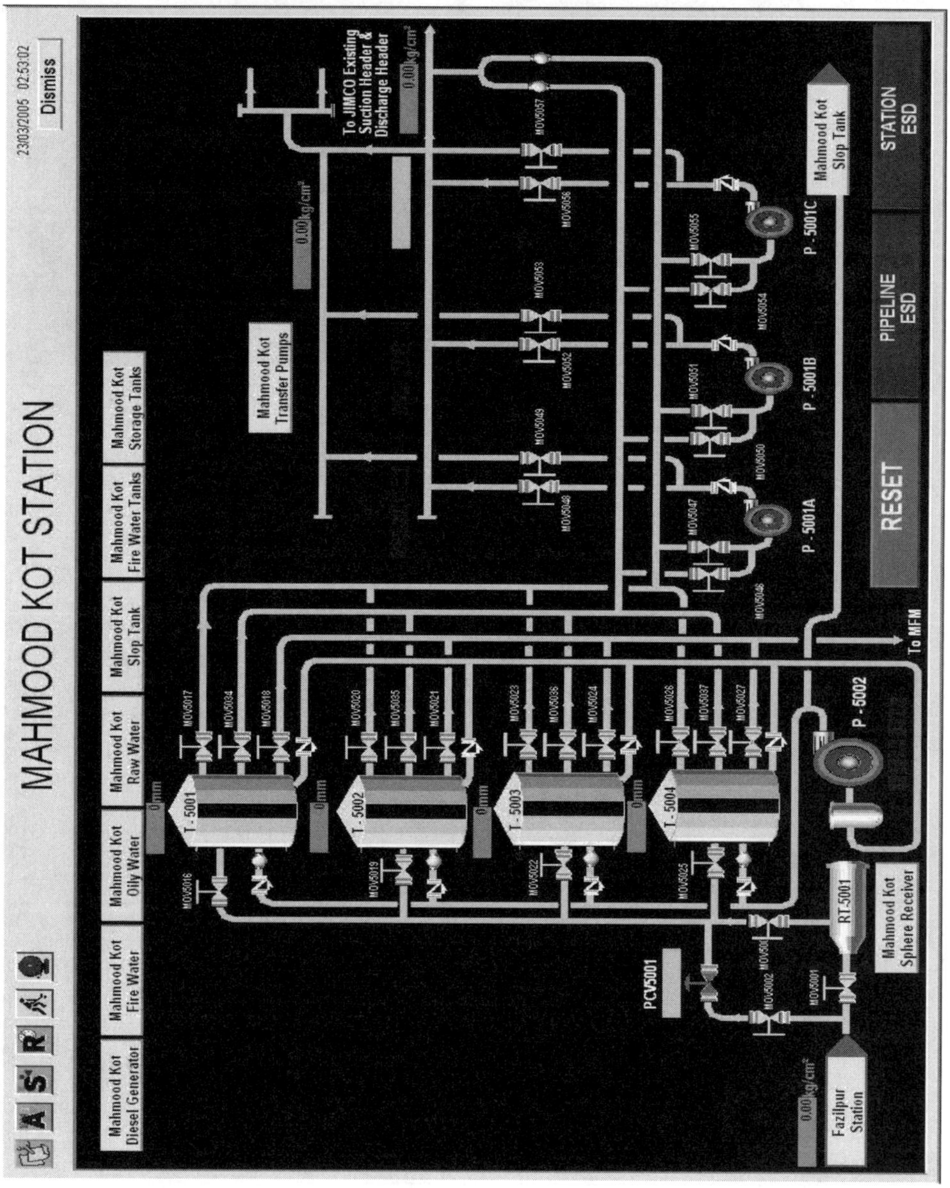

Figure 10 Tanks and Booster Pumps (Courtesy of Telvent)

2. **Hydraulic Profiles Display** – Hydraulic profiles of pressure, flow, density, and temperature help the operators to understand the current pipeline states.

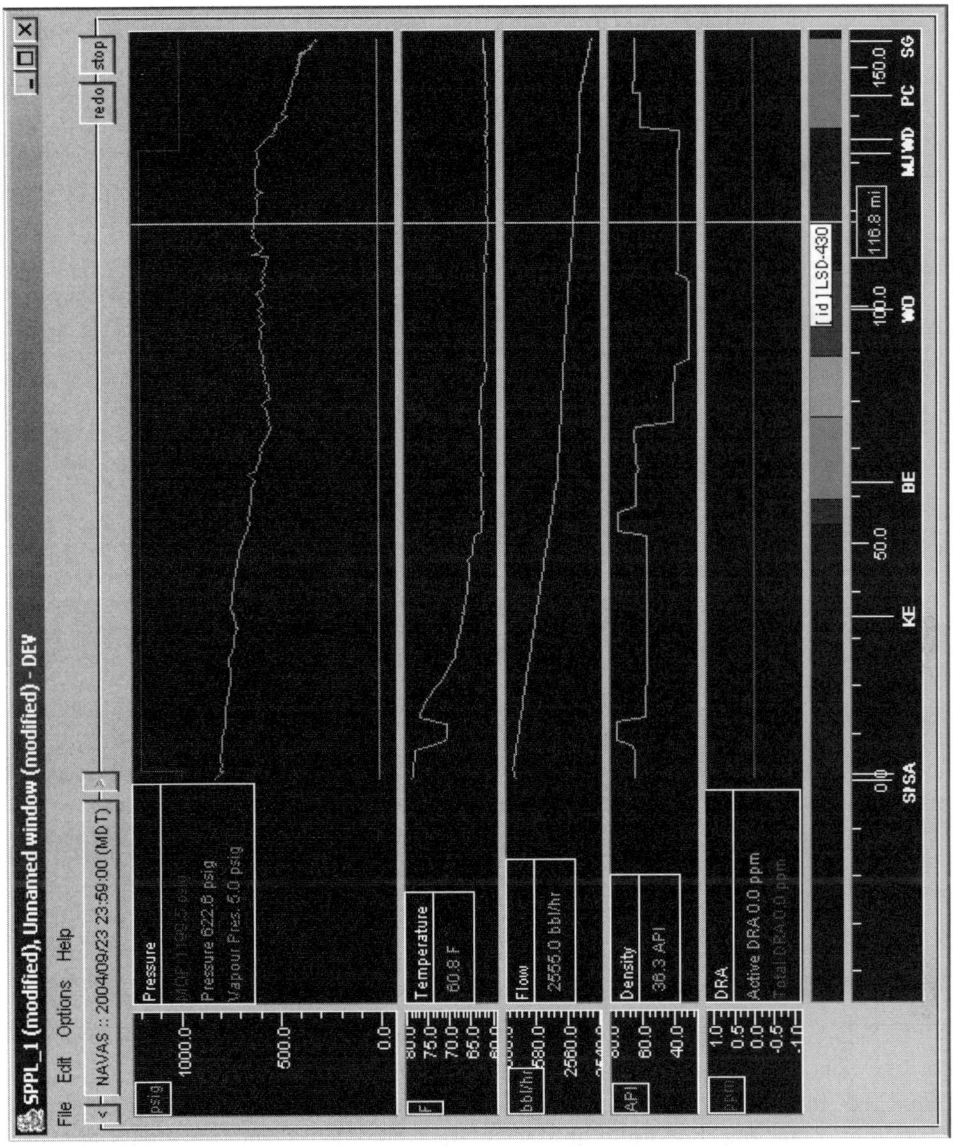

Figure 11 Hydraulic Profiles (Courtesy of CriticalControl)

3. Trend Display

Data trending capability is one of the most important functions of any SCADA system, because it helps the dispatchers and operations staff to identify potential problems before they arise and to diagnose alarm conditions. Data trending is to display any analog values over time at a specific location or locations, which are stored in the historical database. Data trending displays are in graphical format due to large amount of data.

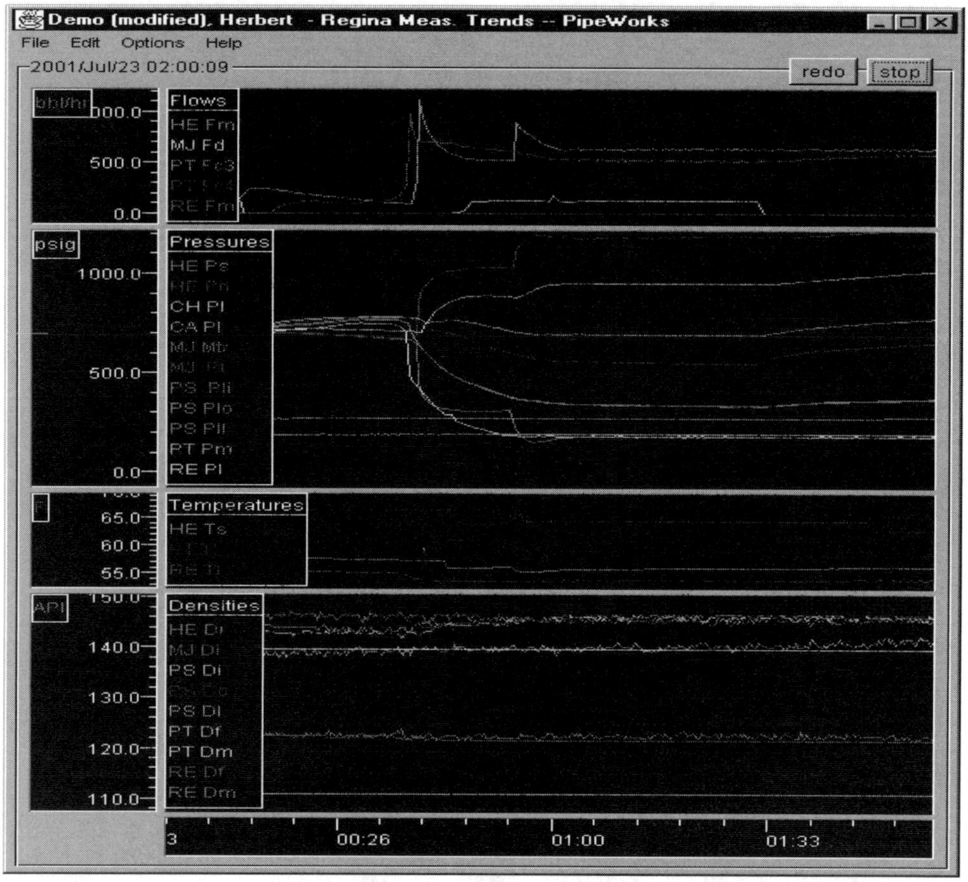

Figure 12 Trend Display (Courtesy of CriticalControl)

1.6.2 Reporting and Logging

All SCADA systems have some method of reporting capability. This will typically consist of both standard reports generated automatically by the system and user-requested reports. These reports are generated from the SCADA databases containing real time, historical and calculated data. The standard reports are of a predefined structure, while the user-requested reports meet the user's specific needs. Examples of standard reports include operating summary reports and billing reports, and those of user-requested reports command/alarm log sorted by station.

Reports are created with a structured query report writer. The report generating software usually comes with the database management system. Some systems will allow for third party software to access values for reporting, which can give the user more flexibility to create reports for use by other business units in their company. For example, the system may allow data to be exported to templates for

Excel or Word. Another option is to include the ability to publish reports to internal and/or external Web sites for view-only users.

The types of reports usually found on a pipeline SCADA system would include some of the following:

1. Operating Reports:
    - Shift or daily operating summary reports
    - Product movement report
    - Alarm summary report
    - System availability, communication and reliability report
    - Emergency scan report, containing operating data during emergency conditions

    Government regulators may require pipeline companies to submit regulatory reports. Normal operation reports may need to be submitted regularly, but emergency reports are mandatory in the event of emergency conditions.

2. System Administration Reports:

    The SCADA system provides system administration tools to configure and maintain the system. An example display is shown in Figure 13.

As shown in the figure, the tools also allow the SCADA users to access various logs.

- Command log, containing a record of all commands issued by the operator
- Alarm log, containing all generated and acknowledged alarm messages for tracking operational problems
- Database maintenance log for recording commands used to change any SCADA database
- System log for recording the SCADA system performance including error data such as the start/stop time, abnormal running time, etc.
- Communication log for recording the statistics of the communications with the RTUs such as the number of attempts, the number and types of error, etc.

The number, content, and style of reports will vary widely depending on the pipeline type, the business requirements, and the regulatory environment. It is important that the SCADA system provides an easy to use, flexible reporting package that does not require programming changes to create and implement reports.

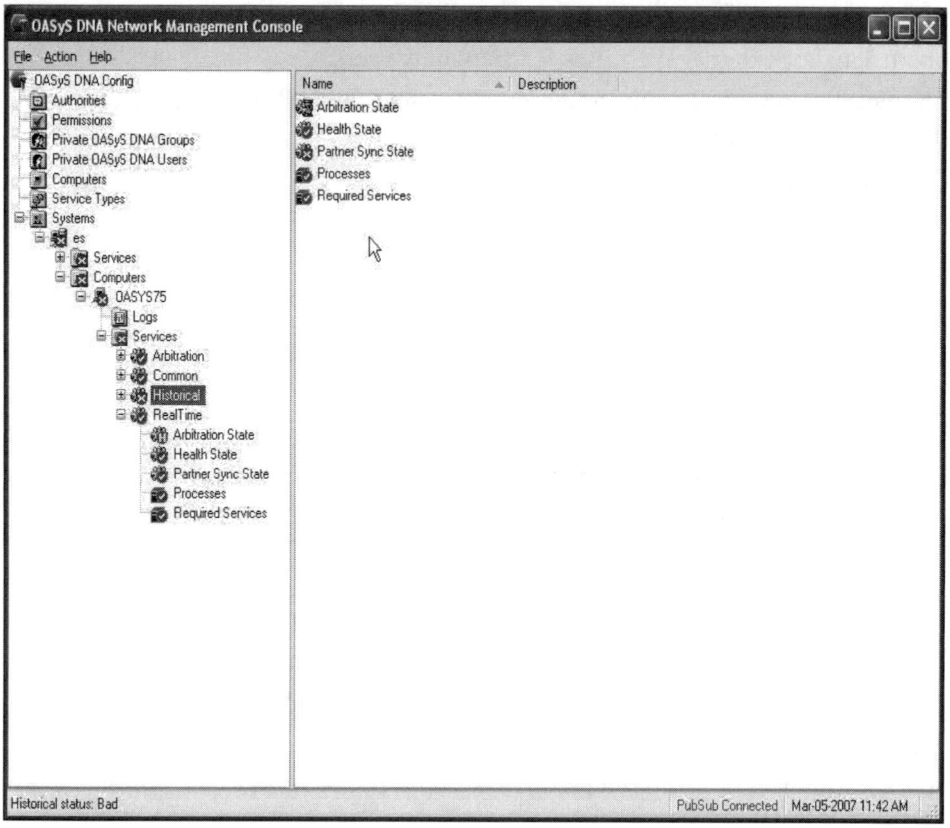

Figure 13 System Administration Tools (Courtesy of Telvent)

1.7 Alarm Processing

Alarm conditions are expected during the course of pipeline system operation. The alarm processing function can help to identify potential alarm conditions before actual alarm conditions occur. Examples of potential alarms include high-pressure violation, high temperature violation at a compressor discharge, leak detection, etc.

The alarm processing function should be able to limit the number of alarms to those that are important. If the number of alarms is too large, the operator's attention is consumed reviewing and acknowledging alarms instead of monitoring and controlling the pipeline system. In general, alarms are prioritized according to their critical nature in order to give the operator an indication of which alarms need to be attended to first. Emergency alarms require the operators to take immediate action to correct the condition, while communication alarms may require them to contact supporting staff immediately. Warning alarms are not usually critical, requiring preventive measure without immediate action. The severity of alarms should be configured to be one of multiple levels of severity (for example, high, medium, or low) for all alarm generating points. Alarms are usually color coded, requiring a different color for each level of alarm. In addition, an audible signal should be generated for high-level alarms.

1.7.1 Alarm Types

1.7.1.1 Analogue

Analogue alarms are generated when a current value for an analogue point reaches a limit pre-defined in the data base attribute for that point. This will typically include the following:

- High-High (or Alarm) means that the point has reached its maximum allowable value. This will generally mean that it is close to or has reached a point where local automatic protection systems may be initiating action.
- High (or High Warning) means that the point has reached a warning level. If remedial action is not taken, the point may reach High-High. The trending system will allow an operator to display such a point to see how long it has taken the point to get to the warning level.
- Low-Low (or Alarm) similar to High-High but for a lower limit
- Low (or Low Warning) similar to High but for a lower limit
- Rate of Change: The slope of a trend line has exceeded a pre-defined limit. This means the process value is changing more rapidly than would be expected.

1.7.1.2 Discrete

Discrete alarms are generated upon a change of state of the data base point. These can represent:

- Change from normal to abnormal such as a high temperature alarm on a compressor station outlet.
- A change of status that was not the result of an operator control action. For example, a valve closes or a pump shuts down with no initiation from the operator.

All such alarms will be reported and logged, as will any change of status of a point. This will provide not only a record of all abnormal events but will also show when equipment was acted upon by an operator.

1.7.2 Alarm Handling

A basic alarm management scheme consists of detecting the alarm and reporting

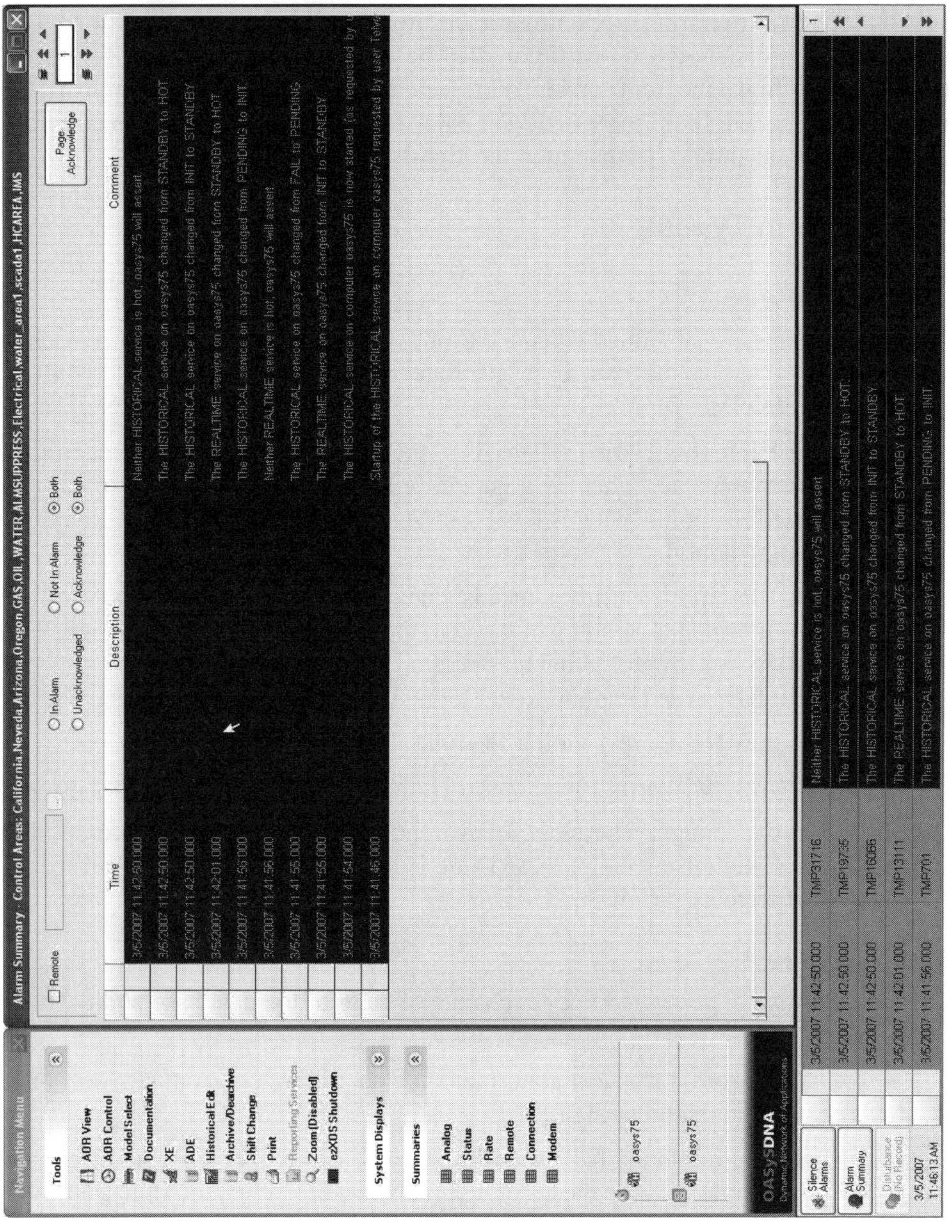

Figure 14 Current Alarm Summary (Courtesy of Telvent)

the alarm to the operator. An alarm management system will also log and provide an audit trail of each alarm. This will include the time that the alarm was reported, when it was acknowledged by the operator and when the alarming point returned to normal. This information along with the database log will provide key information for post-event analysis.

In any system upset, there will be an initiating event followed by secondary indications or alarms. For example, a control valve may fail causing pressure to rise, which may then cause pressure relief valves to operate and flow rates to exceed expected values. Some SCADA systems may incorporate some form of artificial intelligence to process alarms automatically to advise the operator of what the potential root cause may be.

The SCADA database will have the ability to assign various levels of alarm severity to individual points to provide an easy means of reporting high priority alarms to an operator. In an emergency condition, it is important to not overload an operator and allow concentration on priority items.

1.7.3 Alarm Message

The alarm message includes the date and time of the alarm, the point that caused the alarm, the severity of the alarm denoted by color and an audible signal, and the state of the point. The message is displayed in the alarm window and in the tabular summary of alarms. The alarm window lists all unacknowledged alarms, which should be made available on the screen at all times.

Alarms are always logged in an event summary, including not only all the information in the alarm message but also the time when the alarm was acknowledged and by whom.

The operators should be able to easily monitor alarm messages and quickly respond to the messages. Therefore, messages should be made readily available to the operator. Figure 14 shows an example display of the current alarm summary. The current alarm summary is mainly used for monitoring and acknowledging the messages, while the alarm history summary is mainly used for reviewing the alarm status and pipeline system operation.

1.8 Remote Terminal Unit (RTU)

1.8.1 Overview

There is no simple definition of an RTU; it can be seen as a device or perhaps better viewed as a set of functions. The primary purpose of an RTU is to act as a data collection point and to manage the interface between the SCADA system and a field location. In its simplest form a RTU will gather analogue data and discrete status indications for transmission back to the host. In turn, it will receive commands from SCADA, and translate and initiate the appropriate control functions locally.

Primary functions provided by an RTU include:

- acting as a data concentrator
- providing a local controller

- providing protocol gateways, such as allowing SCADA to communicate with other devices at the RTU site
- providing a flow computer
- allowing for local data logging

In the past, SCADA vendors supplied RTUs, which tended to be proprietary. Now some RTUs incorporate standard protocols and can utilize non-proprietary hardware and software. RTUs can also receive digital information from local systems via a network connection or serial link. An RTU may connect to one or more local PLCs, a flow computer, a metering skid controller, or other computerized systems. In order to meet these requirements RTUs need to be able to handle a wide range of standard communication protocols.

In a situation where there is a local control system such as a DCS or PLC (e.g. at a pump station) the RTU functions as a simple interface and/or protocol converter. In many SCADA upgrade projects, the master system is replaced but the existing RTUs may be retained, thus requiring that the newer SCADA host/master (which may be from a different vendor than the older RTU) be able to communicate to the RTU.

1.8.2 RTU Hardware Architecture

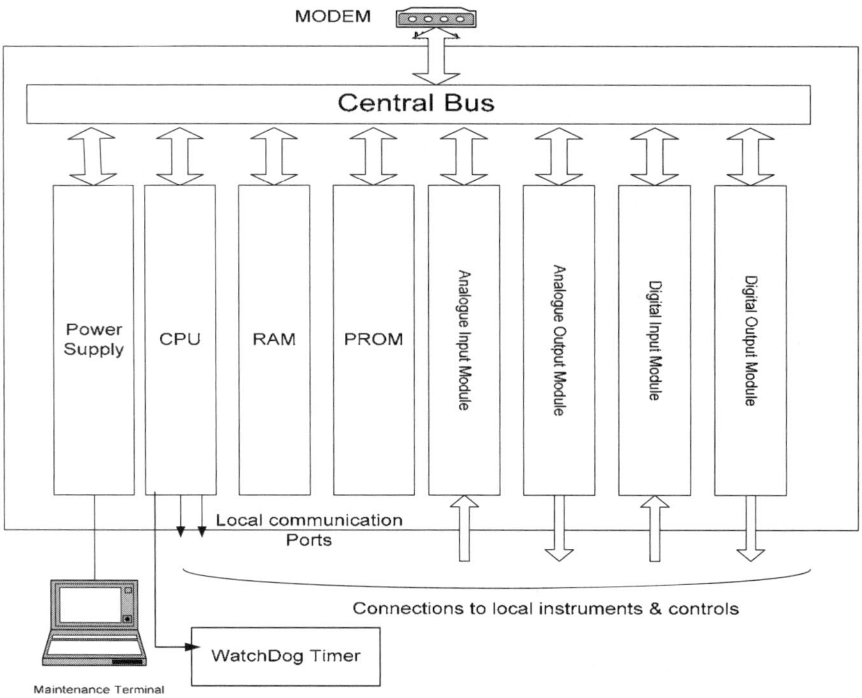

Figure 15 Typical RTU Architecture

There are two basic types of RTUs: a small, single board RTU that contains all components including I/O used for applications with limited I/O requirements and larger RTUs utilizing PLC components configured to provide communication with a SCADA host and extensive local control and monitoring capability.

Most RTUs consist of a microcomputer, I/O termination equipment, and I/O circuitry, communications circuitry, local interface, and a power supply. In spite of the environmentally hard conditions in which an RTU is typically mounted, most users expect the RTU to operate unattended and virtually trouble-free for many years. To achieve this industrial ruggedness, special consideration in design is required. Figure 15 shows a typical RTU architecture.

1.8.2.1 Microcomputer

The microcomputer is the heart of the SCADA RTU, controlling all processing, data input/output control, communications, etc. A general microcomputer configuration will have a processor, memory, real-time clock and a watchdog timer.

(a) Processor

The processor oversees and controls all functions of the RTU. It is generally 8- or 16-bit, with preference in recent years for 32-bit as the demand for processing power increases.

(b) Memory

The RTU has both random access memory (RAM) and read only memory (ROM). RAM memory can be both written to and read from by the processor, and provides a storage location for dynamic RTU data such as the RTU database. RAM is volatile in the event of power failure, and is therefore generally provided with on-board battery backup (typically a lithium battery, or such). Many RTUs will only support a few kilobytes of RAM memory, while a complex RTU, supporting extensive applications programs, may be configured with 1 megabyte or more.

ROM memory is loaded at the factory and cannot be changed by the processor. As such, ROM provides a storage location for the RTU executable program code such as input/output tasks, process control, calculation routines, communications, and operator interface. This ROM is sometimes referred to as "firmware." It can also contain RTU configuration information such as the RTU address, number and type of I/O points, alarm thresholds, engineering units, etc. ROM memory is non-volatile and does not require battery backup. The amount of memory in an RTU and the ratio of RAM to ROM are dependent on the RTU database size and the amount of program code.

A feature available from many RTU vendors is the ability to download the RTU database definition and user applications program code over the communication link from the host SCADA system. The RTU database is built up at the host using menu-driven routines and downloaded to the RTU. EPROM (Erasable Programmable Read Only Memory) may be employed (or battery-backed up RAM) in place of ROM to allow for remote firmware updates which eliminates the need for technicians to travel to each RTU location to implement such updates or upgrades.

(c) Real-time Clock

The real-time clock is typically provided by a crystal oscillator and is used for time tagging of events and real-time process control. The RTU real-time clock is frequently synchronized with the host computer in order to maintain system-wide time synchronization.

(d) Watchdog Timer

The watchdog timer is a timing mechanism that expects to be reset by the CPU at regular intervals. Failure of the CPU to reset the watchdog circuit will indicate RTU failure. The watchdog circuit will timeout and perform some specific functions, such as annunciation, disable I/O power, signal a backup RTU through a set of contacts, etc. The intent of the watchdog timer is to identify an RTU failure and minimize the effect.

1.8.2.2 Input/Output (I/O) Circuitry

Typically, an RTU will support standard signal levels found in an industrial environment. These include analog inputs and outputs, as well as discrete and pulse inputs and outputs.

In some installations, it will be required by local conditions and wiring codes to provide intrinsically safe (IS) barriers to the terminal blocks for the field wiring. This is especially important for installation locations potentially containing explosive atmospheres.

1.8.2.3 Communications

By the nature of the SCADA system, all RTUs must communicate with the host computer. In addition, there is frequently a requirement for serial communication between the RTU and other devices, such as smart transmitters, flow computers, programmable logic controllers, and personal computers. Therefore, there will be at least one, but possibly many, serial communication ports in an RTU. RS-232-C and RS-422/423/485 are often used for these ports, as they are widely accepted standards for short-range, point-to-point communications.

Long haul communications, such as the link to the control center, are typically handled with modems. Many different modems are used, depending on the transmission media and the data rate. Auto-dial modems are available for use on the public switched telephone network (PSTN), and broadband modems are available for high-speed data communications via media such as satellite or microwave.

Some applications may have a network connection between the SCADA host and the RTUs. For example if there is a fibre optic cable installed along with the pipeline, this enables WAN connections to a RTU.

1.8.2.4 Operator Interface

Typically, the RTU operator interface is an ASCII serial device, such as a monochrome CRT or a dedicated low power single-line ASCII terminal. Alternately, it is now more common to enable laptop computers to be connected and to act as the local operator interface.

The operator interface is usually very simple and intended to provide limited operator interaction with the RTU. Generally, the local operator at the RTU can call up the status of alarms and value of analog inputs, tune control loops, drive outputs, perform diagnostic tests and change database definitions.

1.8.2.5 Mass Storage Media

An RTU is rarely equipped with mass storage devices such as disk drives or tapes. This is due to the need for industrial ruggedness in the system. Disk and tape drives will rarely withstand the operating environment that is required for RTUs.

1.8.2.6 RTU Power Requirements

RTU manufacturers will generally supply an RTU with whatever power requirements are specified by the user. Common choices are 120/240 VAC and 24 VDC. The choice determinant is the available power at the user's site.

Often commercial AC power on-site is subject to transient or frequent failure. To avoid such problems, RTUs are often powered by an uninterruptible power supply (UPS) for the RTU, consisting of a battery charger and batteries.

Large multi-board RTUs with extensive processing and I/O support capabilities will easily draw several hundred watts. Many RTUs are installed in remote locations nowhere near any source of commercial power and will utilize solar, wind or fuel cell power sources to charge a battery system. For this reason, many RTUs are designed utilizing technologies that will limit the power requirements of the RTU.

1.8.3 RTU Software

The RTU RAM memory provides a storage location for the RTU database, which includes all I/O points, constants and other points for flow calculations and control. The RTU ROM memory holds the executable code for scanning, transmitting and controlling.

The RTU database is normally small. It may include the following information:

- I/O address associated with the signal being scanned,

- raw value of the point
- range of the input
- factor for converting the raw input into engineering units
- converted engineering unit value
- alarm limit and status

1.8.4 Control

In the early days of RTUs, control logic was implemented by a combination of discrete electronic logic circuits and relays or high current electronic switches for output to field devices. In the 1960's and 70's microprocessor-based controllers were introduced that allowed control engineers to change the logic of a control system by programming and not re-wiring relays. Today programmable logic controllers (PLCs) are a de facto standard for industrial control. A pipeline system will now likely contain a PLC acting as an RTU, a PLC doing local control at a station and perhaps a PLC that is part of a turbine control panel.

With the growth of computers, intelligent electronic devices/instruments, PLCs, etc., each of which can be seen to supply a portion of traditional RTU functionality, the line between RTU/SCADA and local control is somewhat blurred. At a small remote site, the RTU will likely be capable of only limited direct control of equipment, such as the opening and closing of valves; or the starting of a sump pump. At larger sites, the RTU may simply interface to a local control system that in turn will be responsible for the direct control of equipment.

In between, there is a complete spectrum of RTUs with varying degrees of control capability, especially if they are using a PLC platform. Generally, unless the station where the RTU is installed is a large one, such as a pump or compressor, it would be typical for the RTU to provide whatever local control is required. In the larger stations, there will be a dedicated station control system (See Chapter 3) and the RTU will be acting as the interface between the station control and the SCADA system.

1.9 Security

1.9.1 Internal Security

A SCADA system will provide for user password access and the ability to configure specific levels of access for each user. For example, there may be users who may access the SCADA system but are allowed only the ability to read some pre-configured reports. The system manager's accounts are at a higher level of access, and should be password protected. In addition, only those who are directly responsible for the database are allowed to maintain the database with password protection. The operating system may enable the SCADA system administrator to assign an access level to a user that will dictate which workstations, interfaces, and displays a user may access, and which operations they may perform.

As well, the system may limit access to a specific workstation regardless of the user who is logging on at that workstation. Such a limitation could ensure for example, that a workstation in an engineering area could never be used as an operating terminal without changing the system access.

Many older SCADA systems were not designed with information security in mind. This omission has led to systems with unsecured data transmission. Most of the older SCADA systems transmit both data and control commands in unencrypted clear text. This allows potential attackers to easily intercept and issue unauthorized commands to critical control equipment.

Furthermore, the lack of authentication in the overall SCADA architecture means that attackers with physical access to the network can gain a foothold to launch denial-of-service or "man-in-the-middle" attacks, both of which can lead to disruption and safety concerns.

1.9.2 Open Public Network Connections

SCADA systems have long been thought to operate in a secure environment because of their closed networks, which are not exposed to external entities. In addition, the communication protocols employed were primarily proprietary and not commonly published. This "security by secrecy" approach has led to a false sense of security that does not stand up to the test of an audit. SCADA networks were initially designed to maximize functionality with little attention paid to security.

Furthermore, the notion that SCADA networks are closed systems, is no longer true. Recent advances, such as Web-based reporting and remote operator access, have driven the requirement to interface with the Internet. This opens up physical access over the public network and subjects SCADA systems to the same potential malicious threats as those that corporate networks face on a regular basis.

1.9.3 Standardization of Technologies

Typically, compliance with industry standards and technologies is regarded as a good thing. However, in the case of newer SCADA systems, recent adoption of commonly used operating systems and standards makes for a more vulnerable target. Newer SCADA systems have begun to use operating systems such as Windows or UNIX variants that are commonplace in corporate networks. While this move offers benefits, it also makes SCADA systems susceptible to numerous attacks related to these operating systems. SCADA systems also face patch management challenges as the vulnerabilities of these operating systems are uncovered.

RTU to host protocols that were typically proprietary in the early days of SCADA are now utilizing industry standard protocols, which may compromise their security.

1.9.4 Securing SCADA from External Systems

The security associated with the SCADA network needs to be designed and assessed by the same policies utilized in other areas of the company. If there are no such clear network security policies in place, then they need to be established before taking specific actions on the SCADA network.

The US Department of Energy has published a list of actions, detailed in the following sections, to increase the security of SCADA networks (9). As with any set of recommendations, the degree to which they are implemented usually depends upon the political will of the organization and the available resources that management is willing to commit in terms of people, time and money.

1.9.4.1 Identify all connections to SCADA networks.

Conduct a thorough risk analysis to access the risk and necessity of each connection to the SCADA network. Develop a comprehensive understanding of all connections to the SCADA network and how well these connections are protected. Identify and evaluate the following types of connections:

- internal local area and wide area networks, including business networks
- internet
- wireless network devices including satellite uplinks
- modem or dial-up connections
- connections to business partners, vendors or regulatory agencies

1.9.4.2 Disconnect unnecessary connections to the SCADA network.

To ensure the highest degree of security of SCADA systems isolate the SCADA network from other network connections to as great a degree as possible. Any connection to another network introduces security risks, particularly if the connection creates a pathway from or to the internet. Although direct connections with other networks may allow important information to be passed efficiently and conveniently, insecure connections are simply not worth the risk. Isolation of the SCADA network must be a primary goal to provide needed protection. Strategies such as utilization of "demilitarized zones" (DMZs) and data warehousing can facilitate the secure transfer of data from the SCADA network to business networks. However, they must be designed and implemented properly to avoid introduction of additional risk through improper configuration.

1.9.4.3 Evaluate and strengthen the security of any remaining connections to the SCADA network.

Conduct penetration testing or vulnerability analysis of any remaining connections to the SCADA network to evaluate the protection posture associated with these pathways. Use this information in conjunction with risk management processes to develop a robust protection strategy for any pathways to the SCADA network. Since the SCADA network is only as secure as its weakest connecting point, it is essential to implement firewalls, intrusion detection systems (IDSs) and other appropriate security measures at each point of entry. Configure firewall rules to prohibit access from and to the SCADA network, and be as specific as possible when permitting approved connections. For example, an Independent System Operator (ISO) should not be granted "blanket" network access simply because there is a need for a connection to certain components of the SCADA system. Strategically place ISDs at each entry point to alert security personnel of potential breaches of network security. Organization management must understand and accept responsibility for risks associated with any connection to the SCADA network.

1.9.4.4 Harden SCADA networks by removing or disabling unnecessary services.

SCADA control servers built on commercial or open-source operating systems can be exposed to attack through default network services. To the greatest degree possible, remove or disable unused services and network daemons to reduce the risk of direct attack. This is particularly important when SCADA networks are interconnected with other networks. Do not permit a service or feature on a SCADA network unless a thorough risk assessment of the consequences of allowing the service/feature shows that the benefits outweigh the potential for vulnerability exploitation. Work closely with SCADA vendors to identify secure configuration and coordinate any changes to operational systems to ensure that removing or disabling services does not cause downtime, interruption of service or loss of support.

1.9.4.5 Do not rely on proprietary protocols to protect your system.

Some SCADA systems use unique proprietary protocols for communications between field devices and servers. Often the security of a SCADA system is based solely on the secrecy of these protocols. Do not rely on proprietary protocols or factory default configuration settings to protect the SCADA system. Additionally demand that vendors disclose any backdoors or vendor interfaces to your SCADA system and expect them to provide systems that are capable of being secured.

1.9.4.6 Implement the security features provided by device and system vendors.

Older SCADA systems have no security features whatsoever. SCADA system owners must insist that their system vendor implement security features in the form of product patches or upgrades. Some newer SCADA devices are shipped with basic security features but these are usually disabled to ensure ease of installation.

Analyze each SCADA device to determine whether security features are present. Factory default security settings, such as in computer network firewalls, are often set to provide maximum usability, but minimal security. Set all security features to provide the maximum level of security. Allow settings below maximum security only after a thorough risk assessment of the consequences of reducing the security level.

1.9.4.7 Establish strong controls over any medium that is used as a backdoor into the SCADA network.

Where backdoors or vendor connections do exist in SCADA systems, strong authentication must be implemented to ensure secure communications. Modems, wireless and wired networks used for communications and maintenance represent a significant vulnerability to the SCADA network and remote sites. Successful "war dialling" attacks could allow an attacker to bypass all other controls and have direct access to the SCADA network or resources. To minimize the risk of such attacks, disable inbound access and replace it with some type of callback system.

1.9.4.8 Implement internal and external intrusion detection systems and establish 24-hour-a-day incident monitoring.

To be able to respond effectively to cyber attacks, establish an intrusion detection strategy that includes alerting network administrators of malicious network activity originating from internal or external sources. Intrusion detection systems monitoring is essential 24 hours a day. Additionally incident response procedures must be in place to allow for an effective response to any attack. To complement network monitoring enable logging on all system sand audit system logs daily to detect suspicious activity as soon as possible.

1.9.4.9 Perform technical audits of SCADA devices and networks, and any other connected networks to identify security concerns.

Technical audits of SCADA devices and networks are critical to ongoing security effectiveness. Many commercial and open-source security tools are available that allow system administrators to conduct audits of their systems/networks to identify active services, patch level and common vulnerabilities. The use of these tools will not solve systemic problems but will eliminate the "paths of least resistance" that an attacker could exploit. Analyze identified vulnerabilities to determine their significance and take appropriate corrective action.

1.9.4.10 Conduct physical security surveys and assess all remote sites connected to the SCADA network to evaluate their security.

Any location that has a connection to the SCADA network is a target, especially unmanned or unguarded remote sites. Conduct a physical security survey and have inventory access points at each facility that has a connection to the SCADA system. Identify and assess any source of information including remote telephone/computer network/fibre optic cables that could be tapped; radio and microwave links that are exploitable; computer terminals that could be accessed and wireless local area network access points. Identify and eliminate single points of failure. The security of the site must be adequate to detect or prevent unauthorized access. Do not allow "live" network access points at remote unguarded sites simply for conveniences.

1.9.4.11 Establish SCADA "red teams" to identify and evaluate possible attack scenarios.

Use a variety of people who can provide insight into weaknesses of the overall network, SCADA systems, physical systems and security controls. People who work on the system every day have great insight into the vulnerabilities of the SCADA network and should be consulted when identifying potential attack scenarios and possible consequences.

1.10 Corporate Integration

In the early days of SCADA, it was considered a major accomplishment to be able to transfer a file from a SCADA system to another system such as the corporate accounting system. The advent of networking has made it much easier to connect SCADA systems to business systems. This now allows for both physical integration of SCADA and business systems as well as business process integration. Process integration means that SCADA systems are becoming a key part of business processes and their associated applications software.

It is becoming more and more common for pipeline applications to be tightly integrated with SCADA systems and to be part of a higher level Management Information System (MIS). The growth of MIS started in the manufacturing sector as a means of summarizing factory operations data to allow management to monitor the business in real time rather than by using monthly reports. This provides for both proactive business processes as well as the ability to provide better information and thus better service for customers. Similar systems are being installed in the pipeline sector, especially with the recent consolidations of pipelines wherein there are now fewer control centers controlling more pipelines. Figure 16 shows an example of an integrated system.

The need for integration of SCADA systems with corporate IT and business applications has to be identified during the early requirement analysis of the SCADA project. Historically the design and operating philosophies of SCADA and corporate IT systems have differences that must be reconciled as part of the system design. Chapters 4 through 8 discuss various pipeline applications in more detail.

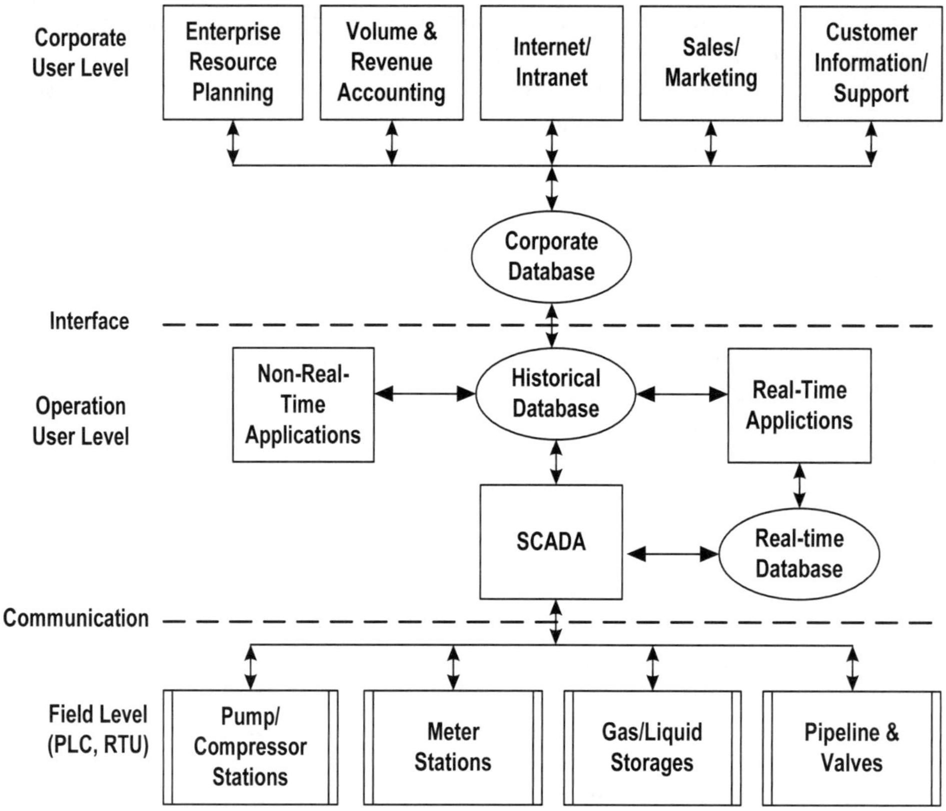

Figure 16 Integrated System

1.11 SCADA Project Implementation and Execution

This section is not intended to be a primer on project management but rather a discussion of specific aspects of managing a SCADA project.

1.11.1 Contracting Strategy

The traditional method of implementing SCADA projects is to use an engineering consulting firm to design the system, issue detailed specifications for bid, evaluate the vendor responses, choose a vendor and manage the vendor throughout the project duration. This contracting strategy (Design-Bid-Build) was developed in the early days of SCADA and automation. These projects usually involved a significant amount of customization, integration of equipment from a variety of vendors and often required the engineering consultant to act as a system integrator.

The current state and capability of SCADA and automation equipment has eliminated many of these issues. Increasingly automation projects are built collaboratively, using what is referred to as co-engineering, or a “Design-Build” approach rather than the older "Design-Bid-Build" contract arrangement. "Design-Build" approach consists of choosing a vendor who will work with your project team, to develop the detailed requirements during the front-end engineering design (FEED) phase. The vendor will then execute detailed engineering and supply the resultant system for the agreed reimbursement. Typically, vendors are paid on a per diem basis for the front-end work and site services. The system supply can be any combination that best fits the needs of the Owner and the particulars of the project. The emphasis of this type of approach is for the project team to concentrate on the performance or “what” aspects of the work and to let the SCADA vendor focus on the "how". However, the SCADA vendor is part of the project team from the front-end engineering design (FEED) through to final commissioning (10).

The Owner must evaluate vendors carefully when using this approach. A vendor must be chosen that has a SCADA system that is compatible with the Owners' project team, having not only the required technical capabilities and personnel, but also a good project track record and cultural fit. The project team would consist of the Owner's representatives, engineering consultants, and technical personnel from the chosen vendor. Some examples of criteria used to evaluate vendors for a "design-build" approach are:

- product quality and functionality
- vendor innovation record
- customer support record
- project management and execution Record
- technical knowledge
- long-term stability and commitment
- local support capability

The advantages of such an approach include:

- The Owner can focus on development of performance requirements rather than detailed design.
- The traditional approach requires a more generic bid specification in order to be vendor neutral. That approach can miss opportunities to make better use of specific technologies. Alternately, such changes have to be made after contract award and will increase the project cost.
- The entity with the best technical knowledge and who understand the capabilities of the technology are the vendors. Involving the vendor in the FEED benefits both the Owner and the Vendor. The Vendor’s knowledge of his system’s capabilities and the Owner’s awareness of his requirements allow the potential of the system to be optimized for the Owner and the Vendor is given a deeper understanding of the Owner’s

operational requirements. Just as significantly, changes and optimization of system design can occur before finalization of specification rather than after, resulting in a potential reduction of project costs. (See Figure 17)

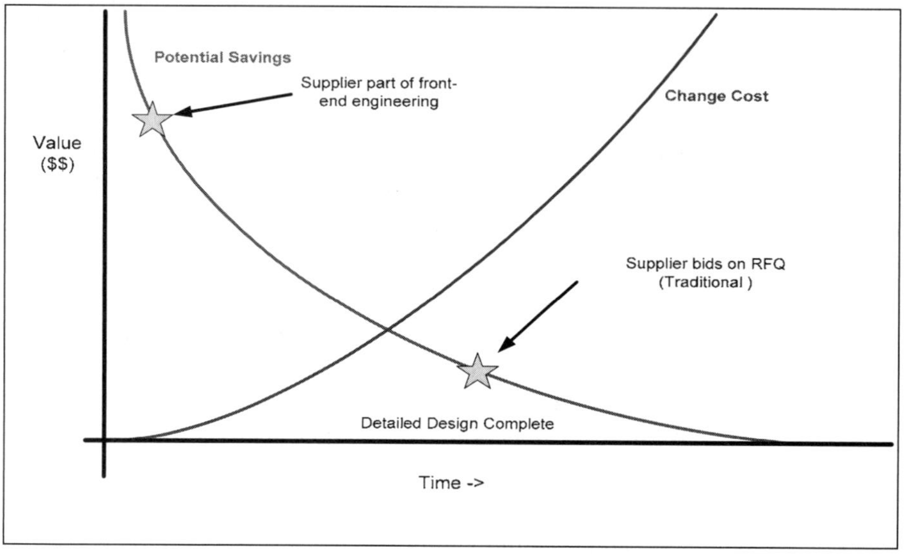

Figure 17 Project Cost Impacts

Although the Owner's focus should be on the functional requirements, it is necessary to understand the technical capabilities offered by suppliers as "off the shelf" in the industry. Restricting the amount of custom software that the SCADA system will require is probably the biggest single action that an Owner can take to reduce costs, risks, and minimize the project timeframe. The Owner's project team needs to work with the various SCADA users to ensure that any custom applications are fully justified, since they will contribute significantly to project cost and risk. For example, in some instances it may be possible to modify a business process to reduce the scope of or eliminate a custom feature.

1.11.2 System Requirements

Thorough planning in the initial phases of a SCADA project, as with any project, are most critical. It is important to develop a clear understanding of why a project is being initiated, who the stakeholders are, what the expected outcomes are, and what the benefits of the SCADA system will be.

In the past SCADA was a basic operational tool to make pipeline operations more efficient and feasible. In today's integrated business environment, SCADA is a key component of many business processes. It is critical to:

- Ensure that all current and potential users of SCADA and SCADA data are involved in the preliminary planning. The project team should include a representative from the operations group.
- Review current business processes to identify possible areas of improvement that may be able to be met by the new SCADA system.
- Review all interfaces to SCADA and the associated business process.
- Identify any new processes that may be required in order to obtain the expected benefits.

If the project is a replacement or upgrade of an existing SCADA system, assess how much of the existing system needs to be replaced. For example, the SCADA host may be replaced but all field RTUs may be retained. A risk identification and assessment should be completed at this stage. Critical risks are identified, quantified, and prioritized. Also, a risk mitigation plan should be developed to minimize the risks and to have a contingency plan in place should any of the identified risks occur.

This requirement and planning phase is a crucial one in a SCADA project. Time invested in this stage will produce benefit throughout the project. Properly executed, this phase will reduce the need for changes after initiation of the project, reduce project risks, and increase the acceptance and usability of the installed system by the operating groups. Finally, all of these outcomes will increase the likelihood of realizing the expected benefits of the project.

The total cost of a SCADA project is a relatively small portion of the overall capital cost of a pipeline project. For this reason, it may not garner the attention and importance that it deserves during a pipeline project. However, it must be remembered that although SCADA is a small portion of the overall pipeline capital cost, the SCADA system will be used every day of the pipeline's operating life and the SCADA system will affect the ability to properly operate and deliver the expected commercial benefits of the pipeline.

1.11.3 Performance Criteria

The system performance can be evaluated in terms of the technical capability and supplier's performance of a SCADA system. Outlined below is a partial list of some key performance criteria:

1. Scanning capability
    - A regular scan rate should be fast enough to satisfy the required response time of the pipeline system. In general, petroleum products with high density require a fast scan rate due to fast transient speeds.
    - A scan update function is needed to refresh real time data at designated locations (most likely all pump/compressor stations) on demand by interrupting regular scanning process.

- A fast scan function is required to poll one location at a faster scan rate than the regular scan rate, particularly vital during an emergency.

2. Accuracy of measured data should be unaffected by the transmission process, with no error being added because of it.
3. For high system reliability, a SCADA system should be nearly 100% available. High SCADA system availability can be achieved with redundancy and regular system backups.
4. The display refresh times should be short during peak and non-peak loading periods.
5. After data has been received from the remote field equipment, alarm response time should be quick.
6. Database and display changes should be possible on-line without interruption in system functioning.
7. A system auditing capability needs to be provided to track database changes and system performance.
8. The display building capability should specify times required to build displays, graphic capabilities displaying pipeline schematics and components, coloring capability for alarms and other displays.
9. The SCADA system should be capable of securing the system and database by assigning various access privileges and restrictions to different groups.
10. SCADA system capability should be easily expanded to accommodate growth of the pipeline system and applications.
11. Third party software interface capability should be specified. Most SCADA systems provide an API (Application Programming Interface) - capability to facilitate easy interface, or use open standards such as SQL (Structured Query Language), ODBC (Open Database Connectivity), OPC (OLE for Process Control), etc.
12. If data exchange with other systems is required, the desired data exchange performance should be specified. These systems may include a backup or distributed control center and shippers.
13. Control commands to supervisory devices should be received at the field devices within a matter of seconds of the operator executing the command. Operator control capability can be enhanced by providing the operator with functions such as time-out of unconfirmed control and command-checking before they are sent.

The SCADA vendor's performance and level of support needs to be included in the performance evaluation criteria.

1.11.4 Roles and Responsibilities

Again, regardless of the contracting strategy employed, it is crucial to have a clear definition of the roles and responsibilities of each of the key parties – the SCADA vendor, the Owner, and the Owner's engineer. The purpose is to ensure that all parties understand their roles and responsibilities and those of every one else. These understandings need to be documented and agreed to before the project execution begins. This will minimize the chances of the project operating on false assumptions. It will help ensure there is no ambiguity as to who is responsible for each task as well as for the problems and issues that arise throughout the life of the project. The responsibility document combined with the Functional Design Specification (see Section 1.11.5) will be a key part of any discussion regarding potential contract change orders. If these documents are clear and complete and have garnered agreement from the concerned parties, they can be referred to with confidence to resolve concerns arising during the project.

The Owner's project team needs to ensure that its key members, including both those from engineering and operations, are full time and do not have other duties that will interfere with their ability to execute the project. Ideally, the operations personnel are seconded from their normal positions for the duration of the project.

1.11.5 Functional Design Specification

This document outlines the functionality requirements of the system. It is not a "how-to" document but rather it is focused on the "what"; it is concerned with functions or outputs of the system, not how they are accomplished. A specification of system performance and functionality will be the guide for the detailed engineering to be completed by the SCADA vendor. It will be the reference document for major design decisions and testing.

The functional design specification will be developed during the front-end engineering phase of the project. The document will address both the functionality of the hardware and of the software for the SCADA system including all interfaces, business process interactions and any application software requirements. Completion of this document and the subsequent approval by all stakeholders is a critical milestone in the project. Since this document becomes the reference document for all subsequent design and technical discussions, it is important that the time and resources necessary are available to ensure its proper preparation.

1.11.6 Testing Plan

In the 1980's contracts routinely specified Factory Acceptance Tests, Commissioning Tests, and Site Acceptance Tests. This was required because the technology was new, expensive and the separation of design and acquisition resulted in a great deal of customization. The modern approach is to use the "design-build" contracts, and pay for performance. A functional test at the end may be all that is required from the perspective of the Owner.

A testing plan is developed to outline testing requirements that will ensure that the system performs as intended and that it is documented throughout the design stage. The specific circumstances of the project will determine the extent of the testing required. All test plans need to be clearly written. Each test procedure needs to be described in detail and include both a description of the expected system response and the pass/fail criteria for each test step. Finally, all test plans need to be fully understood and accepted by all parties.

The test plan may include the following:

1.11.6.1 Factory Acceptance Tests (FAT)

More and more projects are dispensing with the need and expense of a FAT. This is certainly possible if the project is more or less "off the shelf" and does not have a high degree of customization and/or is not a complex system. In addition to the complexity of the system, the vendor's reputation and experience will be considerations in determining the value and need of a FAT. A traditional FAT consisted of a complete installation of all computer hardware and representative RTUs and interfaces to test functionality and response times, etc. A FAT that only tests SCADA functionality may be sufficient if it is coupled with a pre-commissioning SAT.

1.11.6.2 Pre-commissioning Site Acceptance Test (SAT1)

This test will confirm full system functionality prior to initial operation of the SCADA system. This can be especially important on a replacement or upgrade project where it is imperative to ensure that the new system is fully functional before the production SCADA system it is replacing is decommissioned. The plan needs to be carefully thought out and executed to minimize or eliminate interruptions to normal operations.

1.11.6.3 Post-commissioning performance Test (SAT2)

This test is the final test before formal acceptance of the system from the vendor. It will include testing of:

- any functionality that was for any reason not tested during SAT1,
- external interfaces utilizing real operating data, and
- system response and loading test of a fully operational system.

It may also include a longer-term test component to verify availability and reliability conformance.

1.11.7 Installation & Commissioning

If the project is a replacement or upgrade, then particular attention needs to be paid to development of an installation and commissioning plan that minimizes disruption to the existing ongoing operations. This may require the ability to have both the old and new system operational in parallel. Sites would be tested and transferred over to the new system one by one. The old system would only be de-commissioned after completing a full system acceptance test of the new system.

Early in the project, an installation, commissioning and testing plan needs to be developed. This is even more critical if the project is an upgrade or replacement since there needs to be minimal or no interruption of pipeline operation.

1.11.8 Training and Documentation

The vendor is responsible for training the Owner's operators on how to use the SCADA system as well as training the Owner's technical personnel on how to administer and maintain the system. The training of technical personnel can be enhanced by having them involved in the project during system configuration, installation and commissioning.

In addition to thorough training it is important to ensure that the system has been fully documented to ensure ease of on-going operation and maintenance once the vendor is no longer on site. Examples of typical system manuals that the vendor should be expected to supply are outlined below.

1.11.8.1 Hardware Manual

The hardware manual should list all system hardware and configurations incorporating any relevant drawings. As a minimum, this should include:

- Workstation configurations,
- Network configurations,
- Operating system configurations,
- Printer configurations, and
- Third-party hardware configurations.

1.11.8.2 System Management Manual

The System Management Manual will contain a detailed compilation of system management procedures and project installation structures, including the structures of the real-time and historical databases. The System Management Manual should be available in an on-line format (launched from the user interface) in either HTML or PDF format. CD-ROM, printed, or both versions of the user documentation must also be available at Owner's option.

Procedures detailed in this suite of documentation will include the operation of the database editor, the display editor, and the report generator. This documentation should be organized according to the sequence required to build and install the SCADA system. This set of documents will also contain, as necessary, additional modules on installation utilities, customizing computer language(s), and installation administration. The System Management Manual Suite shall also document any specific modules for applications added for this project not included in the baseline documentation that would otherwise be provided by the vendor.

1.11.8.3 Operation and Control Manual

Included in the Installation Management Manual Suite, the Operation and Control Manual will contain a detailed compilation of all installed functions. The Operation and Control manual should be written in non-technical terms and organized for easy access to information. Procedures in this document will explain systematically how varying parameters affects the immediate operation of the SCADA system and its associated specialized applications.

1.11.9 Project Closeout

Project closeout is the final task before the project is officially declared "completed". These would include:

- Formal sign-off between the vendor and the Owner that represents the final acceptance of the SCADA system
- Completion of a final project report
- Archiving and transfer of all project documents and files
- Closeout of any outstanding deficiencies/non-conformities. This will be a combination of accepting some items "as is" and a remedy plan for non-acceptable non-conformities.
- Post-implementation review (PIR)
- Demobilization of project team

A formal project closeout including a post-implementation review (PIR) is something that is rarely undertaken, but should be a mandatory part of all projects. It is important that an assessment be made of how well the system is meeting the organization's needs, as they are now understood. This process may include a post-project review with the vendor as well as an internal review.

The PIR should also review the final system against the initial benefit realization plan created during the requirements phase. With the knowledge and experience gained during the execution of the project, it is beneficial to review the initial benefit realization plan. This review should revise the plan if necessary and ensure that the metrics and methodology is in place for ongoing review of the plan.

This is a final opportunity for the project team to document what they have learned during the project and to identify process and procedures that went well and those that did not. Finally, the team can make recommendations to improve future projects and avoid repeating the same mistakes.

References

(1) Chudiak, G. J. and Yoon, M. "Charting a course in the 90's – From field measurement to management information systems", Proc. of International Pipeline Conference, 1996

(2) Fussell, E., "Wireless Technology a Hot Topic at ISA 2001", Intech, May 13, 2001

(3) The Linux Information Project, December 3, 2005

(4) Mohitpour, M., Szabo, J., and Van Hardeveld, T., "Pipeline Operation and Maintenance" ASME, New York, 2004

(5) NCS, Technical Information Bulletin 04-1, "Supervisory Control and Data Acquisition

(6) Trung, Duong "Modern SCADA Systems for Oil Pipelines", IEEE Paper No.PCIC-95-32, 1995

(7) Ellender, Damon, "Digital Architecture Technology Brings Full Scale Automation to Remote Oil, Gas Fields", The American Oil and Gas Reporter, August 2005

(8) Sang Son, Iannacone, Carmen and Poris, Marc "RTDB: A Real-Time Database Manager for Time Critical Applications", 1991

(9) "21 Steps to improve Cyber Security of SCADA Networks", US Department of Energy, http://www.ea.doe.gov/pdfs/21stepsbooklet.pdf

(10) The Construction Industry Institute, "Reforming Owner, Contractor, Supplier Relationship", Research Summary 130-1, September 1998

2 Measurement Systems

2.1 Introduction

This chapter discusses pipeline measurement systems in the context of centralized automation. It addresses general measurement system characteristics, introduces their measurement devices, and discusses the data required for an automation system. This chapter does not address issues related to the selection and installation of measurement systems. Reference (1) discusses flow measurement subjects extensively with an emphasis on meter selection and installation. This chapter restricts the discussion of measurements to those required for custody transfer. Meter stations are discussed in Chapter 3.

The purpose of a measurement system is to determine a numerical value that corresponds to the variable being measured. Measurements are required for producers, customers and transportation companies. Transportation companies include pipeline, trucking and other transportation media. Pipeline companies charge their shippers for the transportation services based on the measured quantities of the products they have transported, assuming that they satisfy other transportation requirements such as the product quality. Measurements are also required for control and operation of pipelines.

The quantities typically measured for custody transfer and monitoring or controlling facilities are:

- Volume flow rate or accumulated volume
- Mass
- Energy
- Pressure
- Temperature
- Density for liquid or composition for gas
- Quality

What measurement is used to establish custody transfer is dependent on fluids and/or for different regulations; for certain products such as ethylene, mass is measured, and for most liquids it is volume. Natural gas custody transfer in North America is mostly based on volume, but gas transactions in certain areas are based on the energy content of the gas.

These quantities are measured with various instruments using many different techniques. Since flow or volume measurement is most critical for custody transfer, this chapter places more emphasis on the flow or volume measurement than on the other quantities.

2.2 Measurement System and Characteristics

A measurement system consists of four elements:

- sensing element or transducer (primary device) that is mounted internally or externally to the fluid conduit which produces a signal with a defined relationship to the fluid flow in accordance with known physical laws relating the interaction of the fluid to the presence of the primary device
- signal conditioning element (secondary device) that takes the output of the sensing element and converts it into a form more suitable for further processing, such as ampere to voltage conversion and amplification
- signal processing element(secondary device) that converts the output of the signal conditioning element into a form suitable for presentation such as analog to digital conversion
- measured data presentation element (secondary device) that presents the measured value in a form that is easily usable such as on a visual display

A sensing element has certain characteristics that have an effect on overall measurement performance. Measurement characteristics require the understanding of a few definitions:

- The range of a sensing element is the limit over which it runs between the minimum and maximum values of its' input and output such as an input range of 1 to 100 psi for an output of 4 - 20mA.
- The span is the maximum variation in both input and output values such as an output span of 4 to 20mA.
- Hysteresis is the difference in the start and end value of output when input is increased and then returned to the same value.
- Sensitivity is the smallest change in a measured variable that a sensor can properly respond. Modern sensors register such minute changes that it seldom causes a problem for controlling pipeline systems.
- Resolution is defined as the largest change in input that can occur without any corresponding change in output.
- Response time is the time a sensor takes to react to a measured variable whose true values change with time. A short response time is required for controlling process equipment.
- Availability is the mean proportion of time that a sensor or transducer is operating at the agreed level of performance, while unavailability is the mean proportion of time that the equipment is not functioning correctly.
- Calibration is the adjustment of the sensor and/or transducer to improve accuracy and response (e.g., zero level, span, alarm and range).

A sensing element is considered to be linear if measured values establish a linear relationship between the minimum and maximum values. If the measured values deviate from a linear relationship, then the sensor is said to be non-linear. Non-linearity, hysteresis and resolution effects in modern sensors and transducers are so small that it is difficult to exactly quantify each individual error effect. Often, the sensor performance is expressed in terms of error and response to changes. Maintaining operations with a small error is the most important factor in custody transfer, while response characteristics are more important for system control.

2.2.1 Measurement Uncertainty

Measurement uncertainty or errors are inherent in all measurements. The measured numerical value will not be equal to the true value of the variable due to measurement errors. From a custody transfer point of view, measurement uncertainty is critical because it is directly associated with the transaction cost. The pipeline industry deals with measurement uncertainty problems by implementing a technical standard acceptable to all stakeholders.

Measurement uncertainty can be biased and/or random, and change with time and environmental factors such as humidity and temperature. An error bias is the difference between the average and true values. It is directional and must be added or subtracted from the instrument reading. Bias error, if known, can be eliminated by a bias correction process. In practice, it is difficult to determine a true bias error, unless standard equipment such as the equipment at the National Institute of Standard and Technology (NIST) in the U.S. is used.

A random error is called a precision error in the ANSI/ASME PTC 19.1-1985 document. Precision can be improved only by selecting a different measuring device than the one in which the error occurred. Three cases regarding accuracy are illustrated in **Error! Reference source not found.** and are discussed below:

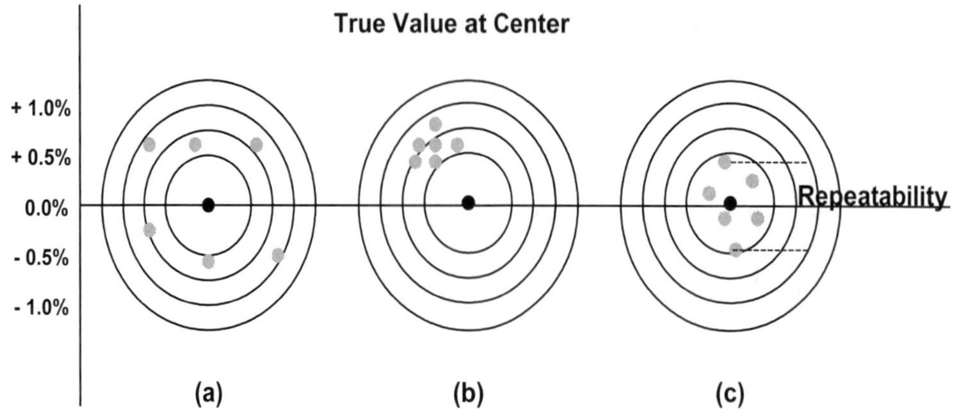

Figure 1: Bias vs. Precision

2.2.6 Transducer/Transmitter

The terms "transducer" and "transmitter" are used interchangeably in connection with instrumentation and measurement but they are not the same. All measuring instruments involve energy transfer and a transducer is an energy conversion device. A transducer is defined as a sensing element capable of transforming values of physical variables into either equivalent electrical signals or a packaged system which includes both sensing and signal conditioning elements and sometimes all four elements listed in Section 2.2. At a minimum it gives an output voltage corresponding to an input variable such as flow rate, temperature and pressure.

A transmitter is a general term for a device that takes the output from a transducer and generates a standardized transmission signal on a transmission medium and is a function only of the measured variable. Like a packaged transducer, a transmitter in a pipeline system amplifies the signal from the sensor and converts it into a more convenient form for transmission.

Certain types of transducers are classified as smart sensors. They contain a dedicated computer which digitizes and linearizes a standardized 4-20mA signal in order to minimize sensor errors. Smart flow transducers combine all of the measured values such as pressure and temperature to correct the flow rate to a reference condition as a way to improve flow measurement accuracy. Smart transducers may have a Transducer Electronic Data Sheet (TEDS), following IEEE 1451.0 standard. The TEDS electronically stores information about the transducer's characteristics and parameters such as type of device, serial number, calibration date, sensitivity, reference frequency, and other data.

2.3 Flow Measurements

A flow meter is a device that measures the rate of flow or quantity of a moving fluid in an open or closed conduit. It usually consists of primary and secondary devices. The secondary devices for flow measurement may include not only pressure, differential pressure, and temperature transducers but also other associated devices such as chart recorders and volume totalizers.

Since volume and flow rates vary with pressure and temperature, the measured volume of a fluid at measured conditions will change with differing pressures and temperatures. Normally, base pressure and temperature conditions are defined for custody transfer in the contract between the parties involved. The correction of measured quantities to base conditions depends on the fluid's properties, particularly density, and thus requires the comparison of pressures and temperatures in order to be calculated. This relationship can be obtained from experimental data or an equation of state, and its accuracy influences the accuracy of the measured value at the base conditions. In North America, API 1101 Volume Correction Factor is often used for hydrocarbon liquids, whereas AGA-8 for natural gas. The flow meters that are popular in the pipeline industry are detailed in this section.

2.3.1 Flow Measurement Devices

The following types of primary flow measuring devices are discussed:

- Differential pressure flow meters such as orifice and venturi meters
- Linear flow meters such as turbine, displacement, ultrasonic flow meters, and Coriolis mass meter.

2.3.1.1 Differential Pressure Flow Meters

All differential pressure flow meters are based on Bernoulli's energy equation. When a flow is constricted, either abruptly or gradually, kinetic energy increases, while potential energy or static pressure is reduced. Flow rate is calculated from the kinetic energy and static pressure differential. Since the flow rate is defined as the multiplication of the flow velocity by the area, it is expressed as a square-root of the measured differential pressure:

$$Q_b = C\sqrt{h_w P_f}$$

where Q_b = flow rate at base conditions
C = discharge coefficient
h_w = differential head or pressure
P_f = absolute static pressure

A differential pressure flow meter registers a pressure differential created by an obstruction in the pipe. The differential pressure transducer measures the pressure differential and determines the pressure drop across the primary device such as the orifice plate. The pressure drop is then converted to a 4-20 mA analog signal. The square root of the signal is proportional to the flow rate.

The turndown ratio or linear range between the minimum and maximum flow rates is limited to 3 to 1 due to the square root relationship of the flow rate to the differential pressure. Since flow rate rather than volume is inferred from differential pressure, a separate flow totalization is required and the accuracy of totalized flow is not well defined. Due to the non-linear relationship of the flow rate to differential pressure, a flow control system requires controller readjustment at different flow rates. Also, pressure loss is permanent and not recoverable for the differential pressure flow meters. Yet, orifice and venturi tube flow meters are popular in the pipeline industry, because they have proven to be reliable and the maintenance cost is low.

1. Orifice Meter

Historically, orifice meters were widely accepted in the pipeline industry, and in terms of installation base they are still most popular. Orifice meter measurement standards such as AGA Report 3 for gas measurement and API MPMS 14.3 for liquid in North America, and ISO 5167-2 in other parts of the world) have been well established. Accuracy is in the order of 1% of flow range.

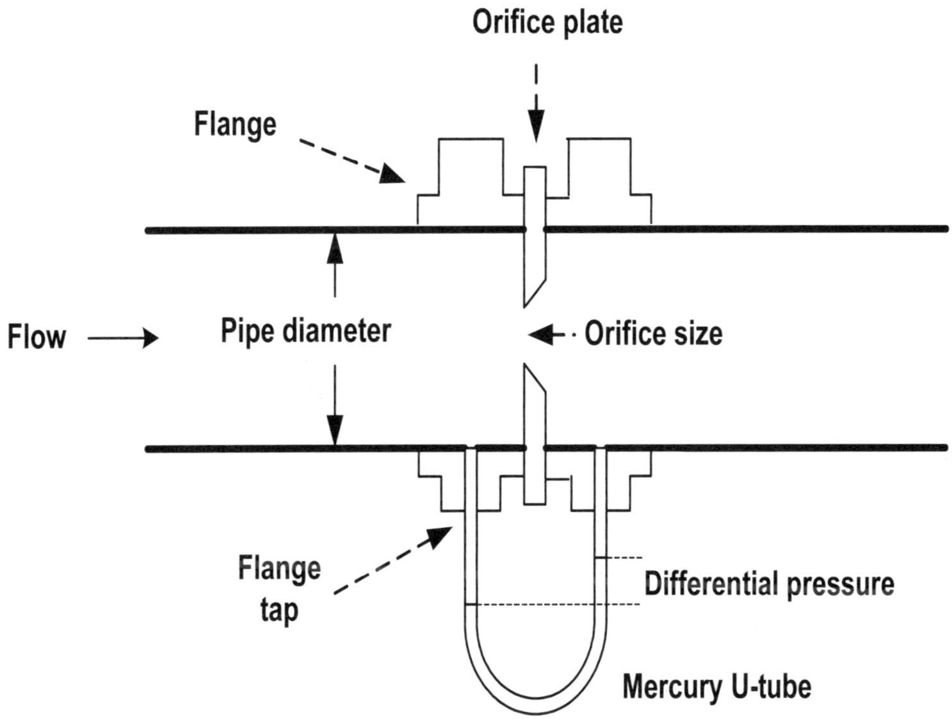

Figure 2: Orifice Meter

A typical orifice meter is shown in Figure 2. The orifice measurement system consists of a meter run, orifice plate, orifice fitting, and upstream and downstream pressure taps. The orifice plate produces the differential pressure, which is measured by a differential pressure gauge such as manometer. The orifice plate is installed inside a pipe. It consists of a hole (normally a circular bore) in a thin circular plate which restricts the flowing fluid. Flange taps are frequently used to obtain greater accuracy and repeatability. The ratio of the orifice plate bore to the inside meter run pipe diameter, normally called the beta ratio, is an important parameter in calculating the flow rate.

The differential pressure across the orifice plate is transferred to a differential pressure transducer or a chart recorder. There are two static pressures: high pressure measured on the upstream side and low pressure measured on the downstream side of the orifice plate. The differential and one of the static pressures are recorded by a flow computer for real time electronic measurement. If an electronic flow measurement is not available, a chart recorder is used to record flows on charts. These charts are removed from the recorder after a certain period for analysis and manually totalize the volume.

The flow rate is determined from the measured differential pressure and orifice discharge coefficient. The orifice discharge coefficient changes with various factors, which are detailed in ANSI API 2430. The orifice coefficient for gas includes the internal energy dissipation during the energy conversion process and includes the following factors (They are expressed here in Imperial Units):

$$C = F_b\, F_r\, Y\, F_{pb}\, F_{tb}\, F_{tf}\, F_g\, G_i\, F_{pv}\, F_a$$

where F_b = basic orifice factor
F_r = Reynolds number factor
Y = expansion factor
F_{pb}= pressure base factor (14.73 psi/contract base pressure)
F_{tb} = temperature base factor (contract base temperature/520°F)
F_{tf} = flowing temperature factor (square root of 520 °F divided by actual flowing temperature in degrees Rankin)
F_g = specific gravity factor (square root of the inverse of specific gravity of the flowing gas
G_i = ideal specific gravity
F_{pv} = compressibility factor (derived from AGA-8 or NX-19)
F_a = orifice thermal expansion factor (this value is normally equal to 1)

In addition to the above factors, the AGA-3 Standard requires the data on flange or pipe tap, upstream or downstream tap location, orifice material, atmospheric pressure, contract base pressure and temperature, orifice size and inside pipe diameters, differential pressure low-flow cutoff, and the pressure ranges that the differential pressure sensors are valid for. Real-time data includes low range and/or high range differential pressure sensor input analog values and status.

Orifice measurement systems are widely used in liquid pipeline applications. The flowing pressure should be kept higher than the vapor pressure of the liquid to prevent vaporization. Since liquids may carry more anomalies and sediment than natural gas, orifice plates and other secondary elements need to be inspected more frequently than do gas orifice meters. The orifice discharge coefficient for liquid is simpler than that for gas because liquid is almost incompressible. The coefficient consists of the following factors:

$$C = F_b\, F_r\, F_{gt}\, F_{sl}\, F_a\, F_m$$

where F_b = basic orifice factor

F_r = Reynolds number factor

F_{gt} = specific gravity factor (temperature dependent)

F_{sl} = seal liquid factor

F_a = orifice thermal expansion factor

A meter station includes one or more meter runs. An orifice meter run consists of both a primary and secondary element. The primary element consists of an orifice plate, a straight pipe of the same diameter called a meter tube, a fitting equipped with tap holes to hold the plate in the tube, and a pressure tap located on each side of the plate. The secondary element consists of transducers that convert values such as pressure differential, static pressure and temperature to an electronic signal. Chart recorders may be used to record the flow rate of the fluid.

An orifice meter has several advantages. It is easy to install, inspect, calibrate and replace if damaged. Orifice plates with different hole sizes are easy to interchange if the measurement of different flow ranges is required. Since there are no moving parts, the complete orifice measurement system is simple and requires minimal maintenance. It doesn't wear out in service. It has been proven in the field and is widely used throughout the natural gas and liquid pipeline industry, even for custody transfer purposes. Its accuracy is within an acceptable range of 0.75% to 1.25% and is repeatable.

However, it has some severe limitations compared to more modern flow meters. It produces a high pressure loss across the orifice plate, which is not recoverable. A small range potential means a large number of parallel meter runs are required in order to measure widely varying flow rates. It is also susceptible to measurement error as a result of liquids in the gas stream and vice versa.

2. Venturi Meter

An orifice plate abruptly changes the flow rate, while a venturi tube changes it gradually, as shown in Figure 3. A venturi meter has a converging section followed by a diverging section. Normally, the pressures are measured at the inlet section where there is no diameter change and at the location with the smallest diameter. The difference between the two pressures is used to calculate the flow rate. The specifications for venturi meters are described in ISO 5167-4.

A venturi meter is similar to an orifice meter in its operation, but can work for "dirtier" fluids due to its' smoothly narrowing tube. Unlike an orifice flow meter, the pressure loss across a venturi tube is low. A venturi meter is not used for measuring gas, but best suited for measuring liquid flow where suspended solids are present. This measurement system and its' installation costs are high. A typical venturi meter is shown below.

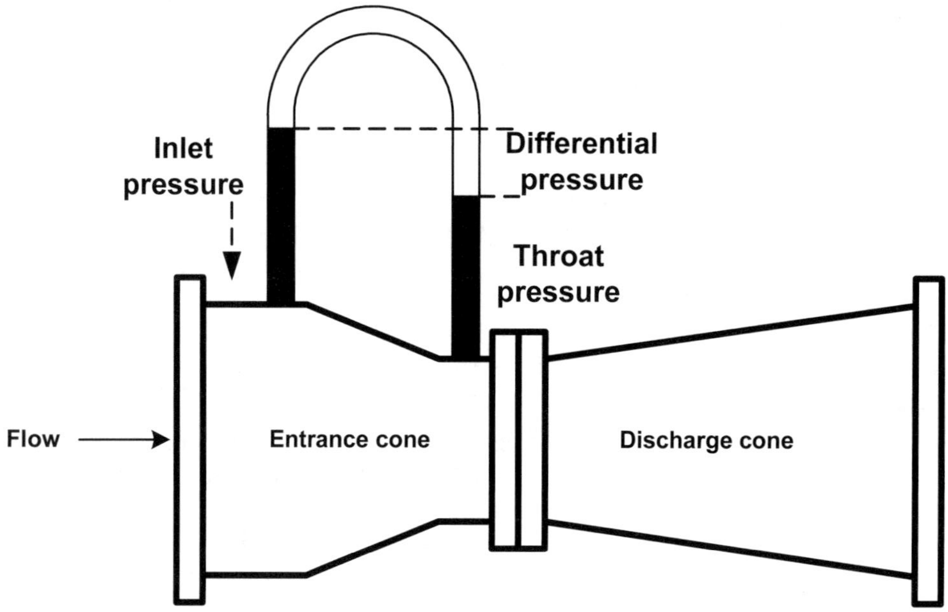

Figure 3: Venturi Meter

2.3.1.2 Linear Flow Meters

For the last thirty years, several linear flow meters have been developed and widely accepted by the pipeline industry. Due to technical advances, they have become more reliable and produce more accurate measurements than when they were first developed. All linear flow meters measure flow volumes directly, based on the principle that the measured volume increases linearly with flow velocity. Turbine, vortex, and ultrasonic flow meters are popularly used, and the applicable flow range is wide (more than 10:1 ratio).

1. Turbine Meter

A turbine meter measures volume directly based on the principle that when a fluid passes over a turbine the fluid makes it rotate proportional to the amount of fluid passing over the turbine at a speed that is proportional to fluid velocity. Turbine rotation is a measure of velocity, which is detected by a non-contacting magnetic detector or by other means.

A turbine metering system consists of a meter run, turbine wheel and housing, bearings, pulse detector, straightening vanes, and pressure and temperature measurement devices. The turbine wheel rotates in the direction of fluid flow. Figure 4 shows the basic construction of a turbine meter. The axis of the turbine coincides with the longitudinal axis of the meter run, which is supported by bearings on both sides of the turbine wheel. These bearings are lubricated by the metered fluid. A permanent magnet embedded in the wheel generates pulses and a small coil mounted on the housing picks them up. Each pulse represents a distinct unit of volume. The total number of pulses integrated for a period of time represents the total volume metered. The straightening vanes provide flow straightening, eliminating the need for long piping upstream and downstream of the turbine meter. A uniform velocity profile is recommended for accurate measurement, but no strict requirements for fully developed flow profiles are required. A pressure tap is located within the turbine meter to obtain a static pressure and a temperature probe on the meter run to obtain a flowing fluid temperature.

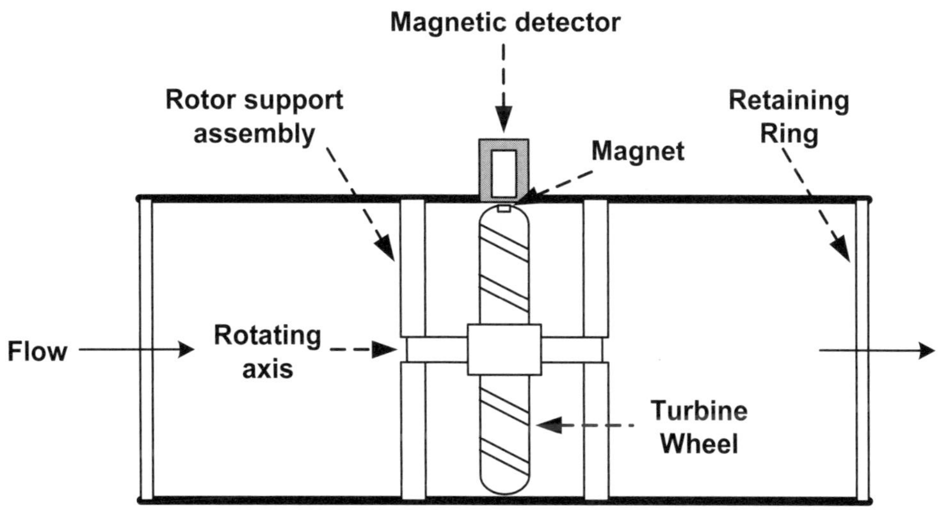

Figure 4: Turbine Meter

The flow rate through a turbine meter is determined using the following equation:

$$Q = V/t$$

where Q = flow rate at flowing conditions
t = time
V = volumetric output over time period t

The volumetric output of the turbine meter is recorded by a revolution counter on the turbine wheel. It is expressed as:

$$V = C/k$$

where C = pulse counts
k = meter factor

The meter factor is expressed as pulses per unit volume. It is unique for each turbine meter and used by a flow computer to calculate the totalized volume through the meter over a given time. The meter factor is a mechanical meter correction factor which accounts for effects such as bearing friction, fluid drag, and many other mechanical and electrical conditions. It is determined by a meter calibration process, using a meter prover under normal flowing conditions.

For custody transfer, the flow rate at flowing conditions should be corrected to the pressure and temperature at base conditions. The flow or volume correction for liquid is simple, but the gas volume correction for turbine meters requires the following equation to find:

$$Q_b = Q\, F_{pm}\, F_{tm}\, F_{pb}\, F_{tb}\, Z$$

where Q_b = flow rate at base conditions
Q = flow rate at flowing conditions
F_{pm} = pressure factor (flowing pressure in absolute/ base pressure)
F_{tm} = temperature factor (flowing temperature in absolute/base temperature)
F_{pb}= pressure base factor (14.73 psi/contract base pressure)
F_{tb} = temperature base factor (contract base temperature/520°R)
Z = gas compressibility factor derived from AGA 8 or NX 19

The data requirements for turbine meters are specified in such standards as AGA-7 and ISO-2715. In addition to the above parameters, the meter factor and real-time data such as pulse counts are required to determine the volume passed through the turbine meter over a specified time period. Therefore, in addition to the measured gas flow or volume, turbine meters require the contract base pressure and temperature, gas composition data or specific gravity, and flowing gas pressure and temperature in order to calculate the net flow.

The liquid volume correction requires an equation of state as specified in API Standard 1101. The liquid volume can be corrected to base conditions using the procedure specified in API MPMS 11.1. Further, there is a minimum operating backpressure level that will prevent cavitation, depending on the characteristic of the specific fluid. A conservative statement of sufficient back pressure necessary when utilizing a turbine meter is given in API Publication 2534.

The liquid volume flowing through a turbine meter is calculated by correcting the raw meter pulses to base pressure and temperature conditions and taking into account the effects of flowing pressure and temperature on the fluid and the meter. The net volume at base conditions is expressed as:

Net volume = (Number of pulses/K-factor) $C_p\, C_t\, M_p\, M_t$

where K- factor is a meter factor obtained from meter proving, pulses/m^3
C_p is pressure correction factor for liquid to base conditions

C_t is temperature correction factor for liquid to base conditions

M_p is pressure correction factor for steel to base conditions

M_t is temperature correction factor for steel to base conditions

C_p and C_t can be determined from the procedures described in API 2534, while M_p and M_t may be obtained from a steel reference manual or the meter manufacturer.

A turbine meter has advantages of high accuracy in the order of 0.25% over the flow range, has a large range of up to 100:1 at high pressure and high flow conditions, of negligible pressure loss across the metering system, and is easy to calibrate and maintain. Turbine meters are most suitable for flow control because of their fast response time to changes. Since turbine meters measure fluid volumes directly, they are known to provide accurate totalized volumes. Because of these qualities, turbine meters are widely accepted for use in custody transfer in pipeline industry.

They do however have certain limitations; they are sensitive to viscosity and their performance is adversely affected by solids or liquids in the gas stream and solid debris in the liquid stream. Therefore, the turbine metering system requires a strainer on the upstream side of the meter run.

2. Positive Displacement (PD) Meter

PD flow measurement consists of a class of devices which measure a specific amount of fluid volume for each cycle. Meters of this design divide the fluid stream into unit volumes and totalize these unit volumes by means of a mechanical counter. The volume displaced during the revolution is multiplied by the number of revolutions to give the accumulated volume passed by the meter.

The method of correcting to base conditions for pressure, temperature, and compressibility/viscosity is the same as that for turbine meters. In North America, the applicable standard for gas is AGA Report 6 (1975), ANSI B109.2 (1980) for diaphragm type PD meters, ANSI B109.3 (1980) for rotary type PD meters, and the standard for liquid petroleum products is ANSI Z11.170 (API Standard 1101). Internationally, ISO 2714 is followed for gas and liquid measurements.

The measurement parameters required for the PD meters are pressure, temperature and density. If the fluid is a homogeneous single product, a proper equation of state, together with the measured pressure and temperature, is used to correct the measured volume to the base conditions.

There are several types of PD meters. A rotary meter belongs to the PD meter class. The fluid flow against the rotating impellers results in a volume of fluid being alternately trapped and discharged in a complete revolution of these impellers. The rotary pistons self-start when the gas flow begins. The rotary movement is transmitted by the magnetic clutch to the totalizer, which adds the

Rotary Meter Diagram

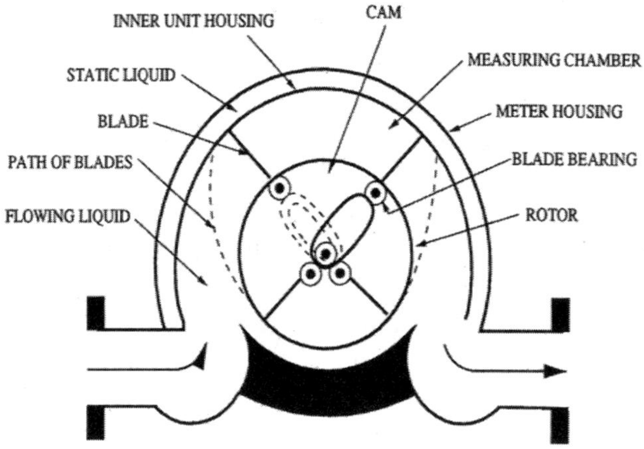

ROTATING VANE METER (DOUBLE CASE)

Rotary Vane Meter

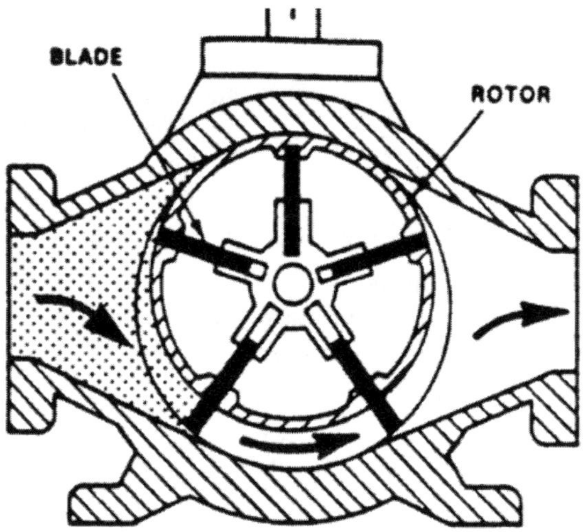

Lobed Impeller Meter

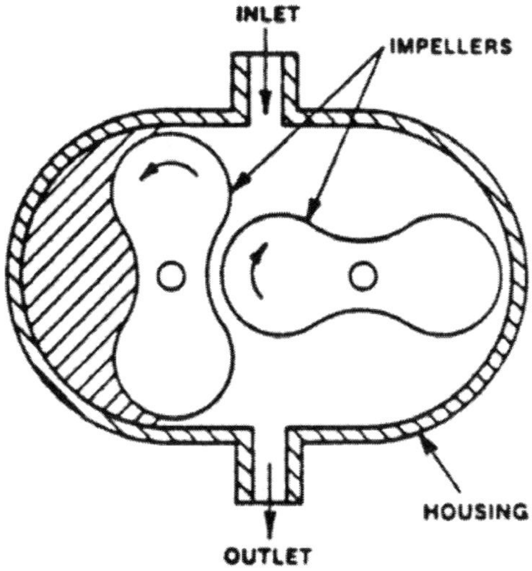

Figure 5 Three Types of PD Meters

number of rotations and indicates the totalized volume. Figure 5 shows the diagrams for a rotary meter, rotary vane meter and lobed impeller meter.

The main advantages of PD meters are:

- Wide applicable range (about 10:1)
- High accuracy (0.5% error)
- Minimum viscosity effects (good for heavy crude measurement)
- Good for low flow rates
- Simple calibration
- No special piping requirement

However, a PD meter can only be used for clean fluids and is expensive to maintain because of its many moving parts. Also, a PD meter with large sizes is relatively expensive.

3. Ultrasonic Flow Meter

Ultrasonic flow meters use acoustic waves of a frequency greater than 20 kHz to measure flow velocity and subsequently flow rates. They operate either on transit time/frequency or on the Doppler effect. The transducers send acoustic waves to the receivers, and acoustic waves propagate upstream and downstream of the flow direction. The range of the flow meter is 20:1 while its accuracy for a multi-path system is better than ±1.0%.

Recently, multiple beams have been used to increase accuracy and repeatability. Multi-path ultrasonic flow meters use more than one pair of sending and receiving transducers to determine flow rates. The transducers send and receive a signal alternately through the same path. Flow rate is determined by averaging the values obtained by the different paths, resulting in greater accuracy and reliability than provided by single-path meters. The applicable standards for gas flow measurement in North America are AGA Report No. 9 and ASME MFC-5M which may also be used for liquid measurements

Ultrasonic flow meters can be classified in terms of the mounting options:

- Insertion flow meters are inserted perpendicular to the flow path, with ultrasonic transducers being in direct contact with the flowing fluid.
- Clamp-on flow meters are clamped on existing pipes. Clamp-on flow meters tend to be less accurate than insertion types, but installation cost is low.

They can also be inserted between two pieces of flanged pipes or threaded into pipes. The ultrasonic transducers can be mounted in one of two modes. The upstream and downstream ultrasonic transducers can be installed on opposite sides of the pipe (diagonal mode) or on the same side (reflect mode).

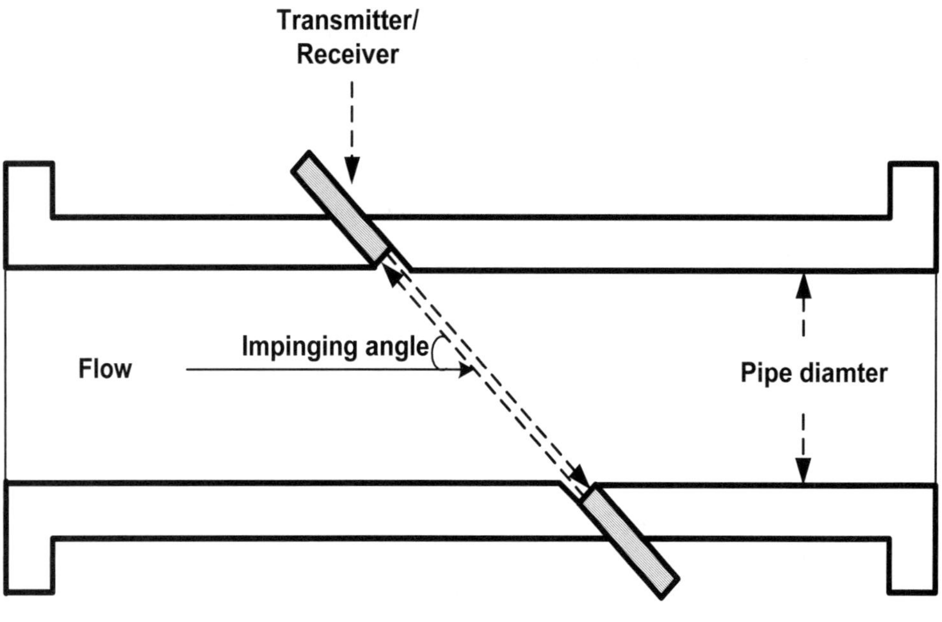

Figure 6: Transit Time Ultrasonic Meter

Transit time ultrasonic flow meters have two ultrasonic transducers facing each other. Two transducers are used with one transducer located upstream of the other as shown Figure 6. Each transducer alternately transmits and receives acoustic waves, and acts as the acoustic wave transmitter and receiver. A pulse traveling with the flow arrives sooner than one traveling against the flow and this time difference is related to the flow speed in the meter. Ultrasonic flow meters measure the difference in travel time to determine flow velocity.

The electronics unit will measure internally, the time it takes for signals to transmit from one transducer to another. At zero flow, there is no difference in time. When flow is introduced, time for the transmission of a signal from the downstream transducer to the upstream transducer will take longer than the upstream to the downstream. Hence we will see a time differential which has a relationship with the velocity of the fluid being measured. Knowing the internal diameter of the pipe, we can now calculate a volumetric flow for the liquid. The flow velocity is calculated as follows:

$$v = D\,(1/t_d - 1/t_u)\,/\sin(2\theta) = D\,\Delta f/\sin(2\theta)$$

where v = velocity of flowing fluid
D = inside pipe diameter
θ = incident angle of acoustic wave
t_d = transit time of downstream pulse
t_u = transit time of upstream pulse
Δf = frequency difference between upstream and downstream pulses

This equation shows that the fluid velocity is directly proportional to the difference between upstream and downstream transit times.

A Doppler ultrasonic flow meter uses the fact that fluid flow causes sound frequency shifts which are proportional to the fluid velocity. Doppler meters also send an ultrasonic signal across a pipe, but the signal is reflected off moving particles in the flow, instead of being sent to a receiver on the other side. The moving particles are assumed to be travelling at the same speed as the flow. A receiver measures the frequency of the reflected signal, and the meter calculates flow by determining the frequency shift of the detected frequency from the generated frequency. Doppler ultrasonic flow meters require the presence of particles in the flow which deflect the ultrasonic signal. Because of this they are used mainly for slurries and liquids with impurities but their accuracy is poor and only applicable to liquids.

The components of ultrasonic flow meters are the transducer and signal processor. The transducer consists of a crystal oscillator transmitter and receiver. The transducer converts the transducer signal to a 4-20 mA analog output signal. The data required to determine the flow rate and volume include static parameters such as flow meter diameter and signal incident angle as well as the dynamic variables such as transit time, pressure, temperature and density.

Ultrasonic transit time flow meters offer the promise of high accuracy, low cost, wide flow range, low pressure drop, and low maintenance because of the lack of moving parts. However, they do not work well for liquids with suspended solid particles or air gaps or for gas with suspended liquids. Doppler ultrasonic flow meters can be used for liquids with bubbles or suspended solid particles.

4. Mass Meters

Mass flow measurement can be made in two different ways: direct and indirect mass measurement. The first approach employs a direct reading of mass flow. The indirect approach measures volume flow rate and density separately at the same conditions and calculates mass flow rate by multiplying the two quantities. This section discusses the direct mass flow meter.

Direct mass flow measurement has the advantage of being unaffected by pressure or temperature, which means no correction to base conditions has to be made for these values. Therefore, the total mass or weight that has passed through a mass flow meter is simpler to determine than volume and the cost is lower because no additional instruments such as a densitometer are required.

Both Coriolis and thermal flow meters measure mass flow rate directly. Coriolis mass meters have been widely used in liquid pipelines recently due to technological advances, and are proven to be a viable option even for use in natural gas custody transfer. AGA Report 11 is a standard applicable to the natural gas pipeline industry and describes the specification for Coriolis mass flow meters. API MPMS 5.6 and 14.7 as well as ISO 10790 standards cover Coriolis mass flow meters for liquid applications. Since thermal mass meters are not reported to be in popular use, this section is concerned with Coriolis mass meters only.

Coriolis force is generated on a fluid element when it moves through a rotating frame. The Coriolis force is proportional to the mass flow rate being produced in the direction perpendicular to the direction of the fluid velocity and rotational vector. The Coriolis mass flow meter measures the force generated by the fluid as it moves through a vibrating U-shaped tube, a rotating frame. The meter induces up and down vibrations of the tube through which the fluid passes. The meter analyzes the frequency and amplitude changes, which represent the mass flow rate. Various designs of the meter are available in the market.

The Coriolis meters contain single or dual vibrating tubes, which are usually bent in a U-shape. The tube vibrates in an alternating upward and downward motion. The vibration frequency is about 80 hertz with a uniform high-low displacement of about 2 mm. The fluid to be measured travels through the tubes, which impart the Coriolis force on the fluid perpendicular to the vibration and the direction of the fluid flow.

As fluid moves from the inlet to the tip of the U-tube, the vertical velocity increases because the displacement is getting larger toward the tip. As a result, the fluid is being accelerated as it moves toward the tip. Similarly, as the fluid moves from the tip to the outlet of the tube, the vertical velocity decreases and thus the fluid is decelerated. The acceleration and deceleration forces oppose each other, creating torque and thus twisting the tube. The angle of twist is linearly proportional to the mass rate of the flowing fluid. When the tube reverses vibration direction, the Coriolis force also reverses direction and so does the direction of the twist. Sensors such as optical pickups and magnetic coils may be used to detect this alternating motion and the magnitude of the twist. A typical Coriolis mass meter is shown in Figure 7.

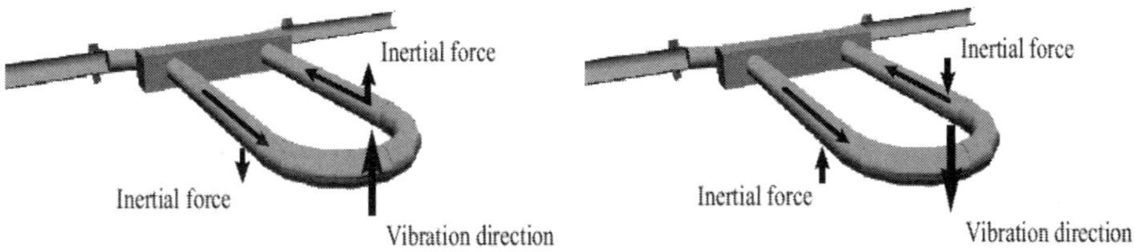

Figure 7: Coriolis Mass Meter

The frequency output is expressed as a pulse scaling factor (PSF), representing the number of pulses for a given mass flow rate. The factor defines the relationship between the mass flow rate and frequency. Coriolis meters have the ability to totalize the mass, by complying with the API MPMS 21.2.

Coriolis mass meters are known to be very accurate and their measurements repeatable and have accuracies of the order of 0.5%, independent of flow profile and composition, and do not require ancillary measurement equipment such as pressure and temperature to determine the mass flow rate. The meters have low maintenance requirements. They are ideal for relatively low flow rate and for custody transfer in mass rather than volume such as for ethylene. They are even suitable for liquid flow measurements with a small amount of gas or vice versa. Even though the meter generally costs more than other types of meters, it doesn't require flow conditioners and thus overall cost is comparable to others. The pressure drop can be high however and thus it may not be suitable for measuring large mass flow rate without its use resulting in excessive pressure drop; this limitation is a factor of the size of Coriolis mass meters, which are not greater than 10 cm or 4 inches.

2.3.2 Flow Computers

Flow was predominantly recorded in the past by chart recorders, particularly in production areas with low flow rates. Lately, due to the rapid development of computer and communication technologies, flow computers are widely used in the pipeline industry. Flow computers not only collect measured flow and other data, calculate volumes, correct flow rates to base conditions, and store all measured and calculated data, but also provide the flow information rapidly on a real-time basis.

The hardware structure of flow computers is similar to personal computers (PC). It consists of a microprocessor-based CPU, RAM, ROM, disk drives, and serial I/O communications ports such as RS232. Such structures need to be rugged due to the severe or even hazardous environments in which they operate. Flow computers are interfaced with flow meters and other measuring devices through their transducers and have programmable capabilities necessary for various applications. Also, they have the capabilities to upload the flow computer data to a host SCADA system and provide measurement data security by setting the security code and authorization.

Unlike PCs, flow computers work in real-time and are dedicated to applications related to flow measurements. Often, the operating systems of flow computers are vendor specific, and most software capabilities are stored in ROM. Generalized measurement related application software can be developed or downloaded onto the flow computers, and specialized application software can also be provided for specific tasks.

Flow computers do not have flexible display capabilities. Their screens are small and a keypad is used to set up and view parameters. Typically, modern flow computers require the following displays, mostly menu-driven, to enter and access the necessary parameters and data:

- The flow computer configuration data such as unit ID, location, elevation, base pressure and temperature, I/O configuration, etc.
- Meter specific parameters such as meter type, meter factor, etc.
- Gas or liquid product property parameters such as API gravity, AGA-8, volume correction factor, etc.
- Communication and modem connection parameters
- Security

- Calibration
- Alarm parameters such as analog and digital input alarm, rate/volume alarm, etc.
- Diagnostics such as analog I/O, digital I/O, calculation, communications, etc.

In general, flow computers perform flow measurement and process calculations, monitor transducer inputs (both analog and digital inputs) in real-time, produce and store multiple measured and calculated output including reports, and can serve as a remote terminal unit (RTU).

Flow computers are able to provide all measurement related functions. They not only read and monitor all inputs of flow and/or volume, temperature, and pressure for most flow meters but also differential pressure from differential pressure meters and pulse input from turbine and positive displacement meters. Input also includes fluid properties such as liquid density or gas gravity, viscosity, and quality and chromatograph data.

For custody transfer, flow computers need to be able to process the flow and other measurement data as specified in the standards appropriate to the measured fluids. They should be able to correct volumes to base conditions and totalize volumes from meter run totals and station totals for each product. For liquid applications, flow computers monitor and store batch operation data, which include batch ID, volume, and batch lifting and delivery times. For gas applications, energy or heating value calculations are required.

In addition to these features, recently developed flow computers can validate input data by using two identical but independent computers and thus correctly provide the intended outputs. Two flow computers are required for continuous comparisons of measurement parameters on a real-time basis.

When a flow computer is used for proving meters, it not only controls the meter prover and calculates the meter factor during the proving time, but also uses the meter factor and K-factor to determine accurate volume. The K-factor is the number of pulses per unit volume and the meter factor a correction applied multiplicatively to the K-factor.

Most flow computers are able to display limited outputs and produce various reports. The minimum required reports generated by a flow computer may include volume totals and quality, batch, alarm, and audit trail reports. They can be directly accessed from the flow computer or uploaded to the host and accessed from the SCADA database.

A good flow computer will give accurate measurement results. It needs to be rugged, economical to install and maintain, and easy to configure and understand. The benefits of a flow computer are:

- enhanced accuracy over chart recorders and integrators

- accuracy achieved by taking into account flow conditions
- instant availability of the required data
- simplified calibration
- high reliability

Flow computers are economical to operate because of the significant savings they afford in labor costs. Some flow computers can control meter proving functions.

2.3.3 Quality of Fluids

2.3.3.1 Quality of Natural Gas

The quality of natural gas is important in gas measurement because it determines not only the accuracy of the volume and heating value of the gas but also correlates to the amount of potential contamination from liquids or particulates, the presence of which may result in measurement problems. Normally, gas pipeline companies are responsible for monitoring and ensuring the quality of the gas being delivered. Gas quality management is detailed in Chapter 4.

2.3.3.2 Quality of Liquid

The quality of liquid is defined differently for different liquids. For example, gasoline is specified for its octane value and diesel for its sulfur contents and cetane number. Contaminants for certain pure products like ethylene are strictly limited to very small amounts of impurities. The quality of liquid is further discussed in Chapter 5.

The following are important factors for most petroleum liquids:

- Basic sediment and water (BS & W) – The amount of BS & W should be limited within the specified percentage in order to avoid various measurement and operation problems including meter accuracy and pipe erosion.
- Air content – Air has to be removed to avoid cavitation problems.
- Transmix – A transmix occurs as a result of the mixing of two adjacent products in a batch operation. Transmixes have to be handled as off-spec products and may be collected in a slop tank or refined again to meet the required specifications.

2.4 Pressure Measurement

Pressure is the force exerted on a unit area. The pressure unit is the kilo-pascal (kPa) in the SI system, and the pound force per square inch (psi) in the Imperial Unit system. There are several pressure related terms (2):

- Absolute pressure – The absolute pressure is the pressure above absolute zero pressure, where a perfect vacuum exists and thus no pressure force is exerted.

- Atmospheric pressure – The atmospheric pressure is the pressure exerted by the atmosphere above absolute zero. The standard atmospheric pressure is referenced to the pressure at sea level, where the pressure is 101.325 kPa or 14.696 psia.
- Gauge pressure – The gauge pressure is the pressure reading referenced from the atmospheric pressure. The absolute pressure is obtained by adding the atmospheric pressure to the measured gauge pressure. Normally, the measured pressure is expressed as gauge pressure.
- Differential pressure – Differential pressure is the difference between two pressures, often measured as the difference in heights of a liquid column in a manometer.
- Static pressure – Static pressure is the pressure of a fluid at rest or in motion. If the fluid is moving, the static pressure is only the pressure component perpendicular to the flowing direction.
- Dynamic pressure – The dynamic pressure is the pressure caused by the kinetic energy of the flow parallel to the flowing direction.
- Total pressure – The total pressure is the sum of the static and dynamic pressures.

A pipeline system is a combination of pressure vessels consisting of the pipe and equipment such as pumps and compressors. Thus pressure is the most important measure of a pipeline state, requiring frequent measurements. Pressures are used for pipeline system control and correcting flow rate to base conditions. For differential pressure flow meters, two pressures or differential pressure measurements are required to calculate flow rate.

Piezoelectric pressure sensors are popular for pipeline applications. They are based on the principle that an external pressure exerted on piezoelectric crystals causes elastic deformation, which is converted into an electric signal. The electric signal is conditioned through an electronic circuit. Materials used for the pressure sensing element are quartz, barium titanate, or tourmaline crystals.

Piezoelectric pressure sensors measure dynamic pressure and generally are not suitable for static pressure measurements. They are accurate (in the order of 1%) and response time is fast. The range is up to 20,000 psi. The sensors are easy to install and use, and their ruggedness is suitable for most pipeline applications. Piezoelectric pressure sensors are available in a variety of configurations for installation with different types of pressure measurement devices. A typical piezoelectric pressure transducer is shown in Figure 8.

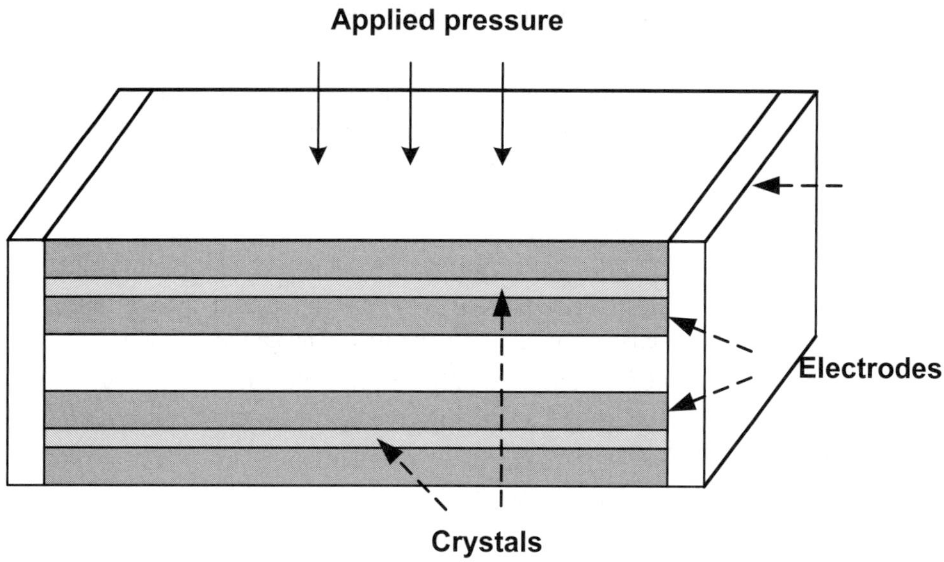

Figure 8: Peizoelectric Sensor

2.5 Temperature Measurement

Temperature is a measure of the thermal energy of the fluid in a pipeline system. The common SI unit for temperature measurement is the Celsius scale, where 0° is the freezing point of water and 100° the boiling point of water under atmospheric pressure. In the Fahrenheit scale, the freezing point is 32^oF and the boiling temperature 212^oF. The Fahrenheit to Celsius scale conversion formula is

$$^oC = 5 \times (^oF - 32) / 9$$

Both the Celsius and Fahrenheit scales are relative because they choose the freezing and boiling points of water arbitrarily. Quite often, it is necessary to use an absolute temperature scale instead of relative scales. An absolute scale finds its zero point at the lowest temperature that is attainable with any substance according to the laws of thermodynamics. The absolute scale in SI units is called the Kelvin scale, and in Imperial units it is the Rankin scale. An absolute zero on the Kelvin scale is -273.15^oC and that on the Rankin scale is -460^oF.

Temperature is used for flow calculation and correction. It is used for compressor discharge temperature control, but seldom used for a liquid pipeline system control. Therefore, temperature measurements are not widely available for most liquid pipeline systems, unless the systems transport heavy crude, whose viscosity is strongly dependent on temperature and thus requires temperature control.

Resistance temperature detectors (RTD) are popular for pipeline applications, because they are simple and produce accurate measurements under normal pipeline operating conditions. Not only is the RTD one of the most accurate temperature sensors for pipeline applications, but it provides excellent stability and repeatability. RTDs are relatively immune to electrical noise and therefore well suited for temperature measurement in industrial environments. Their sensitivity is in the order of 0.02°C and applicable range between -150 °C and 600 °C, depending on the RTD materials.

RTDs operate on the thermoelectric effect that certain materials exhibit linear changes in electrical resistance in response to temperature changes over certain temperature ranges. This electrical resistance property of certain materials is reproducible to a high degree of accuracy. Resistance element materials commonly used for RTD are platinum, nickel, copper and tungsten. Platinum is used for wide temperature range, but since copper is cheaper than platinum it may be sufficient for a range up to 120 °C.

In practice, two RTD elements with different conductivity form a thermocouple. A thermocouple is a sensor for measuring temperature and the temperature difference between two points but not for measuring absolute temperature. A thermocouple joins two elements, one at each end of the conductor to make an electric current flow continuously. One of the ends is the measuring junction and the other is the reference junction. When the temperature changes at the junction of the two metals, an electric current flows in the thermocouple and produces a voltage that can be correlated back to the temperature. Because a thermocouple measures in wide temperature ranges and can be relatively rugged, thermocouples are also widely used in industry.

Often, a thermocouple can be installed on the pipe surface with or without insulation. Sometimes a thermocouple cannot come into direct contact with the measuring fluid because the environment may be corrosive, erosive, or vibrating. To protect the thermocouple, a thermowell is used. It is a tube into which a thermocouple is inserted. Thermowells allow the replacement of measuring elements from the measuring position.

2.6 Density Measurement

Due to increasing prices, custody transfer of certain hydrocarbon liquids by mass measurement is growing in importance because it is more accurate than by volume. Such liquids are pure products such as ethylene, propylene and high vapor pressure (HVP) products such as ethane and propane. However, direct mass measurements for large mass flow rates are not practical, because the application of Coriolis meters is limited to small diameter pipes. Therefore, the majority of mass flow measurements are determined by measurement of volume flow rate and density.

Density can be measured either directly or calculated from pressure and temperature using an appropriate equation of state if composition data are available. It has been common practice to measure the fluid density in the laboratory and calculate densities at other operating pressures and temperatures in pipelines where the operating conditions are well defined. However, calculation of density sometimes can result in inaccurate values for certain liquids or be difficult for liquids such as ethylene and ethane when these products are in their supercritical state and where their density is very sensitive to small changes in temperature or pressure. Therefore, direct measurement of the density is preferred for most hydrocarbons including pure hydrocarbons and liquids of unknown composition or mixtures.

A choice of densitometers, each using a different method is available for measuring density: densitometer measuring the mass and dividing it by the known volume, dielectric property, or variation of vibration frequency with density. Since a Coriolis meter is capable of measuring density, it can be used as a stand-alone densitometer. Density of fluid may be determined by a pycnometer, which is a vessel of known volume that is filled with a fluid and weighed. The density is calculated from the fluid mass and vessel volume.

The vibration frequency type densitometer is widely used in the petroleum industry. The sensing element of a vibration frequency densitometer is immersed in the product and vibrates at its natural frequency creating its resonant frequency. As the density of the product changes, the vibrating mass changes and so does the resonant frequency. The various resonant frequencies are correlated by a calibration process to obtain the density.

The density of natural gas may be measured in the same way using a pycnometer or vibration frequency densitometer. It has been reported that the uncertainty of these densitometers can be as low as 0.1% if the instruments are calibrated accurately (5). If the gas compositions are known, gas density can be accurately calculated from AGA-8 or an ISO equivalent equation. If only gas specific gravity and heating value are known, NX-19 or its equivalent equations can be used to calculate the density of natural gas with reasonable accuracy. The API MPMS 14.6 addresses installation and calibration of density measurement. These densitometers are now frequently used to measure fluid density on-line under the conditions of operating pressure and temperature.

2.7 Chromatograph

The determination of fluid composition is very important in establishing what the flowing and thermal properties of a fluid are. Density and viscosity directly affect hydraulics and heating value. These quantities and other thermodynamic variables can be calculated from the composition. The most popular method of determination of the composition in use in the pipeline industry is gas chromatography.

A gas chromatograph is an instrument that determines the components of a gas mixture. Gas chromatographs can be operated either off-line in laboratory conditions or on-line in field conditions. Better accuracy can be achieved in laboratory conditions than in on-line conditions.

References

(1) ANSI/ASME PTC19.1-1985, "Measurement Uncertainty"

(2) R. W. Miller, "Flow Measurement Engineering Handbook", 3rd Ed. McGraw-Hill, New York, N.Y., 1996

(3) American Gas Association, AGA Report No. 11, "Measurement of Natural Gas by Coriolis Meter"

(4) Nored, M.G., et al, "Gas Metering Payback", Flow Control, Feb. 2002

(5) Jaeschke, M. and Hinze, H. M., "Using densitometers in gas metering", Hydrocarbon Processing, June 1987

3 Station Automation

3.1 Introduction

This chapter presents an overview of the key aspects of the automation of major pipeline facilities for both gas and liquid pipelines. This is not a design primer but rather an introduction to the major considerations and characteristics associated with the automation of such facilities.

A typical station automation system consists of several components; station control, unit control, driver control, storage control, equipment control, and/or meter station control. The detailed control elements of pumps, compressors, meter station and their auxiliaries are discussed only when knowledge of them is required to understand the automation of a station. Mohitpour et al (1) discuss the control of pumps, compressors, and auxiliaries in detail.

The term "station" is used to mean a major pipeline facility that has some combination of equipment, measurement, and automation. Pipeline "stations" can vary from a relatively simple block valve site to a complex multi-unit pumping/compression station. A station can be operated locally as well as be interfaced to a SCADA system to enable remote control from a central control center.

There are many similarities between the automation hardware and operator interface hardware for a compressor station and a pump station, but the specific control requirements are quite different. A turbine-driven compressor is discussed as it is used for the compressor station and an electric-motor-driven pump reviewed as used in a pump station. This will give the reader an opportunity to review the unique design features of each, as there are some significant differences in the control systems and interfaces between the two types of drivers.

3.2 Design Considerations

As with any project, it is important to take time with all stakeholders at the beginning of the project to confirm their requirements, identify design requirements and constraints, and agree on what the expected benefits for the automation system are. The station automation system is the starting point for any business process that relies on obtaining field information at the SCADA or corporate level.

It is now generally accepted practise that stations are automated and operated under remote control from a central SCADA control center. Only under abnormal conditions or during some maintenance tasks will the station be under local control. Some stations may be completely unmanned whereas others will have maintenance staff on site but who will not normally be in control of the station equipment.

A properly designed remote control system will provide the ability to:

- Monitor all equipment associated with the station including station auxiliary systems
- Have two-way communication between the station and the host
- Monitor starting and stopping sequence of the drivers and compressor or pump units
- Control and monitor sequencing of station valves
- Initiate an emergency shutdown of the station or unit

These extra system control capabilities can meet the following objectives for station operation:

- Operate the station safely and reliably, while maintaining cost efficiency
- Allow constant monitoring of critical components of the station
- Shorten response time to potential problems
- Eliminate mundane tasks for the station operators

Unplanned outages can cost a pipeline company tens of thousands of dollars per day. The station automation system must be reliable, robust and have a high level of availability in order to minimize business interruptions and maintain a safe environment for personnel and equipment. It must also be able to transfer control from "remote" to local in the event of an emergency or an abnormal situation.

Smaller stations such as meter stations or valve control stations will typically not have a two tiered control system and will be implemented using a RTU with control capability or will utilize a PLC providing both local control and RTU functionality. For larger pump or compressor stations, economics and required functionality will be key factors in choosing between a DCS or PLC based design for the station control system.

Pipeline system control requires the selection of a control strategy. The strategy depends on the type of fluid (gas vs. liquid), type of prime mover (fixed vs. variable speed), type of controlling station (meter station, compressor station, pump station, backpressure controller, etc.), and location of a controlling station and pipeline system (delivery junction, steep terrain, permafrost zone, etc). The control variables are pressures, flows and temperature.

Since the hydraulic effect of fluid density is not significant for gas pipelines, the discharge pressure is the primary control variable at compressor stations. On the other hand, the discharge temperature can exceed the maximum temperature level due to high compression. This situation requires a temperature control to turn on a gas cooler in order to reduce the discharge temperature to the tolerance temperature. Temperature control is not required for most liquid pipelines except when the fluid is heavy crude with high viscosity or when the pipeline runs along a permafrost zone.

For a pump with a constant speed prime mover, the pump head is fixed for a given flow and thus the discharge pressure can be reduced by throttling the flowing fluid. The throttling action is performed by a pressure control valve, installed downstream of the pumps. The pressures discharged from pumps or compressors with variable speed drivers are controlled by the speed of the drivers with maximum power override.

Flow is the primary control variable at meter stations, but a delivery meter station should also control and maintain the required minimum pressure or contract pressure. A centrifugal compressor requires flow control to avoid a surge condition. This is accomplished by increasing the flow rate. The flow can be increased by recycling part of the discharged flow back into the compressor inlet. A pump does not require flow control as long as the flow is within the pump's capacity.

Side stream delivery may disrupt the main line pressure. To avoid potential pressure disruptions, the main line pressure is controlled by holding the delivery pressure. If a liquid pipeline runs along a terrain with a steep elevation drop, the pressure around the peak elevation point drops below the vaporization point, creating a slack flow condition, unless the high pressure downstream of the peak point is allowed. A backpressure controller is needed at a location downstream from the peak point to avoid this condition, if the pipe strength is lower than the backpressure.

Pipeline system control is accomplished by means of a set point mechanism. In other words, the dispatcher sets pressure, flow or temperature at the desired level and the control system responds to reach the set point. Since pressure is the primary control variable, several pressure set points are discussed below. The controlling pressures, that can be monitored and changed by the dispatchers through the SCADA system, are:

- Suction set point: the desired suction pressure at the station. During normal operation, the suction pressure is equal to or higher than the suction set point. The control system doesn't function properly if the suction pressure is less than the set point, unless the pressure measurement is erroneous. For liquid pipelines, the suction pressure control with discharge pressure override is commonly used to maintain the pressure above the vapour pressure and at the same time the pressure below the maximum allowable operating pressure (MAOP). Normally, the minimum suction set point is higher than the station trip pressure below which the station automatically shuts down.
- Discharge set point: the desired discharge pressure at the station. The discharge set point is the pressure that the station control system tries to maintain as a maximum value. No control action takes place if the discharge pressure is below the discharge set point. For a pump with a constant speed driver, a control valve is used to control the discharge

pressure. The discharge pressure is equal to or lower than the pump casing pressure, and the difference between the casing and discharge pressures is called the throttle pressure. Normally, the maximum discharge set point at a station is lower, say by 100 kPa, than the maximum operating pressure, in order to avoid an accidental station shut-down.

- Holding pressure: the holding pressure is set to maintain a desired main line pressure at the junction where a side-stream delivery may take place.
- Delivery pressure: the holding pressure at a delivery location without a pump station.

3.3 Station Control System Architecture

Although the details of the actual control will be quite different between a compressor station and a pump station, the architecture will be similar. Figure 1 shows the typical station architecture for a multi-unit pump/compressor station, typically with at least three types of control modules within a station. First, there is a unit control system in a self-contained control system usually supplied by the pump/compressor vendor that controls all aspects of the driver-pump/compressor unit. It will control the start-up and shutdown sequencing, and the dedicated unit auxiliary equipment. In addition, it will maintain the unit within operating limits and maintain the set point provided from the station controller.

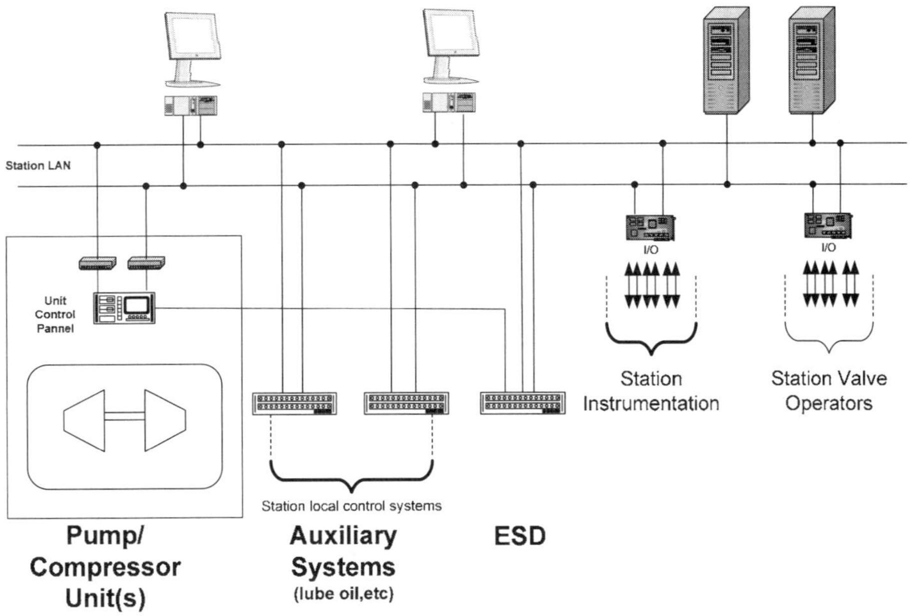

Figure 1 - Typical Station Control Architecture

Second, there are auxiliary control systems for smaller equipment used for specific processes such as lube oil conditions, station air compressing, and fuel gas conditioning. These may be locally controlled or have only local instrumentation for monitoring purposes. At this "level", there are also dedicated control systems such as a Fire/Gas detection system and ESD controllers.

All equipment required to support the pump/compressor units and auxiliaries would be monitored and controlled via the station control system. All other station equipment systems such as heating and ventilation, security monitoring, etc. would also be monitored and controlled from the station control system.

The control system hardware for a pump station is almost identical to that for a compressor station. The number of I/O points for the unit and station at the compressor station may be slightly greater. However, for both station types all the key functions, such as serial communications, LAN communication, operator interface, and SCADA interface, would be provided in a similar manner.

The machine monitoring equipment will also be similar. However, because gas turbine-driven compressors rotate at higher speeds than electrically driven pumps, they require more instrumentation and monitoring than the latter.

3.4 Control Solutions

3.4.1 DCS vs. SCADA

In any discussion of station automation, and to a lesser extent, SCADA systems, the idea arises that a SCADA system is really a distributed control system (DCS) or it and a pump station control can be implemented using a DCS. Before addressing these questions we need to understand what a DCS is and what are the differences between a DCS and a SCADA system.

The goals of DCS and SCADA are quite different. It is possible for a single system to perform both DCS and SCADA functions, but few have been designed with this in mind, and therefore they usually fall short somewhere.

A DCS is process oriented. It looks at the controlled process (the chemical plant or thermal power plant) as the center of the universe, and it presents data to the operators as part of its job. SCADA is data-gathering oriented; the control center and operators are the center of its universe and the remote equipment is merely there to collect the data - though it may also do some very complex process control.

DCS systems were developed to automate process control systems. These systems are characterized by having many closed loop control elements controlling an analogue process in real time. The key differences and characteristics of DCS and SCADA are:

- A DCS normally does not have remotely (i.e. off-site) located components and is always connected to its data source. Redundancy is usually handled by parallel equipment, not by the diffusion of information around a distributed database.
- SCADA needs to have secure data and control over a potentially unreliable and slow communication medium, and needs to maintain a database of 'last known good values' for prompt operator display. It frequently needs to do event processing and data quality validation. Redundancy is usually handled in a distributed manner.
- When the DCS operator wants to see information, he usually makes a request directly to the field I/O and gets a response. Field events can directly interrupt the system and the operator is advised automatically of their occurrence.
- The majority of operations, such as start/stop commands and alarm detection of a SCADA system are digital. They also gather/poll analogue readings but do not implement closed loop control; humans determine if set points need to be adjusted. A DCS is process control oriented and therefore is designed to be able to implement many control loops as well as standard operator initiated start/stop commands.
- A DCS does not poll data but rather needs to be able to process a high number of transactions at a high speed in order to implement multiple real time closed loop control.

3.4.2 Programmable Logic Controllers (PLCs)

PLCs are a control system that consists of a programmable microprocessor unit, communication modules, and input/output modules for connection to field devices. They were first developed to provide a flexible and economic replacement for the traditional relay-based control systems. Their functionality and capabilities have expanded and PLCs are now used as RTUs on SCADA systems, the heart of local control for field equipment (pump drivers, lube oil systems). They can also be networked to provide a complete control system for a complex station (See Chapter 1).

At the same time, the architecture of DCSs has evolved and now they are no longer only economic for large installations and can be a solution choice for larger pump and compressor stations. DCSs would certainly be considered for installations where there is a station and an associated processing facility or a refinery that would utilize a DCS for its control.

The traditional boundaries between various control system solution options have become blurred due to the flexibility of today's automation equipment. For small systems, the control system will generally be implemented using a PLC. As the facility gets larger and more complex, choosing between installing a control system using networked PLCs or a DCS system requires an experienced automation designer to work closely with the end user to ensure the operating requirements are met while at the same time the design dovetails with corporate business information gathering and processing.

3.5 Interfaces

3.5.1 Equipment Interfaces

A station will contain a number of different equipment types, some of which will have their own control systems requiring an interface to the station control system. These systems can range from a complex unit control system to a relatively simple closed-loop controller. A station control system must be able to handle the range of interface "standards" that exist at the plant level.

Similar to the growth and maturation of the SCADA industry, the field instrumentation and control industry continues to evolve and mature. There have been attempts to standardize the interface protocols for field applications. The more popular protocols encountered include OPC, FOUNDATION™ Field Bus and Modbus.

Another significant change in the process industry has been the growth of intelligent electronic devices (IEDs). Whereas traditional instruments and control loops utilized an analogue connection (4-20 mA loops or a voltage output) for transmitting their values to a control system or RTU, IEDs use digital communication. This has the advantage of eliminating the A/D conversion at the control system, reducing noise impact on the field wiring, and allowing for direct communication with each field device. This final feature allows for remote calibration checks, upload of data to the instrument and the direct interrogation of the device via the internet.

3.5.2 RTU/SCADA connection

The interface between a station control system and a SCADA system will vary depending on the nature of the SCADA system and the age of the technology used in the station controls.

In older systems, a separate RTU was installed that was then hard-wired to the station control system to connect station digital and analogue I/O. An improvement to this arrangement was enabling a serial connection to be made between the two devices to exchange data and control commands. The station controller would typically have a protocol converter installed in order to communicate between the RTU and the station controller.

Modern systems are likely to utilize a WAN to connect the SCADA system to the station control system and have a common interface protocol. This interface may be set up so that even if a station is in "local" mode, status and process values could still be transmitted to SCADA for monitoring and logging.

3.6 Common Station Control

This section deals with topics that are common to both pump station and compressor station controls. These stations may also contain some form of storage, which is discussed in Section 4.9.

This section deals with topics that are common to both pump station and compressor station controls. If a station has more than one pump or compressor unit, it is operated in one of two modes: parallel operation and series operation. The primary purpose of operating units in parallel is to allow a wider range of flow than would be possible with a single unit for systems with widely varying flows. Parallel operation is shown in Figure 2, where more than one unit can be operated at the same time. When two or more units operate in parallel, all units have the common suction and discharge pressures.

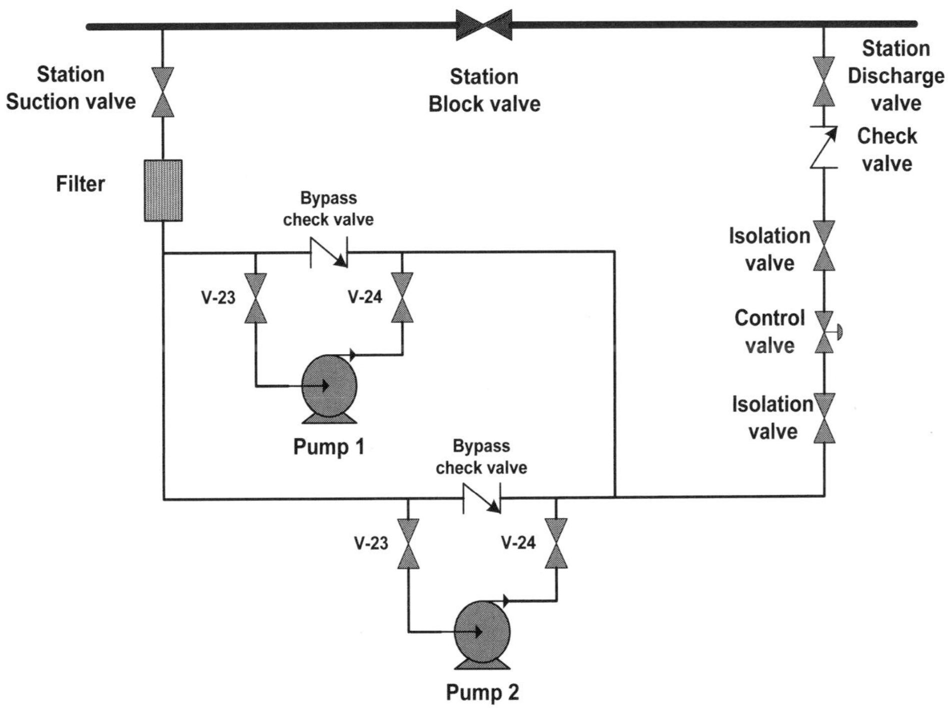

Figure 2 Parallel Operation

The main reason for operating units in series is to increase the pumping head from what would be possible with a single unit. Series operation is shown in Figure 3. In series operation, the flow through all of the units is equal and the discharge of one pump feeds the suction of the next unit.

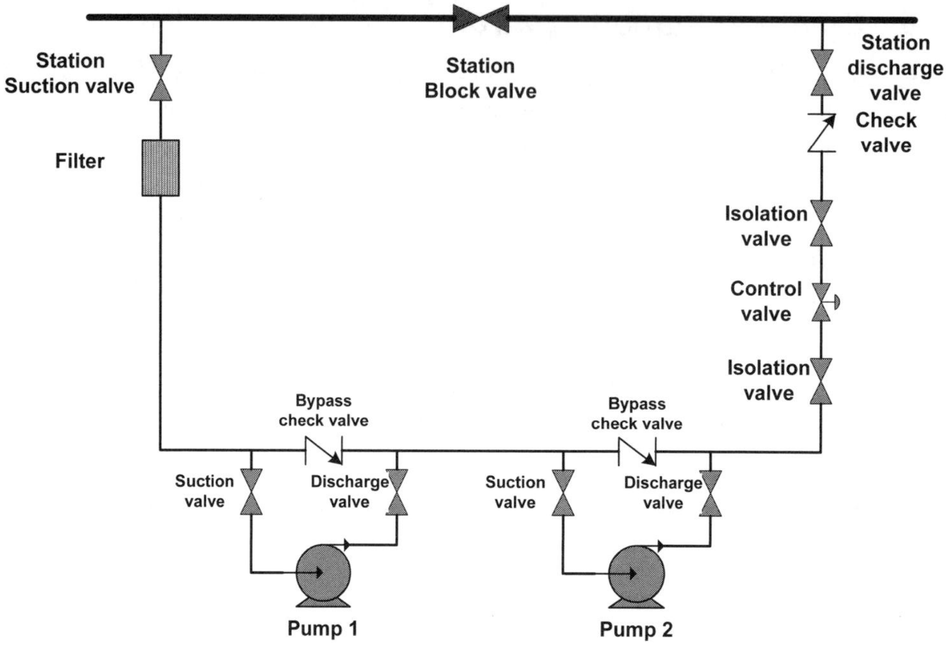

Figure 3 Series Operation

3.6.1 Station Operation

From the perspective of the overall operation of a pipeline, a pump or compressor station can be viewed as a "black box" that maintains product flow by offsetting pressure losses in the pipeline. The pipeline operator may only be interested in setting pressures at the various stations and not be concerned with the control of the individual units. In this situation, the station control system would receive station set points rather than individual unit set points from the SCADA system. It would then determine how many units should be operating and the set points for each unit.

An alternate control scheme is to include the station control system within the SCADA system. The system operator would then be initiating start/stop commands and relaying them to individual units as well as sending them the required set points.

The station control system has overall control responsibility for the station. This control includes all equipment not under the direct control of a unit control system. The station control ensures that the station operates within the parameters for the station and mainline piping (above minimum inlet pressure, below maximum allowable operating pressure, etc). In addition, it determines the required set points for the operating units based on the required station set points received from the pipeline operator via the SCADA system. The individual set points sent to each unit will be determined based on a load sharing strategy. This will vary depending on the type of units installed and the overall pipeline operating strategy. They may include such strategies as:

- Base Loading: one or more units may be operated at a constant load while other more efficient units used to compensate for small changes.
- Optimum load sharing: set points for each unit are determined based on knowing the individual unit operating curves and allocating load to minimize overall energy consumption.

With the increase in computing capability, it is now more common for pipeline companies wanting to optimize their pipeline operations to consider having a system optimizer that would optimize pumping (or compression) usage on the entire pipeline. This is discussed in more detail in Chapter 6.

3.6.2 Control Modes

In control design practice, there are two main methods of failure control. One is called "fail-safe" and the other is called "status quo" design. The fail-safe design ensures that all equipment moves to a predefined safe state in the event of a failure of the control system element. This failure could be on an open circuit, a processor shutdown, or a power failure. In order to accomplish this, most circuits are normally energized.

The status quo design ensures that the loss of a signal will not cause a shutdown. Safe status quo designs usually have redundant paths for tripping in case there is a failure of one of the tripping devices.

Generally, there are three major levels of monitoring and station control in the hierarchy of automated pipeline stations, namely:

Local: In this mode, command control is limited to the local device or the skid control panel. For example in the case of a compressor station in "local" mode an operator would have to be at the engine control panel in order to initiate any control commands to the gas turbine.

Remote: In this mode, command control of all local devices and skids is passed to the station control system. This allows a local operator to control the complete station and all auxiliary equipment from a single location at the station. If the station control system is in "remote" mode, then in effect all control is from the SCADA control center.

Control center/SCADA: In this mode, command control of the station is passed to the central control center via the SCADA system. No local control is possible. This is essentially the "remote" mode for the station control system. Process values and status may still be sent to SCADA for monitoring and logging purposes.

It is important to realize that the control levels described here affect the state of operator control. In all modes, the local device is always being controlled by its control equipment. The change of mode describes from where, for example, set points or commands to the local controller will originate.

3.6.3 Shutdown Modes

A typical arrangement for station controls is to have different levels or severity of shutdowns such as:

Normal Shutdown: This will shutdown the equipment through a normal shutdown sequence. The unit can be restarted normally. This would be initiated by an operator command or may be required if process conditions exceed limits. Once process conditions have been restored, the unit can be restarted.

Shutdown Lockout: This is activated to stop a unit for a serious problem such as loss of lube oil, etc. "Lockout" means the unit cannot be restarted until manually reset locally. This ensures that the site is visited by a technician/operator, who must evaluate the situation before the unit can be restarted. This can apply to individual units or the complete station. Wherever possible, the shutdown will follow normal shutdown procedures to minimize hydraulic disturbances.

Emergency Shutdown (ESD): This condition requires immediate shutdown of all units and will initiate a hydraulic isolation of the station. In a natural gas pipeline, this will also result in the activation of associated blow-down valves. Following an emergency shutdown, all controls will be in a lockout state and require local resetting.

3.6.4 Station Valves

The station valve control logic is included as a part of the station control system. Typically, each valve is able to be in either local or remote mode but will normally be in remote mode. The control logic uses valve position indication to interlock valve operation. This logic ensures that valves are opened and closed in the proper sequences for putting a pump or compressor station on line, for launching or receiving a pipeline inspection tool, for bringing additional units on line and for launching and receiving batches.

Some pump/compressor stations may also be terminal stations with interconnections to other pipelines or tank farms. These also require valve control logic to ensure the proper operation and flow of product to the correct destination.

The valving control logic incorporates interlocks with motor-operated valves to ensure proper sequencing and to avoid damage to equipment. Some sequencing scenarios that the control system contains include:

- Scraper launching and receiving
- Station start up and shutdown
- Station by-passing
- Batch receiving and batch launching

In addition, there may be some control logic required to help minimize or reduce pipeline surges (transients) depending on the results of the pipeline hydraulic studies (2).

3.6.5 Station Auxiliary Systems

The station control system controls and monitors the functioning of all station auxiliary systems common to the operation of all units. These systems include some or all of the following equipment, depending on the specific station requirements:

- Auxiliary (emergency) electrical generator
- DC Battery charger(s)
- Inverter
- Security system
- Boiler (if required)
- Air-conditioning
- Commercial AC power monitor
- Ground fault detection
- Starting Air System(s) (for gas turbine driver)
- Mainline scrubbers (for compressor station)
- Fuel system (for non-electric drivers)
- Vent fans and louvers
- Inlet air filter system

- Central lube oil conditioning (filter/cooling)
- Fire and gas detection system
- Station Emergency Shutdown Device (ESD)
- Gas cooler

Generally, the design of the station control system allows for the complete control of the station to be from a local control room, with the option of passing control to the pipeline controller via the SCADA system. This would allow the station to be operated remotely and be unattended.

3.6.6 Emergency Shutdown System

The purpose of an Emergency Shutdown System (ESD) is to provide a fail-safe independent control system that can shut down a station and isolate it in the event of a pipeline rupture, station piping rupture or a fire at the station.

From a design perspective, ESD systems should be hardened against the explosive forces and fire associated with this type of system failure. Indeed, to be fail-safe, the ESD feature should be capable of automatically isolating the flow of product to an accident site until it has been verified that it is safe to reactivate normal operations. The ESD system overrides any operating signals from the station or local controls and its' design therefore, needs to meet the requirements of both the regulatory regime and the owner's own design philosophies and criteria.

The ESD is the last line of defence to shutdown a station and must be able to perform its function even if the station has lost normal power supply, has lost the ability to communicate with SCADA or in the case of local control system failure. Normal designs of an Emergency Shutdown Controller (ESD) provide for them to be independent of the station controller itself. It should also be possible to test the ESD system on a regular basis without interrupting normal operations. They will typically include redundancy control capability to ensure that no single point of failure in the ESD system will disable the capability to properly detect and execute an ESD action.

A station ESD system has associated shutdown valves to isolate the station. If ESD valves close too quickly a pressure transient can be generated that could damage facilities. Hydraulic studies are usually done to determine ESD valve closure times in order to limit pressure transients along the pipeline from the station.

3.6.7 Condition Monitoring

Condition Monitoring (CM) of the pipeline rotating equipment is designed to monitor and diagnose potential failure. The type of data required for use in such a system includes:

- Vibration
- Oil analysis

- Performance monitoring
- Monitoring of parameters that can have long-term detrimental effects if they are outside of their appropriate operating range (e.g. bearing temperatures, gas temperature.)

The results of special techniques such as ultrasonic and thermography are also used for CM purposes.

A host of software support these techniques, which are undergoing continual development. Some of the systems that host this software are stand-alone and others are being integrated with control and information systems (1).

This type of system is now much easier to install at a station due to the standardization of interface protocols and networking. A stand-alone system can be configured to connect to the station control LAN and to exchange data with the control system.

3.7 Pump Station Control

A pump can be driven by an electrical motor, gas turbine or a diesel engine. This section will consider pump station operation using a constant speed electric pump. The objective is to illustrate the different requirements for the control of an electrical drive versus a gas turbine driver described in the Gas Compressor Station section. The other type of electric motor drive used for pump applications is a variable speed drive that controls the motor speed and thus changes the pump outlet pressure.

Constant speed electric motors provide a cost-effective solution for base load applications where electrical power is available and reliable. They have the advantage of low maintenance costs and are simple to operate. Variable frequency drive motors are becoming more popular, despite being more complex than a constant speed motor because capacity can be controlled without the disadvantage of the pressure loss incurred by a discharge control valve (1).

Electric motor drivers are receiving renewed interest, especially for compressor applications, due to more stringent environmental requirements.

3.7.1 Control Strategy

The pump control strategy must incorporate the following criteria:

- Pump suction pressure must be above the minimum Net Positive Suction Head (NPSH) for the pump in order to prevent cavitation of the pump.
- Pump discharge pressure must be below the maximum allowable operating pressure (MAOP) of the station discharge piping to avoid pipe and associated equipment damage.

- Station discharge pressure must be below the MAOP for the pipeline to avoid damage and to ensure the pipeline is operating within the acceptable limits approved by the regulatory agency.
- Station suction pressure must be above the minimum allowable operating pressure to meet contractual requirements and in the case of liquid lines to avoid "slack line flow" or column separation upstream of the pump station.
- Driver power must be kept within acceptable limits to avoid tripping of the driver.

A pump driven by a constant speed electric motor driver requires a discharge control valve to control pump throughput; the system controlling this valve must have a station suction pressure (or station discharge pressure) control loop. Set points for maximum station discharge pressure, minimum station suction pressure, and maximum motor power are set on the controller. The controller will satisfy the set point for station discharge until the suction pressure or driver power limits are reached then these will override the discharge pressure set point.

Pressure switches are set to provide a trip signal in the event of controller failure. The final backup is a pressure relief valve in the event of a complete control system failure.

For a pump station, that contains both fixed-speed and variable speed motors, the control strategy is to run the fixed-speed units at a base load with minimal throttling and utilize the variable speed unit(s) to adjust for the required station set points.

3.7.2 Station Electrical Control

A pump station using electric motor drivers requires a reliable source of electricity. This may be supplied from a commercial source or generated at the station. Economic and reliability considerations usually determine the choice of power source.

The electrical supply usually will have high voltage feeders, voltage reduction equipment, and be a multi-bus operation with its' associated transfer equipment. All the bus and equipment protection required to support such a system is normally provided with the electrical equipment. Controls for this equipment may be incorporated into stand-alone control equipment or they may be part of the station control system.

The electrical protection is always contained in stand-alone, specialized equipment that will protect against:

- Over and under voltage
- Over and under frequency
- Over current and short circuits
- Ground fault

- Voltage imbalance
- Phase reversal
- Transformer gas and high temperature

The electrical supply control system monitors the electrical system and sends the following information back to the station control system:

- Voltage and current values
- Real power, power factor
- Electrical energy consumption
- Circuit breaker and disconnect position
- Frequency

3.7.3 Driver Unit Controls

The unit control for a constant speed motor is not a complicated process system when compared to a gas-fired turbine; the controls in the latter are all incorporated into the controls of an electrical circuit breaker or a start-controller for the motor.

The typical feedback signals to the start-controller are the following:

- Circuit breaker closed and open
- Circuit breaker control circuits healthy
- Electrical protection (varies, depending on the motor)
- Status of local/manual switch

It should be noted that there should always be a method of controlling the motor circuit breaker locally in case the control system is not functional and it is necessary to shut down the motor.

Larger motors also have other interlocks associated with lube oil and vibration. However, the pump/motor is usually bought as an integral package from the vendor and the lube oil and vibration systems are set up as systems common to both parts.

The lube oil systems may be very simple bath types, or complete circulation types similar to those found on compressor units. In the latter case, the controls will have minimum oil pressure interlocks and backup lube oil pumps.

3.7.4 Pump Unit Control

With a constant-speed unit, there are no controls associated directly with the pump other than the lube oil and vibration monitoring systems, which are usually integrated as part of the motor-pump unit.

One item of control that must be carefully considered during the design and operation of this unit is the minimum flow requirements of the pump. Typically, the pump manufacturer will place a minimum flow requirement of 40% of design flow for pumps associated with the pipeline industry. For most of the time, this does not limit operations but care must be taken during the start-up of the line.

Typically, at the inlet station, some method of recirculation is provided so that the inlet pumps can be brought on-line safely. When the pipeline is running at a large base flow, this recirculation valve may be manual, but for lines where the product is stopped and started, it may be controlled automatically by the unit control logic so that it is open until the minimum flow requirement down the line has been established. The goal is to have the pump operate at or near the most efficient point - labelled Best Efficiency Point (BEP) on Figure 4.

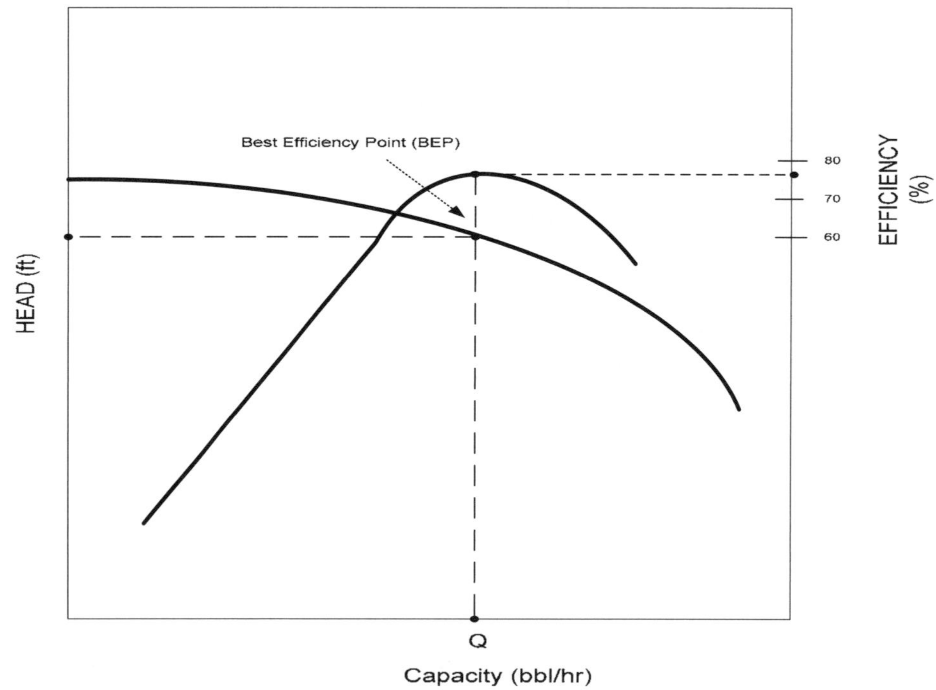

Figure 4 - Constant Speed Pump Curve

3.8 Compressor Station Control

This section discusses the typical control arrangement of a gas turbine driver and compressor set. The specific design application is natural gas transmission. The unit controls discussed are associated with the automatic control and sequencing of a turbine/compressor unit.

The unit control hardware and software are usually a physically and functionally self-contained package, separate from the station control system and typically supplied by the unit vendor. The functional and physical separation of unit and station controls allows local unit operation during station control system outages and simplifies maintenance and troubleshooting. Early in the design process, it is important to ensure compatibility between unit and station control systems and of both with SCADA to minimize duplication and reduce the costs associated with interfacing and integration.

The unit controls have two basic modes of operation - local and remote. In local mode, the unit controls are independent of all external control systems, except for the station emergency shutdown circuit (ESD). In remote mode, the unit will be controlled from the station control by a speed set point, a start and stop signal and an ESD signal.

3.8.1 Control Strategy

The compressor control strategy must incorporate the following criteria:

- Compressor discharge pressure must be below the maximum allowable operating pressure (MAOP) of the station discharge piping to avoid damage.
- Station discharge pressure must be below the MAOP for the pipeline to avoid damage and to ensure the pipeline is operating within the acceptable limits approved by the regulatory agency.
- Station suction pressure must be above the minimum allowable operating pressure to meet contractual requirements.
- Driver power must be kept within acceptable limits to avoid tripping of the driver.
- The maximum station discharge temperature should be below the pre-defined temperature limit to protect pipe coating, so coolers are installed downstream of compressor discharge.

Similar to a pump station, the main control loop (via the station control system) for a compressor station will typically be based on discharge pressure control or flow control. These loops will adjust unit speed to maintain the control loop set point and will employ overrides to limit unit speed based on a secondary condition such as minimum suction pressure.

3.8.2 Turbine Unit Control

The turbine unit vendor typically provides the local unit control system. This system handles all of the controls associated with start-up sequencing, shutdown sequencing, and normal operation.

The unit control system interfaces to the station ESD and the station control system. The station control system provides start, stop and operating set points to the unit control system. It receives analogue signal information from the unit controller instrumentation that monitors conditions in the unit such as bearing temperatures, vibration and internal temperature of the turbine, lube oil temperature, etc. These signals are also sent to a condition monitoring system if it exists.

The unit control system monitors and controls the turbine, the compressor and various auxiliary equipment, such as the:

- Starter
- Lube system
- Seal system
- Surge control system
- Bleed valves and inlet guide vanes
- Air inlet system
- Unit fire and gas monitoring

In addition to the standard monitoring of bearing temperatures and internal temperatures of the turbine, it monitors the ambient temperature. The exhaust temperature is usually limited to a maximum set point, based on ambient temperature. An ambient temperature bias may be required to ensure that a maximum horsepower rating is not exceeded in cooler ambient temperatures. A backup shutdown trip is provided in case the temperature limit function fails to respond adequately.

Complex temperature control is also carried out during unit start-up. The temperature control loop overrides the speed control loop in order to ensure that safe operating temperatures are not exceeded during this period.

Vibration monitoring is used to stop the machine when a high vibration level on any bearing is detected.

Once normal operating conditions are reached, the maximum speed of the gas turbine is regulated to ensure the temperature limit is not exceeded. Backup mechanical and electronic over-speed devices are usually installed on most machines. Under-speed limits and annunciation may be provided for the turbine and compressor. Turbine under-speed causes a shutdown.

3.8.3 Compressor Unit Control

The compressor described below is coupled directly to a power turbine. The power turbine is not mechanically coupled to the gas generator turbine.

The compressor is capable of operating under a specific set of speed and pipeline conditions. A plot of these conditions and the appropriate operating range is provided in a wheel map as shown in Figure 5. This may be used by the operator when manually operating the compressor, or integrated into the control system algorithms and logic to enable automatic control.

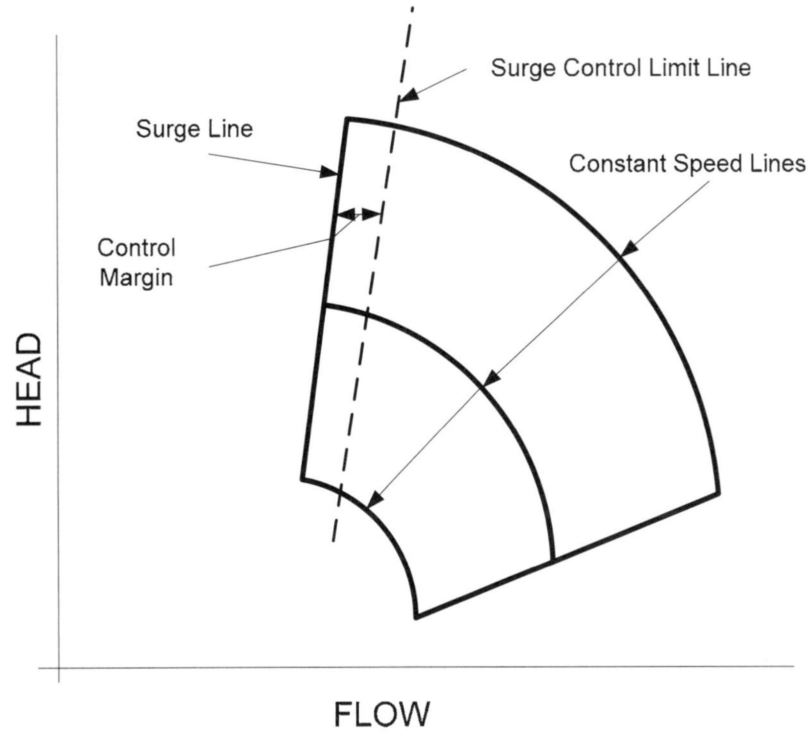

Figure 5 - Compressor Wheel Map

There are some special conditions that the station controls need to monitor for:

3.8.3.1 Choke Operation

A choke condition occurs when there is a low discharge head and a high flow. The compressor is attempting to pump more gas than can enter into the compressor suction. Control of the head developed becomes difficult. If this condition is prolonged, it can be detrimental to the machine. The station control system is responsible for automatically correcting this situation or shutting down the unit. The operators must try to prevent this condition from developing by establishing suitable operating conditions.

3.8.3.2 Surge Control

A surge condition occurs when there is a high head and low flow and can be very damaging to the compressor. Surge occurs when the head differential between the compressor discharge and suction is greater than the head that the compressor is capable of developing at any given speed. This means that at a given flow the existing compressor head is greater than the head that the machine can develop and flow reversal can occur. Surge cycles can continue until the compressor is destroyed, unless pipeline conditions change or corrective action is taken.

The function of a surge control system is to prevent surges by providing a controlled recycling loop around the compressor. The unit recycle valve opening will be set by the surge controller. As the unit approaches the surge line, the recycle valve will open. This decreases the unit discharge pressure and increases the flow through the compressor (but not through the station), which moves the compressor's operating point away from surge. Typically, the surge control system is comprised of a surge controller, a recycle control valve and the required flow and head measurements (See Figure 6). The specific set of head and flow conditions at the surge boundary is called the surge line. All devices in the surge control loop must have very fast response times.

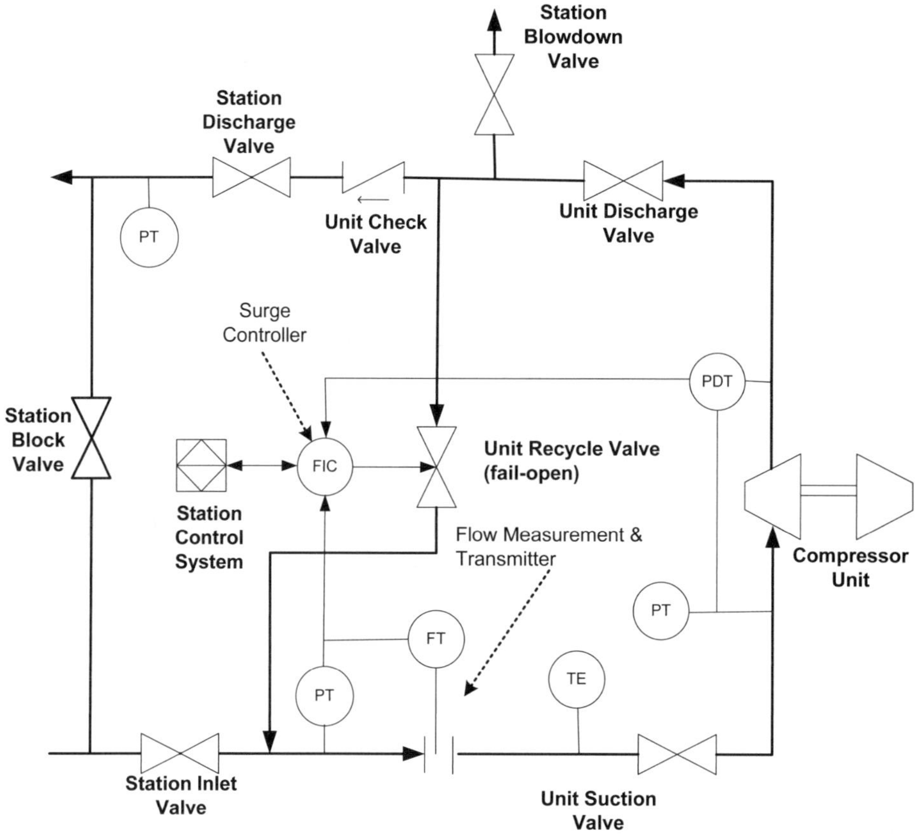

Figure 6 - Compressor Station Surge Control

Operating in a recycle condition is extremely inefficient since a percentage of the flow through the compressor (and hence a portion of the energy used in the compressor) is being recycled. Surge control should be augmented by features in the station controls, which can request an increase in speed whenever the unit is approaching the surge line or is in recycle. This increases the speed of the compressor and the flow, thus moving away from the surge line without having to open the unit recycle valve.

Simple surge controllers usually use a 10% safety margin for the surge control line (110% of midrange flow). More sophisticated controllers are available that can compensate for varying suction pressure and temperature, and can maintain protection at a 5% margin. This allows more turndown of the compressor without requiring recycle, and therefore makes for more efficient operation at lower throughput flows.

Some compressors come supplied with their own self-contained surge systems in order to maximize both operational flexibility and safety, and to minimize unit interactions. However, some designs use the distributed features of a networked architecture to allow the surge control to be located in a seperate module, while still being part of the control system. This avoids having an interface between two devices of different manufacture. A third option is to have the surge control function as part of the station control.

3.9 Meter Station

A meter station is typically located where there are injections or deliveries from a pipeline. Flow metering can be provided for control purposes to supply the operator with operational flow data. Alternately, the meter station may be part of a pipeline leak detection system. Metering stations may also be custody transfer meter, measuring the amount of gas or liquid for commercial purposes.

These meter stations are designed for a high degree of accuracy and a wide range of flow rates. In order to meet both of these requirements, a meter station is usually installed with one or more parallel meter runs, each containing metering devices. The number of parallel meter runs will determine the flow range measurable by the meter station. A multi-run station needs to be remotely controllable, with meter station control logic that will automatically put the required number of meters into operation to meet the meter demand. Flow or pressure regulators may be required to control flow or pressure at the injection site or the delivery station.

Meter stations should be designed as per the requirements specified in the appropriate standards. A typical meter station has the following components as shown in Figure 7

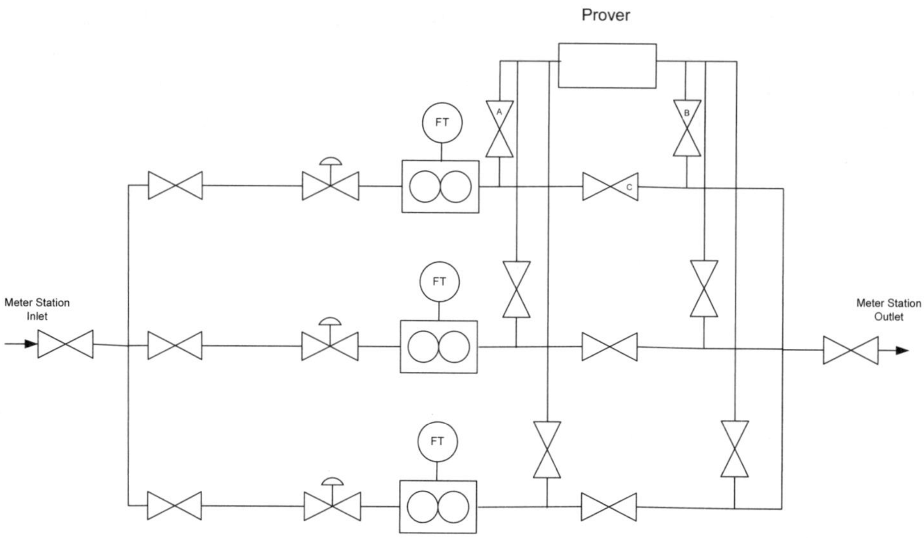

Figure 7 - Simplified Meter Station with Prover

- Headers: Headers are the upstream and downstream pieces of pipe, to which the meter runs and the yard piping in a meter station are connected. Headers allow for the addition or deletion of meter runs at the meter stations. At a multi-run meter station it is recommended that all headers be the same size.
- Meter station valves include check valves to avoid back flow, block and bypass valves to regulate flow direction, and a flow control valve to regulate the flow rate. In addition, a blowdown valve is required for a gas meter station to relieve high pressure. Valves for isolation of each meter are installed so that individual meter elements can be removed for repair without shutting down the meter station. The control valve in each meter run is used by the meter station controller to balance flows between each meter run.
- A pressure regulator maintains a constant downstream pressure regardless of the flow rate in the line.
- A strainer or filter is installed at the turbine and the PD meter station, since the metering is susceptible to damage from solid particles or liquids in the gas stream. A strainer is installed on each meter run.
- A meter prover is used with turbine and PD meters to establish a relationship between the number of counts or revolutions of the meter and the volume flowing through the meter. The number of counts on the meter being proved is related to the volume passing the detectors on the prover.

3.9.1 Meter run

Fluid enters a meter through a meter run. A meter run is defined as the length of straight pipe of the same diameter between the primary measuring element and the nearest upstream and downstream pipe fittings. It consists of a meter and pipes, pressure and temperature measuring devices, valves including a check valve, a strainer and straightening vanes for turbine and orifice meters. The flow range of the meters installed and the volume of fluid flowing through the meter station primarily determine the size and number of meter runs.

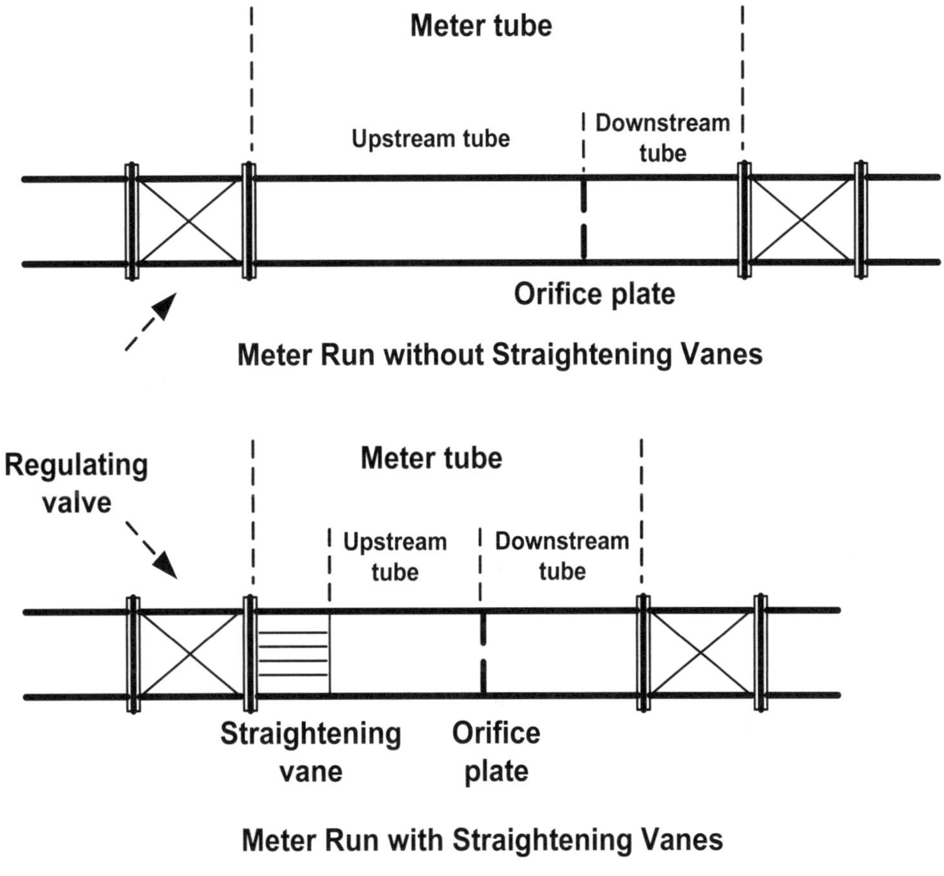

Figure 8: Orifice Meter Run

A typical orifice meter run is shown in Figure 8. The meter tube diameter used in a meter run should be consistent with the size of the orifice plate or other meters such as turbine and PD meters. The meter tube is used to maintain the accuracy of flow measurements. The meter tube should be placed aboveground and connected by a section of pipe installed at a 45° angle to the ground and then another 45° angle pipe at each end of the meter tube at the desired height above the ground. Straightening vanes may be installed in the upstream section of the meter tube to minimize turbulence ahead of the orifice plate. As shown in Figure 8, the upstream tube without straightening vanes should be longer than those with them. The tube lengths depend on the pipe diameter; all the orifice meter run specifications are described in AGA-3 or ISO 5167 standards. The specifications must conform to one of these standards in order to be used for custody transfer meters.

A typical turbine meter run for a turbine meter is shown in Figure 9. Upstream of the turbine meter, a strainer is required to remove debris such as solid particles and straightening vanes. A check valve is required downstream of the meter to prevent back flow into the meter. The tube lengths, which depend on the pipe size, between the strainer, straightening vanes, turbine meter, check valve, and various measurements taps, are specified in turbine meter standards such as AGA-7.

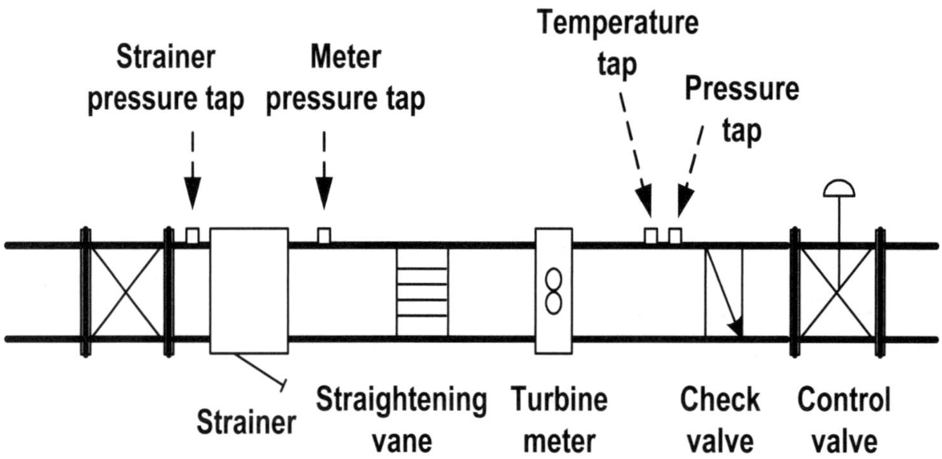

Figure 9: Turbine Meter Run

A typical positive displacement meter run is shown in Figure 10. Unlike a turbine meter run, it requires a bypass valve because a PD meter blocks flow when it stops running.

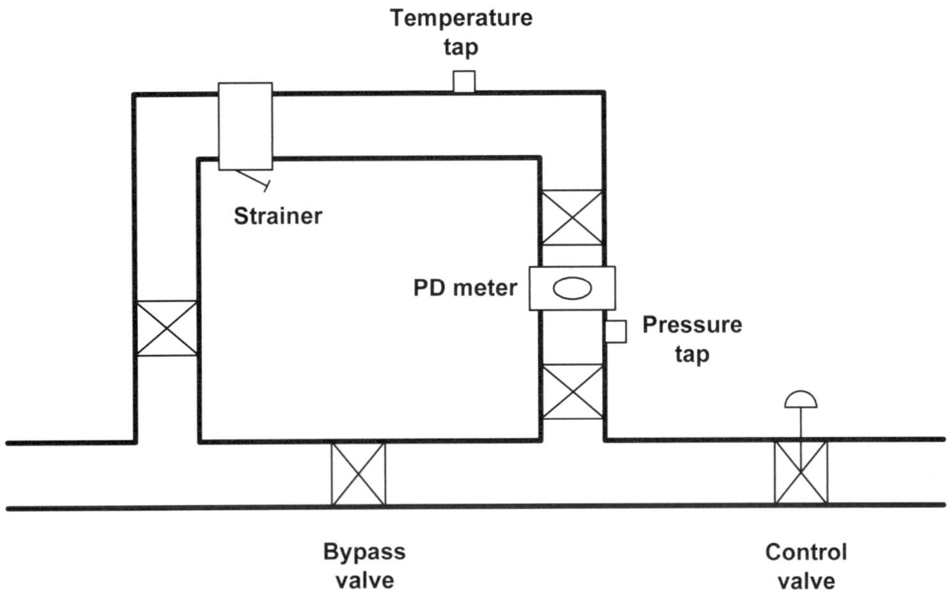

Figure 10: PD Meter Run

3.9.2 Straightening Vanes

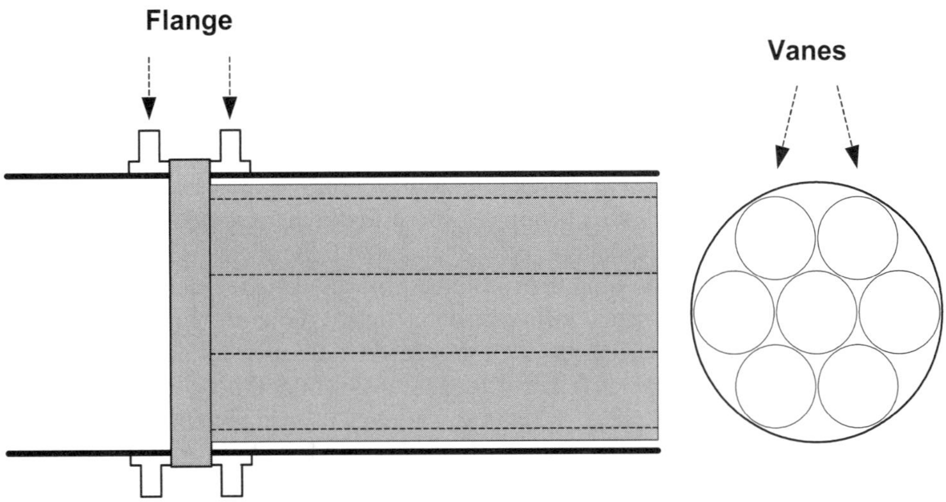

Figure 11: Straightening Vane

Swirls and other turbulences are created when fluids pass over the pipe bends and fittings in a meter run. To reduce turbulence straight lengths of pipe (length equal to 100 pipe-diameters) may be required, while a short length of pipe is sufficient when using a flow conditioner. A flow conditioner such as straightening vanes is used to reduce turbulence before turbulent flow can reach the meter. Therefore, the use of straightening vanes in a meter run serves the following purposes:

- The flow profiles are smoothed, resulting in improved meter accuracy.
- The pipe length required preceding a meter is significantly reduced.

Straightening vanes should conform to the requirements of standards. A flange type straightening vanes is shown below.

3.9.3 Meter Prover

Pipeline companies are responsible for accurately determining the amount of product received into and delivered out of their pipeline systems. This objective can be accomplished through a meter proving process, which uses more accurate metering devices and recording instrumentation to check the calibration of the primary meter.

Proving is a method of checking a measuring device against an accepted standard to determine the accuracy and repeatability of that measuring device. Meters are proven immediately after repair, removal from service for any reason, when changing fluid products being measured, when product viscosity changes, or on demand if meter history indicates that it is required. On liquid pipelines that carry multiple products or products of varying properties, it is important to be able to perform meter proving on demand in order to test the accuracy of the meter station.

When proving or determining the accuracy of meters, proper practices and procedures must be followed. The API Manual of Petroleum Measurement Standards (MPMS) chapters 4 and 5 as well as ISO 7278 provide guidelines on meter proving techniques and standards. A meter prover is used to verify the accuracy of the liquid meters. Essentially, the prover determines the meter factor that is representative of the volume being put through the meter. The API MPMS defines the meter factor as a number obtained by dividing the actual volume of liquid passed through a meter during proving by the volume registered by that meter. The meter factor accounts for non-ideal effects such as bearing friction, fluid drag and mechanical or electrical readout drag. In addition, turbine and PD meters are subject to accuracy variations as a result of temperature, pressure, viscosity, and gravity changes. Therefore, the meter should be proved under the same operating conditions as those that the meter experiences during normal operation.

During proving the meter outlet is diverted through the prover and the measurement of the prover is compared to the meter's measurement. If the difference between the two readings exceeds the allowable tolerance, a new meter factor is calculated and used to correct the meter until it is proven again. This correction factor is called the meter factor and will be the data point sent to the station control and SCADA system. A history of meter factor changes is required to monitor the meter's performance and for billing audit purposes. It is also used to adjust previous meter tickets according to the metering policy of the pipeline company.

The method of determining the meter factor is to put a known volume through the meter and count the number of pulses generated from the test meter. Temperature and pressure need to be stable before running the prover and should be measured and recorded during the proving to correct their effects on volume. The meter factor or the meter K-factor is determined by

Meter K-factor = number of pulses from meter/actual volume

If direct volume readout is obtained from the meter, the meter factor is determined by

Meter factor = corrected prover volume/corrected meter volume

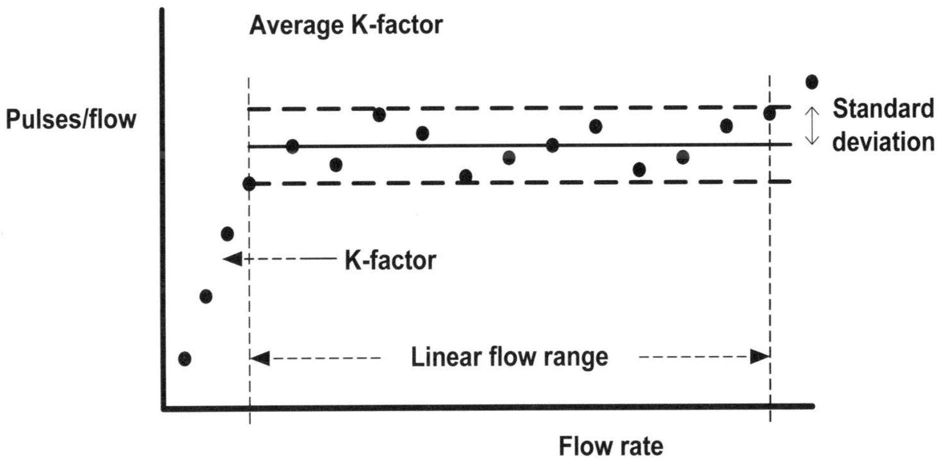

Figure 12: Meter Factor

Since meter factor depends on the fluid properties, different meter factors are required for different fluids. The following data are required for meter proving:

- Data on the test meters and meter prover
- Properties of all the products that are put through the test meter
- Temperature, pressure and density during a proving run
- Meter factor for each product and flow range including their variables to calculate the meter factor such as pulse counts and actual volumes

- Status information of the prover and the proving progress, pump starts/stops, ball launch/detection, etc.

K-factors are determined over various flow rates by conducting several proving runs and a representative K-factor is selected for an applicable linear flow range. The K-factor is acceptable if the average from multiple consecutive proving runs is within 0.05%. A typical K-factor linearity curve is shown above.

Meter proving can be done using a flow computer, RTU or PLC, depending on the telemetry equipment availability. If meter proving is done locally using the flow computer or local PLC, either the local facility is instructed to perform meter proving or it is automatically initiated upon the detection of a batch interface for a batch pipeline. When the local proving is completed, the flow computer or the local PLC uploads the proving report to the host. Alternately, meter proving can be done via the host SCADA. The host SCADA may be able to control the entire proving sequence, from initiating each prover run, gathering all data at the completion of each run, to calculating the meter factor.

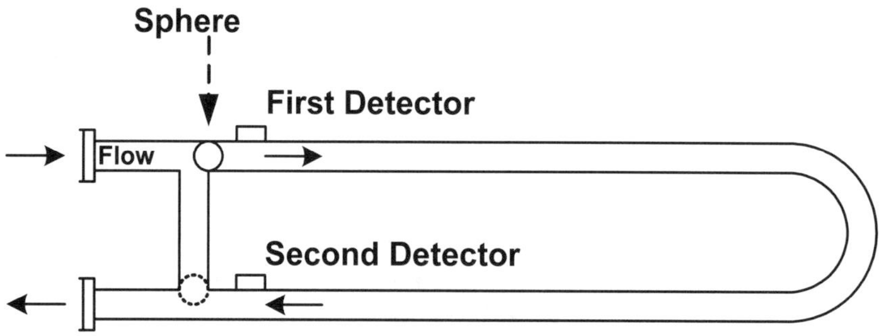

Uni-directional Prover

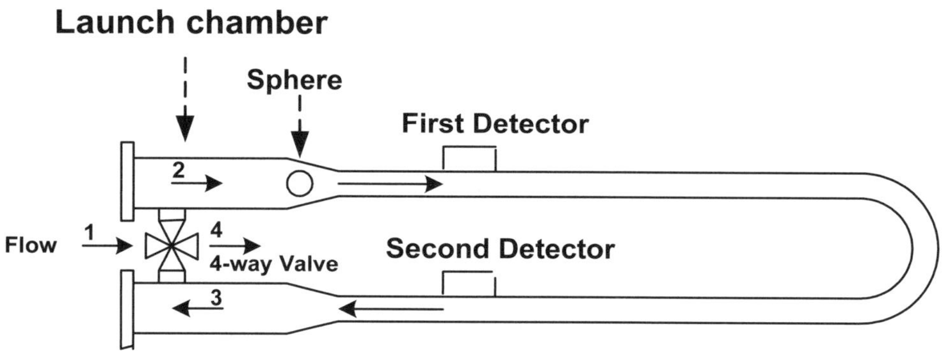

Bi-directional Prover

Figure 13 Meter Provers

There are two types of provers: fixed type and portable type. Fixed type provers are either kept in service continuously or isolated from the flowing stream when not proving. All provers may be unidirectional or bidirectional. The API Manual of Petroleum Measurement Standards (MPMS) details prover types, applications and methods of calibration and corrections. Refer to Figure 13 for unidirectional and bidirectional provers.

In the unidirectional prover, the spheroid always moves along the prover loop in one direction and thus the fluid flows in the same direction. The volume of the prover is the volume displaced by the spheroid from detector D_1 to detector D_2. For the bidirectional prover, the spheroid moves through the loop in one direction and then returns around the loop in the opposite direction. The volume of the prover is the sum of the volume displaced by the spheroid in one direction from detector D_1 to detector D_2 and the volume displaced when the spheroid moves in the other direction from detector D_2 to detector D_1.

The volume of the prover needs to be corrected to base pressure and temperature conditions. The correction is necessary to accurately determine the prover and fluid volumes at base conditions. The pressures and temperatures of both the prover and fluid are measured for volume correction throughout the proving period.

3.10 Storage Operation

3.10.1 Tank Farm Operation

Crude oil and petroleum products, including light hydrocarbons, are often stored in tanks in various locations such as producing areas, refineries, petrochemical plants, and/or distribution centers. Petroleum liquids are stored underground or in aboveground storage tanks. Storage allows for flexible pipeline transportation and efficient transportation management through the existing pipeline system and minimizes supply/delivery disruptions. The stored liquids need to be measured and accounted for accurately in order to keep track of all volume movements including custody transfer when appropriate.

A tank farm refers to a collection of tanks located at refineries, shipping terminals and pipeline terminals. Tank farms at refining operations are used to store various products produced by the refinery and to hold them until they are scheduled for injection into a pipeline for transportation. Similarly, tank farms at shipping terminals hold products until a shipping route has been scheduled. This may be tanker ship, truck, railcar, or another pipeline. Tank farm operations must be measured and controlled. Figure 14 shows the key elements of a simple tank farm.

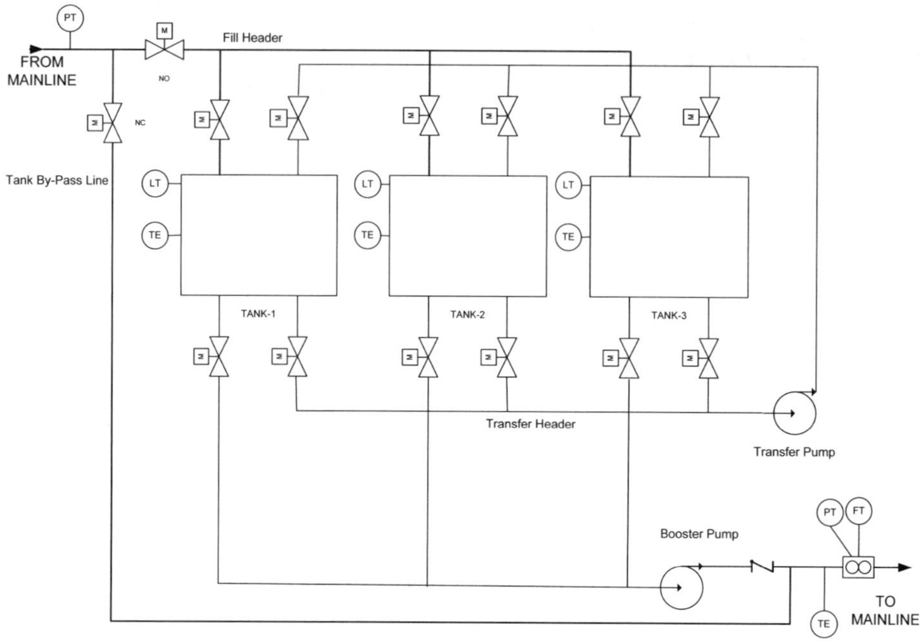

Figure 14 - Simplified Tank Farm

3.10.1.3 Tank Measurement

One way to measure the volume of a stored liquid is to determine the level of the liquid in the tank and then calculate the volume from a capacity or strapping table that relates the level to the corresponding gross volume of liquid in the tank. The strapping table is established during the tank proving process, using a tank prover which has thermometers mounted in the measuring section to accurately measure temperature. The API Standard 2550, Measurement and Calibration of Upright Cylindrical Tanks, describes the strapping procedures. API MPMS Chapter 2 describes the strapping procedures for cylindrical as well as other types of tanks.

Tank level to volume conversion requires that the parameters and strapping table or equation associated with the tank be defined. In addition to the level measurement, the gravity and suspended BS&W content and the temperature of the liquid and ambient temperature near the tank need to be measured to determine the net volume and liquid head stress caused by high hydrostatic pressure on a large tank. The accurate calculation of the volume in the tank requires parameters such as tank roof types (fixed or floating) and the level of free water. The volume conversion can be performed by a field automation system such as PLC and RTU.

Once the tank level has been measured, whether manually or automatically, the level data is converted to a gross volume using a volume conversion process. The process uses either the strapping table data for each individual tank or an incremental table that defines incremental volumes per number of level increments for the tank. The conversion equation associated with the tank can be used for the volume conversion. The gross volume should be corrected for tanks with a floating roof by taking into account the weight of the roof and any snow load.

The level of free water is also required to determine the gross volume of the petroleum product in the tank. This value is converted to its equivalent volume using the volume conversion table and then subtracted from the gross volume to determine the gross volume of the product by assuming that the water is on the bottom of the tank.

A gross volume is converted to a net volume using the density and temperature of the fluid in the tank. The density or API gravity is used to calculate the temperature correction factor, which is detailed in the API Standard 2550. Once the temperature correction factor is determined, it is multiplied by the gross volume to obtain its equivalent net wet volume. If sediment and water (S&W) is present, the value of the S&W content is needed to determine the net dry volume.

3.10.1.4 Tank Control

The purpose of a tank farm control system is to assist the operator in moving product and product inventory. Terminals that handle multiple products (i.e. a batched pipeline) with a large number of tanks and interconnecting pipelines can have quite a complicated routing within the terminal. There will be a significant number of motor-operated valve controls and level monitoring systems.

A tank farm control system can assist the operator by verifying that proposed valve line-ups represent a valid path before he initiates the sequencing and starts the pumps to move the product. This is important, as an error such as the injection of crude oil into a refined product tank would be costly.

A tank farm control system generates and stores product delivery and shipment feed information in business applications such as inventory tracking, billing for product receipts and deliveries, as well as feeding the same information into a pipeline scheduling system.

The tank control process involves several functions. It establishes a tank's maximum level or volume, from which is calculated the volume required to fill and/or avoid over-flow of the tank and it also determines the minimum level or volume from which can be calculated the volume necessary to pump out in order to avoid over-drainage of the tank. The flow rate into and out of a tank is calculated by dividing the volume change by the difference in time between the two. The resulting flow rate can be used to estimate the time required to fill or empty the tank.

A tank control requires alarms and events to be generated in response to various conditions. These may include alarms for when maximum or minimum tank levels have been violated and for abnormal rates of change. Typically, the information contained in a tank report includes such data as tank level and water level, measured and corrected gravity, temperature, gross and net wet volume, S&W volume, net dry volume, and flow rate.

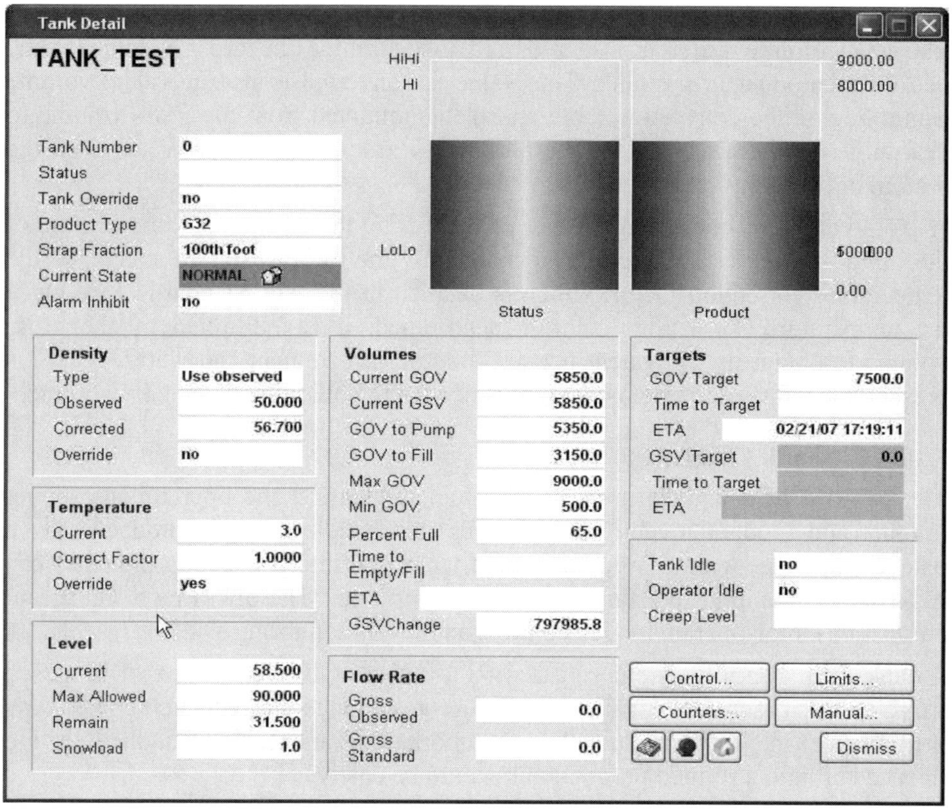

Figure 15 Tank Control (Courtesy of Telvent)

3.10.2 Natural Gas Storage

Natural gas is normally stored underground in order to shave peak demands and is an integral part of an efficient gas inventory management system of a pipeline complex. Underground storage is in rock or consolidated sand formations that have high permeability and porosity. Natural gas storage is usually located close to consuming centers and near the transmission pipeline. The natural gas is injected into the storage during off-peak season, typically during summer, and withdrawn from the storage during peak periods if the line pack is not sufficient to meet the peak demands.

The base inventory and deliverability can be calculated from the reservoir pressure and temperature measurements, together with the gas composition. (See reference (3) for a detailed calculation method). The inventory of a gas storage reservoir is normally made using metered injections and withdrawals. The required measurements includes: well head and flowing pressures, temperatures, and injections and withdrawal flows. A gas chromatograph may be required to measure the quality of gas, unless its' composition is known. In addition, reservoir characteristics are required to estimate deliverability.

References

(1) Mohitpour, M., Szabo, J., and Van Hardeveld, T., "Pipeline Operation and Maintenance" ASME, New York, 2004

(2) Dempsey, J.J and Al-Gouhy, A.H., "Simulation Effectively Sites Surge Relief Facilities on Saudi Pipeline," Oil & Gas Journal, Sept., PP. 92-98, 1993

(3) Tek, M. R., "Underground Storage of Natural Gas," Gulf Pub. Co., Houston, 1987

4 Gas Management System

4.1 Introduction

This chapter discusses the functionality and implementation related issues of a computer based gas management system whose intended use is for managing daily transportation services. The primary focus of this chapter is on gas nomination and volume accounting functions. Also described are related applications such as gas inventory monitoring, gas storage injection/withdrawal and gas load forecasting.

The purpose of a computer based gas management system is to automate the provision of gas transportation with the goal of improving the efficiency and profitability of the service. The gas management system begins with the initial gas contract and carries through to the final invoicing and account settlement. It allows the system users to access the required information quickly and to provide the shippers with accurate information (1). The users may include not only the pipeline company staff and management but also shippers such as producing companies, other transmission companies, distribution companies, and gas marketing companies. A computer based gas management system can help the operator make the most efficient use of the pipeline capacity and facilities and keep more accurate track of the transportation process than a traditional manual system, thus increasing profits.

Historically, pipeline companies in North America provided total gas service including gas supply and transportation. In 1992, the Federal Energy Regulatory Commission (FERC) in the U.S.A issued Order 636, which transformed American gas transmission companies from gas merchants into transporters. In other words, the order required the unbundling of these two business activities. Pipeline companies had to provide gas producers and shippers with equal and open access to transportation services and eliminate the discriminatory contracts that limited access to small volume suppliers. The ultimate objective of the FERC's Order 636 was to provide consumers with access to an adequate and reliable supply of natural gas at a reasonable price. As a means of achieving this objective, the order mandates that transmission pipeline companies open their transmission services to all shippers regardless of the ownership of the gas or its quantity. It was reported (2) that the unbundled service requirement allowed a gas transmission company to set up hourly customer nominations and determine daily balancing and billing, while achieving 99.5% daily billing accuracy.

A major transformation of the gas industry has taken place in North America as a result of Order 636. Gas transmission companies have used automation to improve their operation efficiency and standardized their business processes from contract to invoice to increase their profitability. Similar changes are taking place in the European natural gas industry.

Open access has altered the business demands expected of gas pipeline operators: the number of commercial transactions has increased dramatically, transportation and operational processes have become more complex, gas marketing requires near real-time information and customers need and are requesting more information in a timely fashion. One key change in the business process that has helped to meet these demands is the automation and integration of the marketing, operation and customer portions of the gas pipeline business.

To address business and operation issues effectively, several standards have been developed for the gas industry. In North America, the Gas Industry Standards Board (GISB), an industry interest group, has developed natural gas transportation standards to respond to regulatory and technical changes (3). GISB develops and maintains standards which address gas industry business practices and electronic data interchange protocol. The GISB provides business practices standards for:

- Contracts including short-term sale and purchase of natural gas
- Nominations and capacity release
- Invoicing
- Data interchange such as data syntax, time synchronization with time-stamp (i.e., time of data capture), batch and interactive processing, security, compatibility for effective operation, etc.

The GISB standards have helped the natural gas industry to change a paper-driven business process to an internet based one (4). They provide many benefits for pipeline industry and shippers, including: increased profitability, information transparency, and fast business processing with short response time. The GISB Standards have been adopted by most North American gas transmission companies and shippers. Effective January 1, 2002, GISB became the North American Energy Standards Board (NAESB).

While GISB or NAESB has been developing and maintaining model business practices to promote more efficient and reliable energy services in North America, similar standardization activities are occurring in Europe as the gas industry becomes more open. Several gas transmission companies from eight different countries in Europe have formed the Edig@s Workgroup, which has developed natural gas transportation standards (5). As a result, the business practices, particularly communication among stake holders in North America and Europe have been standardized.

Due to these significant changes in both the North American and European regulatory environments of the natural gas industry, business needs have changed. To meet them gas management systems must have the new standards and regulations incorporated into their design.

4.2 Transportation Service

There are several main stakeholders in the natural gas industry: gas producers, gas gathering and/or processing facility operators, gas marketers, transmission companies, local distribution companies, and end users. Gas producers produce gas from gas wells, which are connected to gathering facilities that include gas gathering lines and processing plants. These facilities are normally owned and operated by the gas producers and connected to transmission lines. The transmission pipeline or midstream company transports the gas through its pipeline networks to local distribution companies, other transmission pipelines, gas storages, and/or end users. Gas marketers arrange for the buying, selling and storing of gas, but they are neither distributors nor end users. Local distribution companies purchase and distribute the gas to such end users as residential and industrial customers. Gas storage is normally used for peak shaving and to minimize supply disruptions. Storage facilities are typically located near end users, so they can deal quickly with unusual circumstances, such as unexpected cold weather.

4.2.1 Gas Transportation Services

Gas transmission pipelines are the link between gas supplies and the markets, and interconnect all of the stakeholders. Three main transportation services are provided by pipeline companies: the receipt of gas from producers and other pipelines, the transportation of gas through its pipeline network, and the delivery of gas to the customers. Pipeline companies may provide other services such as:

- Storing allows customers to store gas at designated storage sites.
- Loaning allows customers to receive gas from the pipeline company and return loaned quantities to it.
- Parking allows customers to store gas in the pipeline system for short periods of time.

There are two types of transportation service; firm and interruptible. Firm service is a guaranteed transportation service and the pipeline company will guarantee that the service will be available during the contract period unless a catastrophic accident occurs. Firm service contracts are generally long-term, say a year. If a customer cannot use his service during the agreed time, contractual terms will allow the pipeline company to release it to other parties. Interruptible contracts infer that the pipeline company can interrupt the transportation service with no economic penalty if the pipeline capacity is not available. Firm service contracts have a higher priority than interruptible services and thus the charge for them is usually higher than for interruptible services.

4.2.2 Gas Transportation Service Process

Once a contract is in place, transportation service processes include three distinct phases: nomination, daily flow operation, and revenue accounting after delivery is completed. Each phase has its own unique tasks and processes. The gas transportation service is summarized in Figure 1, showing the nomination management phase, pipeline system operation and measurement phase, and revenue accounting phase. The pipeline system operation and measurement phase is discussed in other sections.

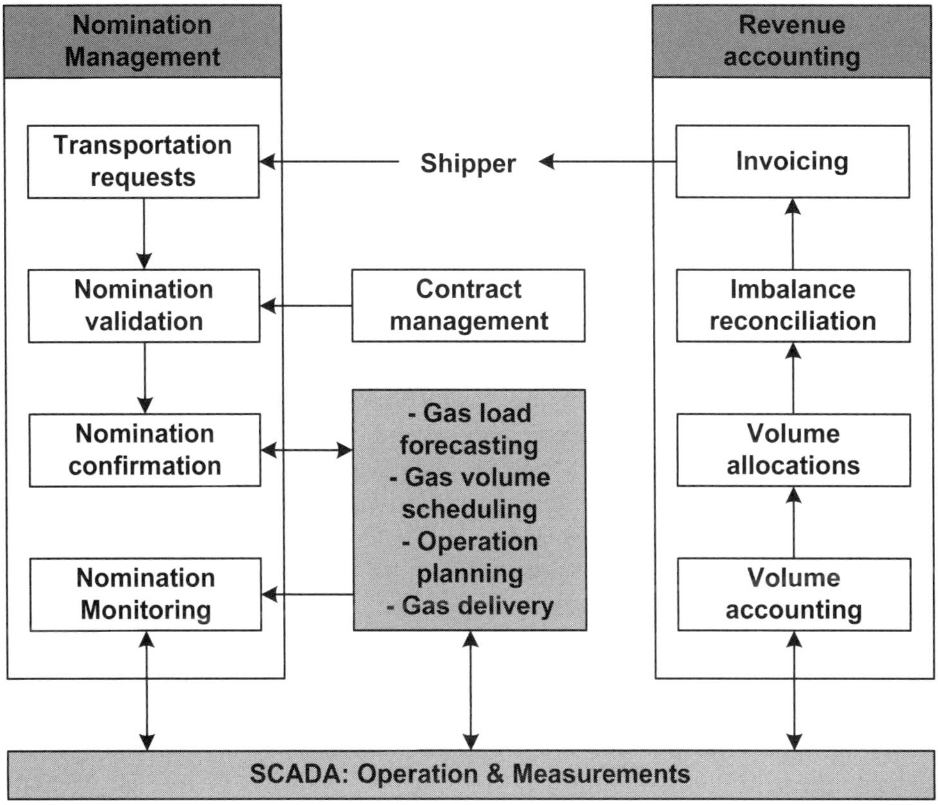

Figure 1 Process of Gas Transportation Service

The contract is a legally binding agreement between a pipeline company and a shipper for transportation, storage, and/or other services. It specifies each service, including the maximum quantity of gas to be delivered each day, the receipt and delivery points of the gas in the pipeline system and the minimum and maximum tariff rates that will apply.

A transportation contract allows a shipper to ship natural gas through the pipeline system for a specified period of time and a storage contract allows a shipper to store gas in storage facilities throughout the pipeline network. A shipper's request to a pipeline company for transportation service typically will include the following data:

- Name of contract
- List of the associated customers
- Type of contract (firm, interruptible, etc)
- Dates the contract takes effect and terminates
- Quantity of daily default maximum
- Custody transfer points

For storage service contract requests, shippers will specify the injection and withdrawal schedules for the contracted storage facilities.

The contract information portion of a computer based integrated management system should include not only the storage of contract information and contract requests but also their maintenance and display functions. The maintenance function includes the contract approval and modifications.

4.2.3 Nomination Management

Nomination management begins with transportation requests called "nominations". Shippers submit nominations to the pipeline company for the next gas day. In North America, NAESB designates a "Gas Day" as a 1 day period (usually a 24 hour day, but could be 23 or 25 hours, depending upon Day Light Saving Time) that begins at 9:00 AM CDT (Central Day Light Saving Time). A given Gas Day (the daily flow) starts with nominations.

The nomination is the process whereby a shipper requests transportation or other services of the pipeline company. Each nomination may include information on the services to be performed: gas volume to be transported or stored, locations of receipts and delivery, lists of customers, etc. Based on the nominations received from all shippers, the pipeline company performs several tasks internally. These include: gas scheduling, nomination confirmation, and the receipt and delivery operations required to fulfill the nomination. The receipt and delivery operations are the only physical operations involved in the transportation service. Gas scheduling is a series of processes that validate nominations for contracted volumes, balance limits, and pipeline capacity rights. Gas scheduling may require gas load forecasting to estimate the gas load at certain delivery points if the locations are sensitive to weather conditions. The validated nominations are checked against the available pipeline system capacity and curtailment volumes to determine if a nomination exceeds the pipeline capacity. Curtailment is a service reduction to a level below the contracted volume due to pipeline capacity limitation. Nominations are accepted or changed at receipt and delivery points through a confirmation process. After all the nominations including receipt and delivery quantities have been confirmed, the pipeline operator physically transports the confirmed volumes from/to the designated points.

4.2.4 Revenue Accounting

The revenue accounting phase includes volume accounting and allocation, volume balancing, billing and invoicing, and account settlement. After nominated volumes are physically received and delivered, the physical volumes are allocated to confirmed nominations based on the contracted allocation method such as proration or ranking. Allocation is a process designed for balancing and billing purposes. Once allocation is completed, the cumulative volume difference (called imbalance) between nominated and allocated volumes and the difference between receipts and deliveries are calculated through a balancing process. Billing and invoicing for the services take place after volume balancing is completed. If a customer disagrees with the invoices and imbalances, the pipeline company has to settle the account using the contract and volume accounting data.

It should be noted that many contracts are now written using energy as a basis rather than volume. This is because the economic value of natural gas is a factor of not only its volume, but also of the energy produced when a unit of gas is burned (also referred to as its heating value or quality). As a result, nomination, allocation and balancing procedures are often performed using energy figures rather than volume. For the purposes of discussing the management of natural gas within industry, the term 'volume' is sometimes used (technically incorrectly) to actually describe energy.

This chapter discusses the volume accounting only, because the other activities in the revenue accounting phase do not belong to engineering disciplines.

4.3 Nomination Management System

The daily nomination is an integral part of a gas pipeline operation because it specifies the transportation request in terms of contractual gas volume for a particular receipt/delivery point. This section describes the components of a nomination management system: the data required, how it is to be entered, and the daily monitoring and display of the nomination function in pipeline system operations. Such a system must also be capable of modifying nominations throughout the gas day.

4.3.1 Nomination Data

Once appropriate contracts are in place, each gas day the shipper provides the pipeline company with daily nominations. These can be modified up to a specified hour. Nominations may include injection to or withdrawal from storage. Daily nominations may include the following data:

- Shipper name

- Associated contract and effective gas day
- Receipt and delivery locations
- Nominated volume and tolerance at each location
- Injection or withdrawal volume from a storage, if applicable
- Maximum daily quantity (MDQ)
- Must take gas volume

After the entered nominations are verified and gas scheduling is completed for the gas day, the pipeline company sends the confirmed quantities for the entered nominations to the shippers. When the gas day is over, the pipeline company determines the allocated volumes from the measurements and sends the volumes and daily invoices to the shippers. If an internet based shipper information system is available, the confirmed and allocated volumes and invoices can be confidentially posted on secured web sites for each shipper.

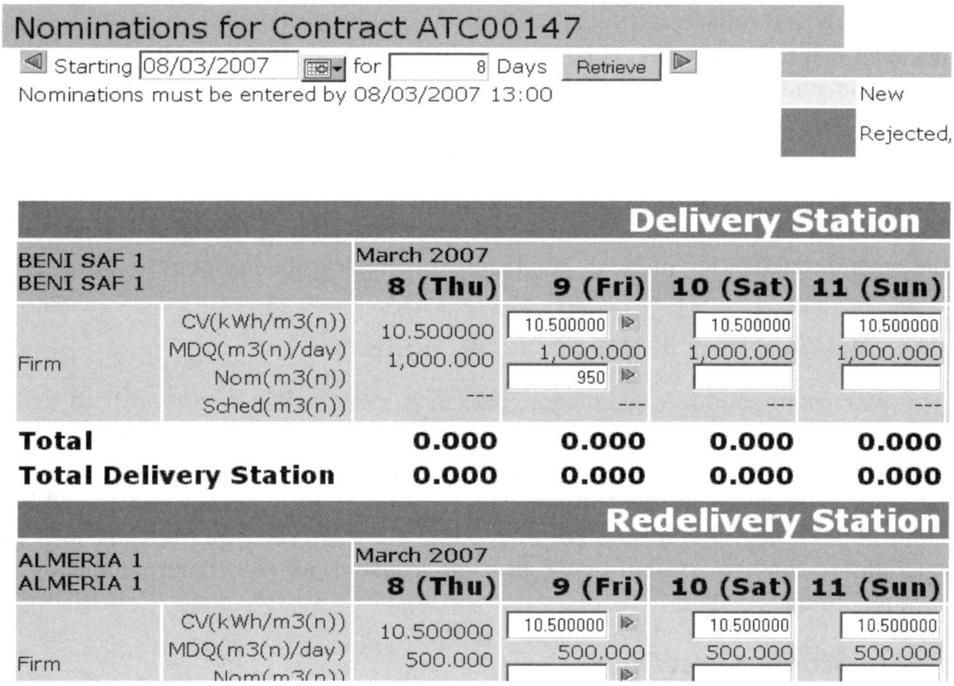

Figure 2 Example of Nomination Display (Courtesy of Telvent)

4.3.2 Nomination Data Management

Depending on the type of nomination management system in place, shippers will provide their nominations by one or more of the following mechanisms:

- a computer-based data entry to a computer based system
- direct link to a computer based system
- email
- phone
- fax

A computer-based nomination system allows the pipeline company and shippers to manage their contract and nomination data easily and expeditiously. The users of the system may be internal, such as operation and management staff and external, such as shippers. The users can view the existing contract and nomination data as well as add, modify and delete data as long as appropriate access privilege has been given. A computer-based system can provide the shippers with nomination data entry and monitoring through the Internet. Having the monitoring information allows the shippers to track and monitor confirmed nomination and allocated gas volumes at supply and/or delivery points. As well, the system can compile and display reports, events, comments, and alarms that occur during the gas day. Shown below is a computer-based nomination system architecture.

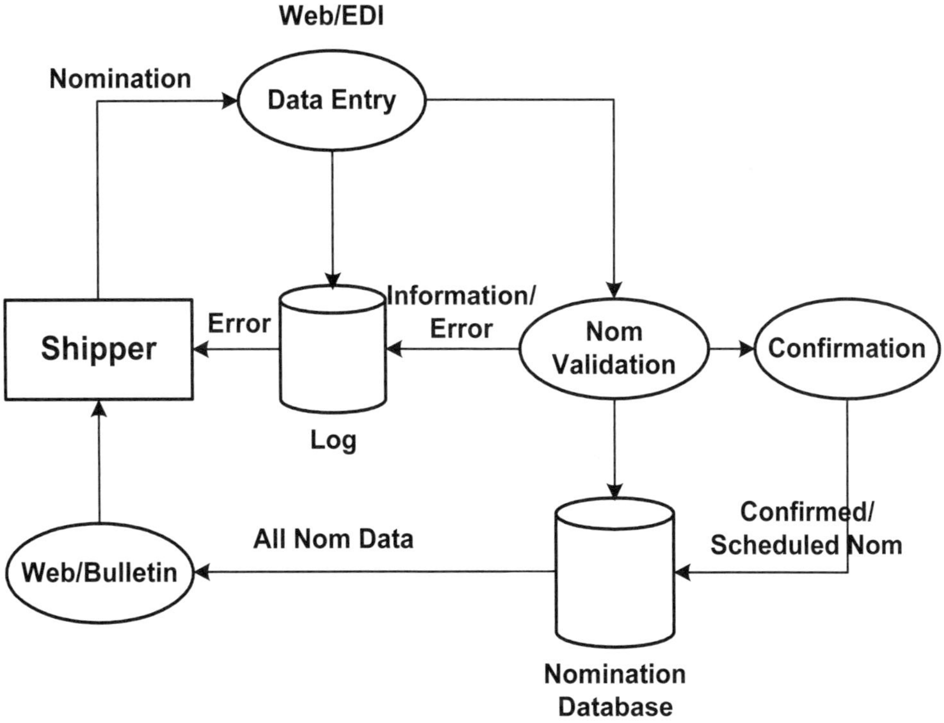

Figure 3 Computer-based Nomination System Architecture

Alarm messages are generated when a certain business process or condition is violated. Such violations might include a nomination that does not match the one contracted for during the current gas day, system communication failure, failure of the nomination to arrive before the nomination deadline or a nomination that has been modified after the deadline. Shippers are allowed to acknowledge their specific alarms.

Typically, a computer-based nomination system can:

- Verify incoming and outgoing data from shippers and the pipeline company.
- Confirm receipts of the shipper information almost instantaneously.
- Monitor nomination status throughout the gas day from the initial nomination through volume allocation and billing.
- Monitor nomination tracking status.
- Control versions of nominations and their modifications.
- Notify of failures, alarms and events almost instantaneously.
- Keep security and confidentiality of transportation service and volume accounting information.

Often nominations must be modified due to pipeline facility failures, changing weather conditions that may necessitate different gas volumes, or supply problems. A computer-based nomination system can handle nomination changes efficiently, allowing both internal and external users to review the changes easily.

4.3.3 Nomination Monitoring Function

The nomination monitoring function tracks and monitors nomination status including gas volumes at supply and delivery points in the pipeline network throughout the gas day. It also maintains the daily nominations, stores the accumulated volume totals up to the hour, calculates the projected volume, and compares the projected volume to nominations to determine nomination imbalances. It can provide the following information:

- Nominated volume
- Remaining take and the hour for the calculation
- Projected end of day (EOD) volume
- Alarms and event messages

Alarms are generated for the operators if the difference between the estimated end of day volume and confirmed nomination exceeds the tolerance specified in the nomination or if the estimated end of day volume exceeds the maximum daily quantity (MDQ). The MDQ is intended to provide the pipeline operators and measurement staff with the gas receipt and delivery information that is required to satisfy the nominations within the contract limits. Shown in Figure 4 is an example of the nomination monitoring display.

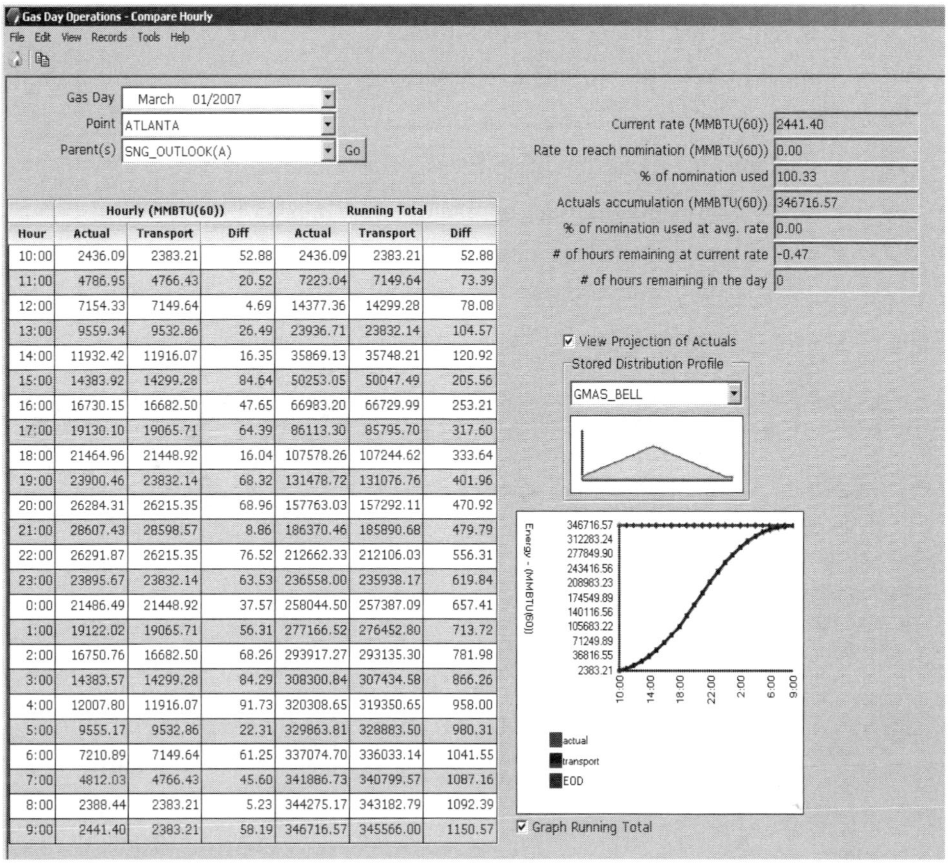

	Hourly (MMBTU(60))			Running Total		
Hour	Actual	Transport	Diff	Actual	Transport	Diff
10:00	2436.09	2383.21	52.88	2436.09	2383.21	52.88
11:00	4786.95	4766.43	20.52	7223.04	7149.64	73.39
12:00	7154.33	7149.64	4.69	14377.36	14299.28	78.08
13:00	9559.34	9532.86	26.49	23936.71	23832.14	104.57
14:00	11932.42	11916.07	16.35	35869.13	35748.21	120.92
15:00	14383.92	14299.28	84.64	50253.05	50047.49	205.56
16:00	16730.15	16682.50	47.65	66983.20	66729.99	253.21
17:00	19130.10	19065.71	64.39	86113.30	85795.70	317.60
18:00	21464.96	21448.92	16.04	107578.26	107244.62	333.64
19:00	23900.46	23832.14	68.32	131478.72	131076.76	401.96
20:00	26284.31	26215.35	68.96	157763.03	157292.11	470.92
21:00	28607.43	28598.57	8.86	186370.46	185890.68	479.79
22:00	26291.87	26215.35	76.52	212662.33	212106.03	556.31
23:00	23895.67	23832.14	63.53	236558.00	235938.17	619.84
0:00	21486.49	21448.92	37.57	258044.50	257387.09	657.41
1:00	19122.02	19065.71	56.31	277166.52	276452.80	713.72
2:00	16750.76	16682.50	68.26	293917.27	293135.30	781.98
3:00	14383.57	14299.28	84.29	308300.84	307434.58	866.26
4:00	12007.80	11916.07	91.73	320308.65	319350.65	958.00
5:00	9555.17	9532.86	22.31	329863.81	328883.50	980.31
6:00	7210.89	7149.64	61.25	337074.70	336033.14	1041.55
7:00	4812.03	4766.43	45.60	341886.73	340799.57	1087.16
8:00	2388.44	2383.21	5.23	344275.17	343182.79	1092.39
9:00	2441.40	2383.21	58.19	346716.57	345566.00	1150.57

Figure 4 Nomination Monitoring Display (Courtesy of Telvent)

4.4 Volume Accounting System

The primary objective of a volume accounting system is to turn measured data into accounting information that meets the requirements for billing/custody transfer. This is necessary because pipeline companies charge the shippers for their gas transportation services based on volume accounting rather than on raw data measured by the host SCADA. Therefore, volume accounting needs to be very accurate, providing the shippers with all the necessary relevant information to verify their billings.

The benefits of a computer based volume accounting system include instantaneous availability of required information, highly accurate measurement and volume accounting data and a commensurate level of confidence in the results for both transporter and shippers, and an economical operation of the volume accounting functions once the necessary system components are installed. Specifically, a computer based volume accounting system can provide the following benefits:

- Efficient measurement services for both internal and external users
- Significant reduction of manual processes
- Reduction of human errors
- Instantaneous processing of all measurement data
- Flexible connection to other applications such as a billing system
- Compliance with GISB standards
- Improved customer services

Figure 5 shows the components of a typical volume accounting system.

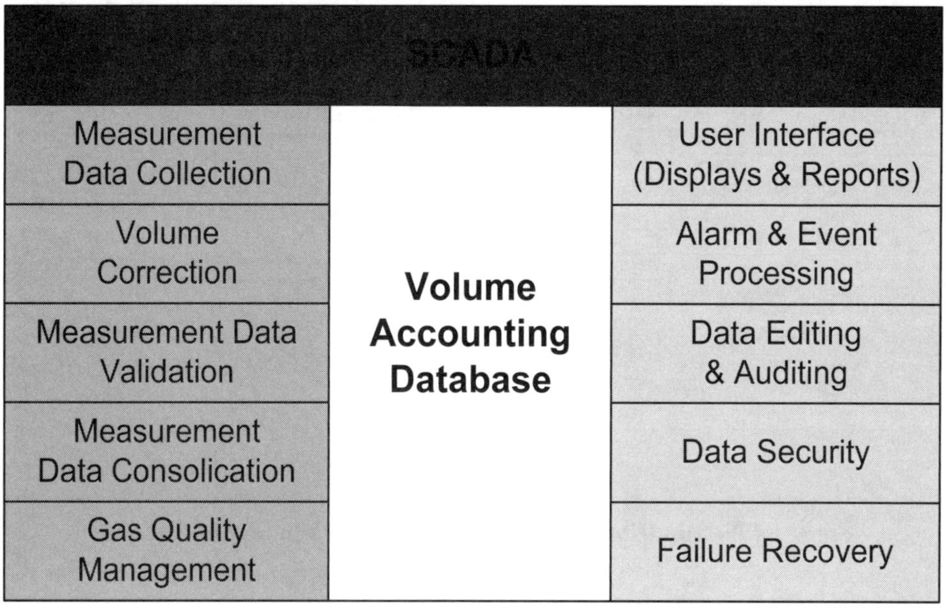

Figure 5: Volume Accounting System

Beginning with the raw measurements received from the host SCADA it takes many steps to produce the required volume accounting information. The accounting process may be required to use specific measurement standards and to apply proper gas quality data. Usually, an accounting system supports the following process:

- Collection of measured data from the host SCADA
- Correction of volume and/or flow data to base conditions
- Validation of measured data

- Consolidation of data and flow totals

In order to support the accounting process, the volume accounting system must provide typical database functions such as database management, auditing, failure recovery, etc. It requires a database separate from SCADA in which to store large amounts of data.

4.4.1 Measured Data Collection

Measured data collection is the first step in the volume accounting process. Field measurement devices supply the raw measured values to remote equipment such as a remote terminal unit (RTU), Programmable Logic Controller (PLC), or a flow computer (FC) at a metering station. The measured values include flows, pressures, temperatures and possibly gas composition. The measured flows are validated and corrected, accumulated into volume, and gas energy is calculated using gas composition. These metering functions are performed either at the remote terminal or at the host SCADA, depending on the capability of the remote terminal and availability of required data. For the purposes of custody transfer, these metering functions are almost exclusively done in flow computers at the metering site, due to the requirement that the flow rate be integrated at a high resolution (at least once a second). This type of resolution is typically not possible at the SCADA host, due to latency in communication with the field device. However, metering done at the SCADA host is usually sufficient for operational purposes.

Each meter station includes one or more remote terminals and measurement devices. The volume accounting system database will store data from each meter station such as station number or name, location, meter run, flow meters, pressure and temperature measurement, and possibly information from a chromatograph. Flow meter information may include static data such as meter identification and type, base pressure and temperature, and possibly ownership.

A volume accounting system may collect flow measurement data both on an hourly and daily basis. Data can be collected automatically, and uploaded and downloaded between the host SCADA and the meter station. Data collection frequency is a function of communication cost effectiveness including such factors as communication resource restrictions and the relative value of metering points. Special contingencies need to be in place to minimize potential loss of data due to communication related problems. The data collection should be time stamped for data validation and flow totals. If the pipeline system crosses several time zones, all time stamps need to be converted to a standard time.

The data collection function should provide the following manual data entries:

- downloading of data to RTU, PLC or FC
- entering of missing data
- editing anomalous data if an automated data validation cannot detect it

- overriding the flow measurement
- entering of enter gas quality data
- editing related measurement parameters

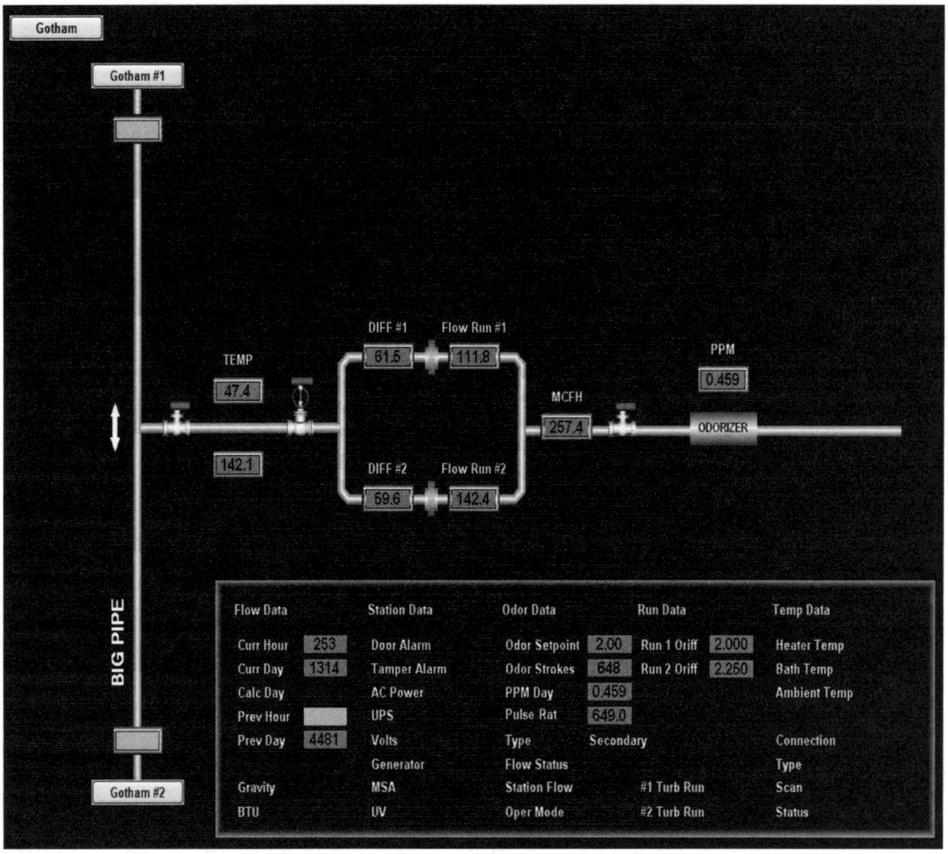

Figure 6 Gas Meter Display (Courtesy of Telvent)

Meter data varies with the means of process and systems. Typical meter data includes the following:

- Time stamp
- Measured flow, differential pressure, pressure and temperature with their measurement status for orifice meters, ultrasonic meters and turbine meters
- Corrected flow and volume with accumulated volume
- Energy or heating value

- Gas composition and quality with quality status
- Meter parameters such as plate size for orifice meters and meter factor for turbine meters

The volume accounting system provides flow accounting functions using the above meter data. It should be able to collect data from the following data collection mechanisms:

- Remote terminals (RTUs, PLCs and FCs) which measure, validate and correct flow
- Remote terminals which measure and/or validate field data and validate and correct flow being done at the host SCADA
- The host SCADA system to enable operation personnel to enter flow measurement data manually. The manual entry system may include a laptop computer and a dedicated data storage terminal.
- Third party data access mechanism in the host SCADA

4.4.2 Metering Capability at Remote Equipment

Most modern remote equipment such as RTUs, FCs and PLCs usually provide flow metering functions. Metering at the remote terminal level is most valuable because measurement data can be retained there in the event of communication system failures; this backup helps ensure data reliability and accuracy. Remote metering functions are possible if the remote terminal has sufficient computing capacity and all required data is available. If a gas chromatograph is not available at the remote location, laboratory tested gas composition data (or data from another upstream gas chromatograph) can be downloaded to the remote terminal from the host SCADA so that flow correction and energy calculation can be performed in the remote terminal. The metering values calculated at the remote terminal should be time-stamped when they are uploaded to the host SCADA.

A remote terminal with metering capability will collect flow and other measurement data frequently from field measurement devices. Even though, for control purposes, raw measured values are sent to the host SCADA at each SCADA cycle, the metering values are uploaded to the host less frequently - normally at hourly and daily intervals - during which the remote terminal performs the metering functions. Typically, a modern remote terminal performs the following key functions:

- Defines meter parameters including instrument specific data such as orifice plate size or meter factor.
- Records flow measurement history, whose quantity, format and frequency depend on the measuring device and communication protocol. Typical flow measurement history includes time stamp, volume, pressure, temperature, energy or heating value, and differential pressure

for an orifice meter and accumulator values for a turbine meter. Some devices can record many other items in their flow measurement history.

- Validates the measured flow, pressure, temperature, and gas composition. In the validation process, the availability and quality of the measured data are checked and the measured values compared with the predefined limits.
- Corrects the raw measured flow and/or volume to a baseline amount. Corrections are made by applying an applicable standard such as AGA-8, NX-19 or ISO-12213 to the validated data on gas composition, pressure and temperature.
- Calculates energy or heating values applying a standard such as AGA-5 or ISO-6976 to the validated gas composition, pressure and temperature data.
- Accumulates gas volumes and heating values on a periodic - normally hourly and daily - basis.
- Stores all measured and calculated values in the terminal's memory, along with the flow history. The memory storage time requirement varies with the amount of data and storage period, but the retention capacity within a remote terminal is relatively short, usually 30 – 60 days.
- Uploads the above metering values to the host SCADA database on an hourly and daily basis. If certain values such as gas composition are not available at a remote terminal, then the terminal must receive them from the host SCADA.

4.4.3 Metering Capability in the Host SCADA

If a remote terminal cannot meter flow or doesn't have the required data, raw measurements are uploaded to the host SCADA, where the metering functions are performed. In other words, the raw measurements are used for both the operation of the pipeline system and for volume accounting.

The data collection and management processes for the host level metering are similar to those for the remote terminal metering. except that in the former they are performed there; raw data collected from the remote terminals is sent to the SCADA host each cycle and is then used at host level metering functions for hourly and daily calculations.

Meter parameters are defined and measurement history is recorded in the SCADA database. Measurement validation, flow correction, and energy or heating value calculations are performed in the host database by applying the required standards such as AGA or ISO standards. Gas volumes and heating values are accumulated on an hourly and daily basis. In addition, these metering values are usually stored in the SCADA database before they are moved to a historical database for long term retention.

4.4.4 Manual Data Entry into the Host SCADA

Authorized operation personnel should be able to enter manual data and override real-time measured data and hourly and daily metering values. This capability is required to ensure data continuity whenever a field measurement device fails, a remote terminal malfunctions, or a communication outage occurs. Manual data must be identified as such to distinguish it from other measured data, and if available, the original data should be kept for audit trails.

Chart-based measurement is still widely used, particularly at small volume measurement sites. Typical chart data include station name or number, chart ID, meter run size with measurement ranges, date and time that chart was put on or taken off, integrator count, and comments. To handle the chart-based measurement data efficiently, the following entries or functions should be possible:

- manually or from 3rd party electronic analysis files
- integration of counts and calculated volume
- recalculation of the gas volume when a meter change is made
- identification of missing or overlapping times
- maintenance and display comments
- inspection reports for meter errors

Typical manually entered or modified data will be:

- flow and/or volume, pressure, temperature, and differential pressure
- orifice meter parameters for orifice meters
- measured volume and accumulator value for turbine meters
- corrected volume and energy values
- meter operating time

Manually entered or modified data must be validated in the same way as other values. Whenever flow related data are modified manually, the following records should be stored:

- name of the person who made any changes
- date and time these changes were made
- comments explaining what was changed and why

4.4.5 Third Party Data

Some meter stations are owned and operated by third parties, i.e an entity that is not the pipeline operating company. In such a case, the third parties should be able to enter their data into the volume accounting system. These third party entry points are treated in the same way as the pipeline owner's remote terminals. The third party data may be sent electronically or manually. Like all measurement data, third party data entry and modification should be time-stamped and separately maintained. The validation of the third party data is required for custody transfer.

Third party metering can also serve as check metering for primary metering owned by the pipeline company. At locations where gas is changing ownership, both parties usually have metering equipment at the site. This affords either party the opportunity to proactively check their own metering against the other party's, and serves as another level of data validation.

4.4.6 Volume or Flow Correction

Volume or flow correction is the process of correcting raw volume or flow measured during flowing conditions to the base condition at which custody transfers occur. This base condition is defined in the contract between the pipeline operator and the shipper. Measured volume or flow can be corrected either by the host SCADA or by the remote measurement devices. The correction process requires the following:

- Pressure and temperature meter readings from flow or volume measuring locations
- Base pressure and temperature
- Gas composition (i.e. its physical properties), such as specific gravity
- Appropriate standards such as AGA or ISO

Volume correction is based on the mass balance principle. The volume or flow rate at the base condition is calculated as follows:

$$V_b = \left(\frac{Z_b}{Z_f}\right)\left(\frac{P_f}{P_b}\right)\left(\frac{T_b}{T_f}\right)V_f$$

or

$$V_b = S^2\left(\frac{P_f}{P_b}\right)\left(\frac{T_b}{T_f}\right)V_f$$

Where:

V_b = Corrected volume at the base condition
V_f = Raw volume measured at the flowing condition
Z_b = Compressibility factor at the flowing condition
Z_f = Compressibility factor at the flowing condition
S = Supercompressibility
P_b = Base pressure
P_f = Pressure measured at the flowing condition
T_b = Base temperature
T_f = Temperature measured at the flowing condition

The supercompressibility or compressibility factors are calculated using either the AGA-8 or NX-19 equation or ISO-12213 Part 2 or Part 3 equation. If the measured values are flows instead of volumes, the volumes are replaced with flows.

When the corrected volumes have been determined, their corresponding heating values can be calculated by applying the AGA-4 or ISO-6975 standards.

To support volume or flow correction, the volume accounting system should be able to record and/or store the:

- history of volume or flow correction, energy or heating value calculation, and volume/energy accumulation
- alarms and events associated with volume correction. For example, an alarm is issued if the volume corrected at a terminal is different from the volume corrected in the host (within some tolerance, these volumes could be different due to differences in integration or averaging of the input values.)
- parameters relevant to volume correction, energy calculation and volume/energy accumulation – time stamp, pressure, temperature, gas composition, volume, energy, AGA parameters, etc.

4.4.7 Flow Measurement Data Validation

The objective of a measurement data validation process is to determine the quality of measured and manually entered data in order to preserve the accuracy of the measured volumes. Erroneous data can be identified and possibly corrected by using accepted validation criteria. Data error or discrepancy can occur due to the following problems:

- Instrument failure – the measurement accuracy may drift, or the flow rate, pressure, temperature or differential pressure remains relatively static over a long period of time. Sometimes, an instrument behaves erratically.
- Flow computer failure – the flow computer fails to function or

incorrectly functions due to configuration error or time synchronization problem

- Communication outage – data cannot be collected during the period of communication outages

The data validation functions include automatic validation testing, editing and auditing, and reporting. Validation tests are usually performed on a per meter run basis. The data types to be validated are measured data, manually entered or modified measured data, and third party data. The validation function may track the changes in measurement parameters such as orifice plate size and meter configuration as well as records all data scrubbing (modification) history.

4.4.7.1 Limit Checking

Limit checking tests the reasonability of measured flow or volume on a per meter basis. Flow or volume measurement is tested against predefined operational limits. This test may include any of the following limit categories:

- Average pressure for hourly and daily volume
- Average temperature for hourly and daily volume
- Average differential pressure for orifice meters
- Gas analysis data
- Heating value

The specified limits include both hourly and daily values, and may be tested separately. If an operational limit violation is detected, the measurement should be flagged as invalid and reviewed by operation staff.

4.4.7.2 Flow Computer Checking

The operation of flow computer should be checked to determine that it functions properly. Flow computer checking involves testing operation status, date and time synchronization, configuration setup, and communication between the flow computer and the host. The event log from each flow computer should be processed to check potential instrument failure and interface status between the instruments and flow computer.

4.4.7.3 Time Stamp Validation Testing

The time stamps of flow measurements are tested against expected time ranges to check their validity and correctness. The following time stamp related problems can be encountered:

- The time stamp is outside of the expected time range. In such a case, the flow measurement should be flagged as erroneous and recorded in the system event log. The operation or measurement staff should investigate it and determine a proper value.

- The SCADA time and remote terminal time are not synchronized. If this occurs, the data polling will be unreliable and the flow measurement data invalid. An investigation of the inconsistencies will be required.
- Measurement data is missing when the time stamp is within the expected time ranges. If hourly or daily flow measurement data is missing, the mistake is detected through the validating process and, the missing data replaced.

When missing or late measurements are replaced, one of the following potential rules can be applied:

- Zero value
- A user defined value
- A value from another measurement device assigned to this device
- The most recent good value for the meter
- The flow measurement value of an identical meter at the same meter station if available

Unless the zero value replacement rule is specified, the second method is used for the meter station with a single meter run. Any measurement that has had a rule applied to it needs to be identified accordingly as an indication of its data quality.

4.4.7.4 Time Series Testing

Testing measured flows or volumes over time can reveal either a frozen value or a rate of change violation. The frozen value checking detects measured flows that have not changed over a specified period of time, while the rate of change violation detects any significant change over a short period of time. Normally, a time series of the hourly and daily volumes is analyzed statistically by examining the trend of volumes with respect to time. Whenever one of these violations is detected, a violation alarm is activated; the violation must be recorded and reviewed by the operations staff and corrective measures taken.

There are several ways of analyzing the time series statistically. One simple approach is that a frozen value is detected if the time series hasn't changed beyond a minimum limit, say two standard deviations of the average rate of change, and a rate of change violation is detected if the time series exhibits a change beyond a maximum limit. The time series can be analyzed and violations detected by applying a statistical testing method.

4.4.7.5 Corrected Flow or Volume Testing

This testing is intended to detect any difference between the corrected flows or volumes reported from remote terminals (flow computer, PLC or RTU) and those values calculated at the host. Such a difference can occur due to the following problems:

- An application of the standards such as AGA-8 fails either at the remote or host level.
- The parameters required by the standards are not configured accurately.

Another corrected volume test includes the comparison of the corrected daily volume against the corrected hourly volumes accumulated over the corresponding gas day. This test will detect not only missing hourly and daily volumes but also any discrepancy between hourly volumes and daily volume.

The differences between the two corrected volumes can be analyzed directly or statistically. Any difference beyond the specified difference limit must activate an alarm and be reviewed by the operations staff to correct the problem.

4.4.7.6 Redundant Testing

If two or more flow meters operate in parallel at a meter station, redundant testing is possible to check the validity and accuracy of the flow meters. This test is based on the assumptions that each flow meter running in parallel at a meter station behaves similarly and the average flow rate of each meter run is the same within the limit of the measurement accuracy.

This validation requires comparison of a specific flow measurement against the flow measurements from the other flow meters. This function can be performed on a real-time hourly and daily basis. Flows with related variables such as temperature and pressure are compared for real-time redundant testing, but hourly and daily volumes from each redundant meter together with station average pressure and temperature are compared for hourly and daily redundant testing. Any meter, where flow or volume differences are greater than a specified limit, should be alarmed and reviewed by the operations staff.

4.4.8 Flow Measurement Data Accumulation

Flow measurement data accumulation involves the totalization and balancing of flow and/or volume measurement data. These accumulated flows and volumes are useful for both operation and measurement purposes. For example, flow totals provide information on peak load and projected flows, which is useful for flow operations to meet the nominated volumes.

Flow totalization is performed for several time periods - hourly, daily and monthly. Hourly, daily and monthly total volumes are normally maintained in the historical database of the host SCADA system, but modern remote terminals have the capacity to store the data. Flow totalization may be performed in the following cases:

- Individual meter stations or groups of meters
- Multiple meter stations

- Multiple supply and delivery points for balancing
- Gas processing plant
- Storage injection and withdrawal points
- Specified regions
- Others as required by the shippers or regulators

After flows are totaled, transportation balancing can be performed on an hourly, daily and/or monthly basis. The following imbalances may be required:

- Supplier totals and imbalances
- Customer totals and imbalances
- System-wide totals and imbalances
- Area send-out totals
- Other balancing required by the applicable regulations

The totaled flow and volume data are time-stamped. It should remain on-line in the historical database for a specified time period (which varies depending on the operational requirements) to make it easily accessible for trending analysis, displays and the preparation of reports. The totaled data is archived for several years as required by the contract and regulation. Figure 7 shows an example of regional monthly totals.

4.4.8.1 Flow Totalization

Flow totalization begins with real-time flow totalization. Real-time flow totalization is performed using real-time flow rates either in remote terminals such as a flow computer and RTU or in the host SCADA system. If the flow rates are totaled in a remote terminal, the amounts are uploaded to the host SCADA on an hourly and daily basis. If the flow rates are sent to the host, the flow totalization is performed in the host. The real-time flow totalization calculations are performed on the data sampling time. Reporting time periods required for the flow totalization are normally:

- Current/previous hour
- Current/previous day
- Current/previous month

The hourly and daily totaled volumes are normally referenced from the starting hour of the gas day. Standard data validation processing or at least limit checking is required to ensure the validity of the totalization process. If a pre-defined condition is violated, an alarm has to be generated and information about the violation stored in the flow totalization database. If real-time flow data is missing due to communication failures, totalized volumes from the affected flow computer are uploaded to the host to recover flow measurement data.

1				**Enerco Gas**					
2				**Region Report**					
3				Period: June 1 2006 to June 30 2006					
4	Region: Eastern								
5	Description: Eastern Delivery								
6									
7									
8									
9		IN		OUT		NET			
10		Volume	Energy	Volume	Energy	Volume		Energy	
11							(%)		(%)
13	Eastern	9,847.4	0	9,625.5	0	221.9	2.3%	0	
14	South Bank	9,847.4	0	9,625.5	0	221.9	2.3%	0	
15	Nickelville	9,847.4	0	9,625.5	0	221.9	2.3%	0	
16	Bigpipe Interconn	9,847.4	0	0.0	0	9,847.4	100.0%	0	
17	Bigpipe City #	9,847.4							
18	SB Industry	0.0	0	524.5	0	-524.5		0	
119	SB Residential	0.0	0	9,101.0	0	-9,101.0		0	

Figure 7 Regional Monthly Totalized Flow (Courtesy of Telvent)

4.4.8.2 Flow Balancing

The volume accounting system uses hourly and daily accumulated measurement data to calculate daily and monthly flow imbalances at all receipt and delivery meter stations, including the flows received from third parties. The flow imbalances at the meter stations can be used to determine account imbalances, customer imbalances, supplier imbalances, and pipeline subsystems, and the entire pipeline system imbalance. Daily flow balancing is difficult at the locations where flow measurements are recorded on charts, due to the logistical challenges of gathering and integrating the chart data in a timely manner.

As the gas day progresses, accumulated volume and energy figures are presented to the operator for each hour that has elapsed since the start of the gas day. To minimize imbalances, the volume accounting system may calculate the average flow rates required to meet the total confirmed nomination requirements based on the current accumulated volume.

4.4.8.3 Peak Load Determination

Information on peak load is important for planning and operating a pipeline system. Peak load is defined as the highest and lowest volumes transported over a specified period for operation and planning purposes. Current operating peak loads and historical peak load values are used for short-term planning. Peak load calculations may start at the beginning of the heating season or the calendar year, say January 1. It may be convenient for the operating staff to retrieve and use the peak load information if the peak load histories and values are time-stamped and stored in a database. The information on peak load may also be required for storage injection and withdrawal operations.

The following peak loads or their variations may be required for controlling flows to satisfy nominations and for operation planning:

- Highest and lowest hourly values and hours in the current gas day
- History of highest hourly and daily values
- History of lowest hourly and daily values
- History of highest three consecutive days and their peak values

4.4.8.4 Gas Day Flow Projection

This function is used for controlling flows to satisfy nominations based on estimated or historical flow profiles. The pipeline operator prepares a plan to reduce or increase receipt or delivery volumes in order to ensure that the gas volumes at receipt and delivery points will fall within nomination limits. The flow projection can be determined by projecting the end of day flow accumulation at each receipt and delivery point based on hourly totalized flows and normalized flow profiles.

A flow profile can be specified for an hourly basis over the gas day and provide the receipt or delivery pattern of gas volumes at specified facilities such as gas processing plants, industrial sites or city gates. If the flow profile is not available, the flow can be projected by using the accumulated volume up to the current point and a constant profile for the remaining operating hours.

4.4.8.5 Nomination Monitoring

A pipeline company needs to monitor volumes at supply and delivery points for shippers throughout the gas day. This function compares actual accumulated volumes against confirmed nominations at the receipt and delivery points. It is intended to balance or minimize imbalances by alerting the operators if the confirmed nominations are not fulfilled or are overrun. The volume data is used in monitoring the gas delivery and supply for clients and projecting the flow rates required to meet a gas nomination. It is assumed that the net gas volume data is available at all receipt and delivery points.

The schedule of nominations for a given gas day is entered into a nomination system either manually or electronically. Nominations are compared to actual flows accumulated during the gas day. This nomination monitoring application uses flow projections at the measurement points to predict the end of gas day value using the flow profiles on the basis of current and accumulated flow information.

Specifically, nomination monitoring takes the following calculations for nomination tracking and monitoring:

- Projected end of day volumes using the flow profile specified in the current gas day nomination and the current accumulated volume at each custody transfer point. The current accumulated volume can be obtained from the volume accounting system or from a flow computer.
- Comparison of the projected end of day volumes against the nominated volume and the MDQ. If the projected volume at the end of the gas day is over or under the nominated volume or MDQ, an alarm is activated to alert the pipeline operator of a problem.
- Flow rates required to satisfy the nomination volumes by the end of the current gas day. The operator may adjust the flow rates if the pipeline capacity limitation has not been violated.
- If the nomination imbalance is larger than the tolerance specified in the current gas day nomination, an alarm alerts the affected shipper that a problem exists.

4.4.9 Gas Quality Management

One of the primary responsibilities of a gas pipeline company is to monitor and ensure the quality of the gas being transported. Gas quality management includes all aspects of determining gas quality from the measurement device through to the proper application of industry standards. Having gas quality data allows operators to identify potential contamination, which could cause measurement and operating problems. Such data can also be used to calculate corrected volume and heating values accurately at remote terminals and/or SCADA host.

4.4.9.1 Definition of Gas Quality

The quality of natural gas is measured in terms of gas composition and other parameters. Gas composition is expressed in terms of mole percentage or fraction of methane, ethane, propane, N-butane, Iso-butane, and trace amount of heavier hydrocarbons. Other important characteristics are:

- Specific gravity – Specific gravity must be known for gas volume correction and flow equation. It is determined by gravitometer or gas chromatograph.
- Heating value – The heating value is a reflection of the energy content of the natural gas and is determined by a calorimeter or gas chromatograph.
- Hydrogen sulfide content – Hydrogen sulfide concentrations must be limited and monitored to secure safety and limit corrosion. They are measured by means of an H_2S analyzer and should be controlled at the source.

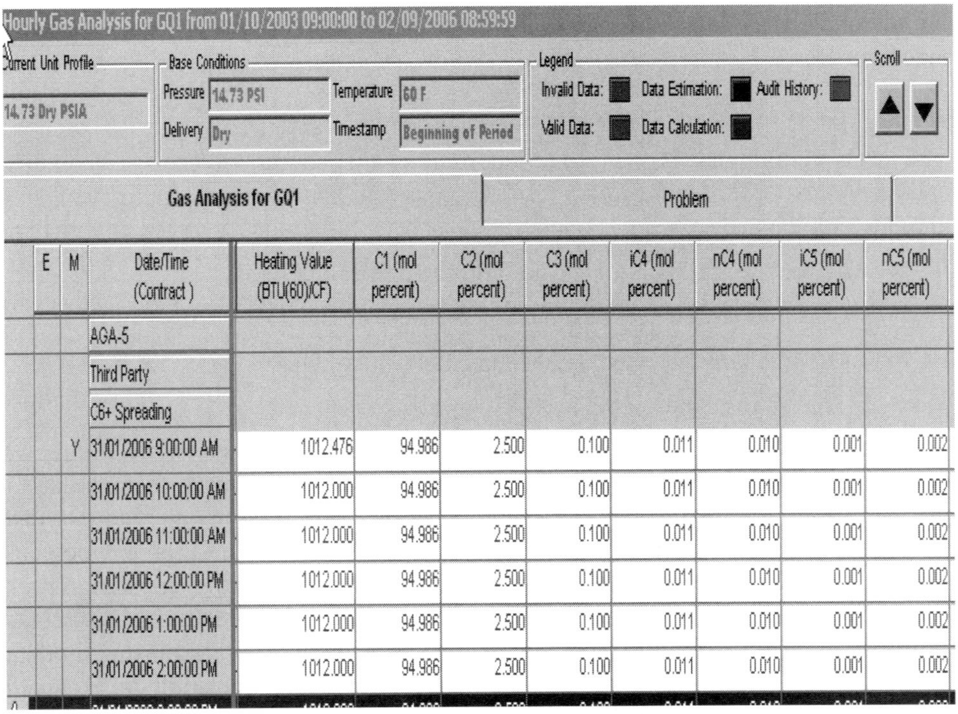

Figure 8 Gas Quality Display (Courtesy of Telvent)

- Hydrocarbon dew point – Dew point indicates the presence of liquefiable hydrocarbons, which include heavier hydrocarbon components. The presence of liquid hydrocarbons in the gas stream can reduce pipeline efficiency and cause internal pipeline corrosion with amines.
- Water vapor content – Water vapor with natural gas in a high pressure pipeline can cause hydrate formation, internal pipeline corrosion, and lower heating value. The content is determined by a moisture analyzer or dew point tester.
- Carbon dioxide content – Carbon dioxide with free water forms carbonic acid, which corrodes steel pipe.
- Liquids and particulates – The presence of impurities such as heavier hydrocarbons and sands may adversely affect the transmission pipeline efficiency and even the accuracy of the flow measurement of gas.
- Sulfur content – Sulfur with free water can form sulfuric acid, which corrodes steel pipe.

There are other contaminants or compounds such as oxygen and nitrogen. Acceptable levels of the above qualities will be specified in transportation agreements. These quality specifications need to be monitored and enforced by the pipeline company to comply with the agreed specifications. Contaminants such as liquids and particulates can cause lower pipeline efficiency and large metering inaccuracy.

4.4.9.2 Determination of Gas Quality

The gas quality management function defines the source of each gas quality characteristic and monitors the resultant gas quality. Typical sources of gas composition data are:

- Chromatograph data measured directly from the field
- Chromatograph data assigned to the field devices
- Off-line laboratory analysis data entered manually
- Third party data

Specific gravity can be measured directly by a gas gravitometer or indirectly determined from gas composition and in this case not requiring measurement. If the chromatograph data is measured directly in the field, it is uploaded to the host SCADA database and then validated. Since all gas flow measurement devices require gas composition data, this function should be able to assign gas composition data to gas measurement devices even where directly measured gas analysis data is not available. It should also be able to make single gas quality data available to multiple gas measurement devices. The gas composition data tracked by a real-time model may be used at downstream measurement points if the tracked composition data is sufficiently accurate and the regulating agency approves the practice (6).

Gas composition data is often obtained from laboratory analysis. Also, composition data may be obtained from a third party measurement. If gas composition data is obtained from lab analysis or a third party it should be time-stamped and manually entered into the database.

The gas quality management function should provide for gas composition data to be downloaded to the field measurement devices from the host SCADA if there is no gas chromatograph installed. Since gas composition does not change frequently, the composition database usually is not refreshed in real-time.

4.4.10 Support Functions

A computer based gas management system will include database support functionality found in corporate database systems. The design of the system should ensure that it is flexible enough to incorporate the individual business processes unique to each pipeline company without the need for revisions. Rather, this should be a configuration exercise and not involve software development. Finally a well designed system will allow for ease of integration into other applications by adhering to industry standard interfacing (such as OPC) and database queries such as SQL.

4.4.10.1 User Interface

User interfaces should be intuitive and graphical. An intuitive user interface helps the user navigate the system more easily than unnecessarily complex ones. A graphic user interface for instance may help simplify the viewing, recording and editing of flow measurement data. Two types of user interface displays are required: static data screens and dynamic data displays. The static data screens normally provide system definition and configuration information such as meter configuration, meter parameters and validation rules. Dynamic data screens are the main operating interface used to display all measured and calculated data such as collected and totalized flow data. At a minimum the following displays are required to maintain and use the volume accounting system properly:

1. Static Data Screens
    - Data on contract and nomination details including client name, location and client role (i.e. supplier or customer)

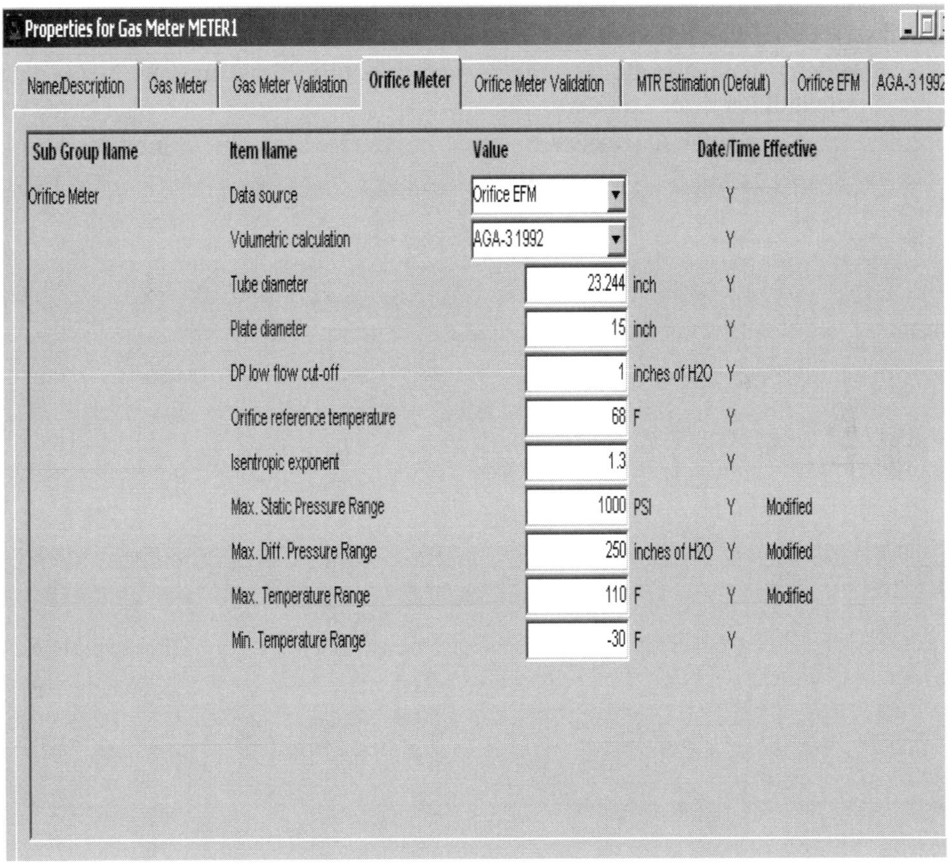

Figure 9 Orifice Meter Information (Courtesy of Telvent)

- Meter configuration data documenting the meter station, name, ID or tag number, type, run information and location as well as the plant name, company name, etc. Chart-based meter information may require separate configuration data entry displays.
- Meter information about the source of gas, its destination, route, flow profile, etc.
- Measurement validation rules and parameters
- Alarm conditions including limits
- Meter parameters including applicable standards for each meter type. Meter types should be of custody quality and may include such meter types as turbine, positive displacement, orifice, ultrasonic, etc.
- Totalized flow parameters such as location, meter station, time period, etc.

2. Dynamic Data Displays of the following information may be included:
   - Nomination data including the contract gas day, volume, tolerance, etc.
   - Meter data for hourly, daily and monthly flow measurement data including corrected flows

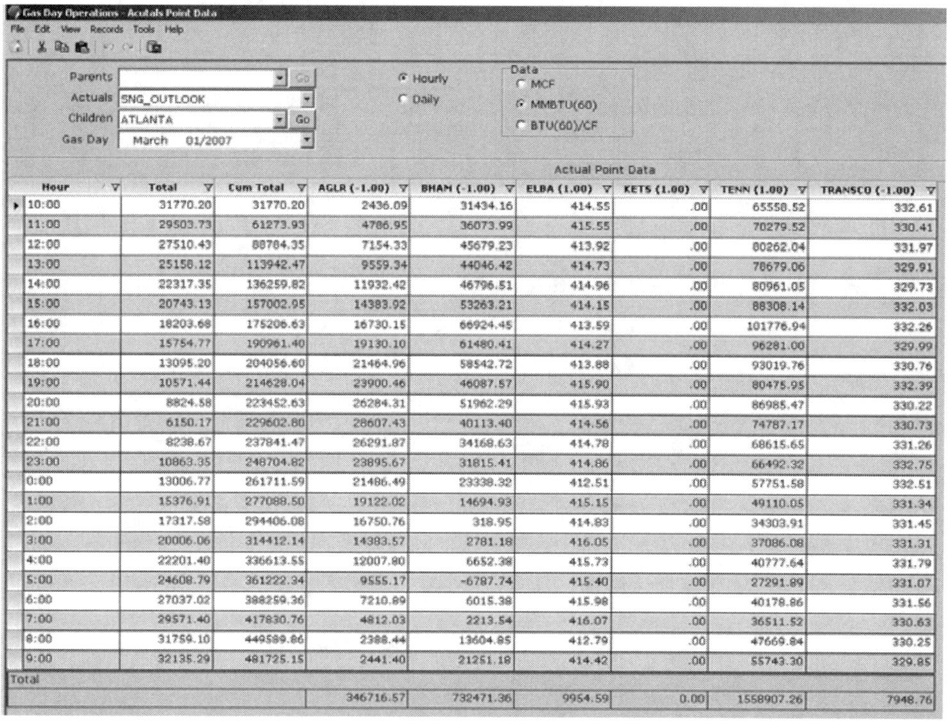

Gas Day Operations - Acutals Point Data

File Edit View Records Tools Help

Parents: | Go

Actuals: SNG_OUTLOOK

Children: ATLANTA | Go

Gas Day: March 01/2007

Hourly / Daily

Data: MCF / MMBTU(60) / BTU(60)/CF

Actual Point Data

Hour	Total	Cum Total	AGLR (-1.00)	BHAM (-1.00)	ELBA (1.00)	KETS (1.00)	TENN (1.00)	TRANSCO (-1.00)
10:00	31770.20	31770.20	2436.09	31434.16	414.55	.00	65558.52	332.61
11:00	29503.73	61273.93	4786.95	36073.99	415.55	.00	70279.52	330.41
12:00	27510.43	88784.35	7154.33	45679.23	413.92	.00	80262.04	331.97
13:00	25158.12	113942.47	9559.34	44046.42	414.73	.00	78679.06	329.91
14:00	22317.35	136259.82	11932.42	46796.51	414.96	.00	80961.05	329.73
15:00	20743.13	157002.95	14383.92	53263.21	414.15	.00	88308.14	332.03
16:00	18203.68	175206.63	16730.15	66924.45	413.59	.00	101776.94	332.26
17:00	15754.77	190961.40	19130.10	61480.41	414.27	.00	96281.00	329.99
18:00	13095.20	204056.60	21464.96	58542.72	413.88	.00	93019.76	330.76
19:00	10571.44	214628.04	23900.46	46087.57	415.90	.00	80475.95	332.39
20:00	8824.58	223452.63	26284.31	51962.29	415.93	.00	86985.47	330.22
21:00	6150.17	229602.80	28607.43	40113.40	414.56	.00	74787.17	330.73
22:00	8238.67	237841.47	26291.87	34168.63	414.78	.00	68615.65	331.26
23:00	10863.35	248704.82	23895.67	31815.41	414.86	.00	66492.32	332.75
0:00	13006.77	261711.59	21486.49	23338.32	412.51	.00	57751.58	332.51
1:00	15376.91	277088.50	19122.02	14694.93	415.15	.00	49110.05	331.34
2:00	17317.58	294406.08	16750.76	318.95	414.83	.00	34303.91	331.45
3:00	20006.06	314412.14	14383.57	2781.18	416.05	.00	37086.08	331.31
4:00	22201.40	336613.55	12007.80	6652.38	415.73	.00	40777.64	331.79
5:00	24608.79	361222.34	9555.17	-6787.74	415.40	.00	27291.89	331.07
6:00	27037.02	388259.36	7210.89	6015.38	415.98	.00	40178.86	331.56
7:00	29571.40	417830.76	4812.03	2213.54	416.07	.00	36511.52	330.63
8:00	31759.10	449589.86	2388.44	13604.85	412.79	.00	47669.84	330.25
9:00	32135.29	481725.15	2441.40	21251.18	414.42	.00	55743.30	329.85
Total			346716.57	732471.36	9954.59	0.00	1558907.26	7948.76

Figure 10 Display of Hourly Totalized Flow (Courtesy of Telvent)

- Gas analysis required for entering and modifying chromatograph and lab data
- Gas validation with audit trails
- Various totalized flows
- Heating values
- Alarms and events
- Peak loads for the current year and current peak heating season
- Nomination status and flow projection relative to the confirmed nomination for the current gas day
- Data trending including flow measurement data, gas analysis data, totalized flows, etc.
- Communications including outage and upload/download history

4.4.10.2 Reports

The reporting system should enable the operation staff to customize reports or select from a collection of defined reports. It may allow the operation staff to select how reports will be distributed and to automate distribution of them via email, web, or fax, or to print the reports and fax them manually. The ability to produce reports of the following types of information is a minimum requirement of a comprehensive reporting system:

- Contracts, including new or revised contracts
- Nominations including confirmed nominations and allocated volumes
- Daily and monthly volumes
- Daily and monthly totalized flows
- Transportation and pipeline system balances including imbalance for each supplier/customer
- Gas analyses including rejected, missing and late analysis
- Meter stations including chart meters
- Alarms including events
- Prior period adjustments if there are adjustments for the prior period
- Communication, including upload and download activities
- Third party input including missing values and third party auditing

Enerco Field Services

Edited Hourly Record Report

December 2005

Report Time: 2/7/06 9:09 AM

Meter: **Buckwheat-12**

Description: Buck Gather

Meter Run No.:

RTU Identifier: BB-1234-AG

PERIOD	TIME	DIFF	PSI ()	TEMP	FLOW HOURS	MCF @ Dry	BTU @ Dry	DTH
01/12/2006	9:00	35.13	119.86	74.88	1.00	3994.80	1083.001	4326.37
01/12/2005	10:00	36.93	121.43	75.03	1.00	3995.33	1083.001	4326.94
01/12/2005	11:00	38.47	122.89	76.12	1.00	3996.66	1083.001	4328.38
01/12/2005	12:00	38.91	124.22	76.85	1.00	3997.31	1083.001	4329.09
01/12/2005	13:00	39.87	125.52	78.59	1.00	3998.40	1083.001	4330.27
01/12/2005	14:00	40.85	126.71	78.65	1.00	3998.44	1083.001	4330.31
01/12/2005	15:00	42.14	128.30	79.95	1.00	3999.03	1083.001	4330.95
01/12/2005	16:00	43.11	129.71	80.65	1.00	3999.15	1083.001	4331.08

Figure 11 Meter Station Report (Courtesy of Telvent)

4.4.10.3 Alarm and Event Processing

An event message is produced whenever certain operational activity takes place such as recovery from a failure or manual data entry. An appropriate alarm message is generated when an error or discrepancy is detected. Alarm messages alert the appropriate parties to take corrective actions. The parties may include the pipeline operators, measurement and billing staff, and shippers. Alarm messages should cover all functions of the gas management system including transportation, storage services, volume accounting and invoicing. Typically, alarms are activated when staff attention is required in the following areas:

- Contracts
- Nominations
- Gas storage including both storage nomination and injection /withdrawal operation
- Transportation imbalances
- Communications
- Measurements
- Allocations
- Invoices

Since the number of alarm messages can be extensive, an alarm filtering system based on certain rules can be useful to identify the alarms appropriate to the problems.

4.4.10.4 Data Editing and Auditing

A gas management system requires an editing function to maintain data accuracy. The editing function allows the system users to modify contracts and nominations, which are subject to frequent change. The ability to conveniently edit flow measurement data is required because it can be affected by events such as communication or measurement device failure, human error, wrong gas composition.

The editing function should serve to view and modify transportation service data such as contract and nomination, and volume accounting data such as measurements, gas quality, metering configurations, etc. Any activities associated with third parties should also be logged. This audit trail capability is a key requirement to meet the needs of commercial invoicing and reporting.

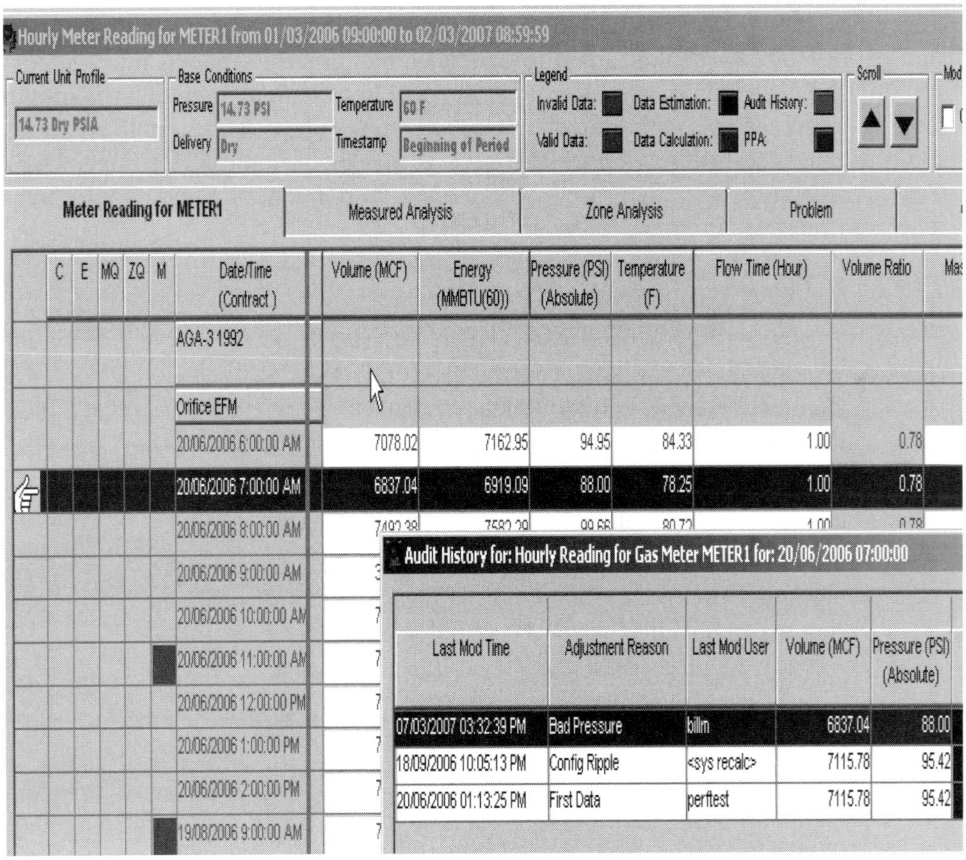

Figure 12 Example Display of Data Auditing (Courtesy of Telvent)

The auditing function maintains logs of transportation and measurement data modification activities. This ensures that the original data is overridden only when modifications to the transportation or measurement data are justified according to the contract. The logs usually include the following information:

- Editor name and signature
- Contract and/or nomination data that were modified
- Measurement data that were modified
- Current and previous values
- Reasons for the modifications
- Modification date

4.4.10.5 Data Security

Information about gas transportation service is essential to gas pipeline companies and their associated business partners including shippers and third parties. Since gas pipeline companies play a central role in providing transportation service, they are responsible for maintaining information security and dissemination. Pipeline companies must provide accurate information expeditiously, confidentially, and in an appropriate businesslike format to the concerned parties. The gas management system used must have access mechanisms capable of both maintaining maximum security in processes and transactions and encrypting them. Unscrupulous parties could use pilfered information to their own ends. For example, if a shipper knew that the pipeline was in danger of not meeting its nominations, they could press for lower transportation fees for a new injection.

4.4.10.6 Databases

Databases of the following operations are the minimum required to support a gas management system efficiently:

- Contract, which contains data such as contractors, contract types, effective dates, custody transfer points, and contract volumes
- Nomination, which contains all the nomination related data such as the initial nomination and confirmed nomination data
- Flow measurement configuration, which defines the gas flow measurement system including the measurement devices, meter runs and stations, applicable standards, and their changes
- Gas quality, which contains gas quality and compositions, composition sources, associated measurement devices, and gas analysis and change history
- Gas measurement, which contains raw and corrected flow or volume data including estimated values, validation of hourly, daily and monthly totalized volume data, shipper data, upload and download records, and validation and recalculation history

- Transportation imbalance, which contains imbalances from suppliers, customers, areas, systems, etc.
- Alarms, which contain measurement related alarms and events data including communication problems throughout the gas transportation and volume accounting processes

All of theses databases need to be traceable and auditable. If the pipeline company maintains any databases for third parties they must be partitioned (physically or logically) into separate entities in order to preserve the data integrity and security.

4.4.10.7 Failure Recovery

The volume accounting system has to be robust enough to preserve the flow measurement data in the event of system failures. Failures can occur at various points: field devices including RTU and FC, communication networks, the SCADA system, and volume accounting databases. The failure recovery process will depend on the sources of failure, the field data collection device, manipulation capability, and duration of failure. The sources of the failures should be alarmed to the operator and logged in a database.

If a primary measurement device such as a meter or FC fails, the host SCADA may be able to detect it and inform the operator of the measurement problem immediately. Possible recovery procedures will vary depending on the capabilities of field devices. The operator may have to take a corrective action immediately in order to avoid a recovery problem, if the failed field size is significant.

If the communication network, SCADA system or volume accounting system (including database) fails, flow measurement data will be uploaded to the host and back to the volume accounting system when the failed component is restored. The flow measurement data uploaded after the failed system is restored should be updated. If the failure duration is long, say longer than an hour, additional measures may need to be taken to obtain corrected hourly volumes and to validate the measured data where only raw data is available from a field device.

4.5 Gas Control Applications

Gas control applications assist the gas operators in providing effective gas control and other users in meeting their business requirements. These applications include functions for gas inventory monitoring, gas storage pool management, and gas load forecasting. This section discusses gas inventory monitoring and load forecasting systems in detail.

4.5.1 Gas Inventory Monitoring System

A gas inventory monitoring system helps the operators to monitor line pack (volume of gas in the pipeline) and manage gas storage pools. The system consists of a line pack component and a storage pool management component. The line pack monitoring function provides estimates of gas volumes in pipeline segments and changes in the line pack, while the storage pool function provides estimates of gas volumes in storage pools and a calculation of their injection/withdrawal from measured flow rate, storage pressure, and temperature.

The inventory monitoring system has to perform the following functions in order to provide complete inventory monitoring capabilities:

- Calculate the line pack volume in each pipeline segment, using known pipe configurations and measured pressure/temperature data.
- Maintain the estimated line pack volume of gas in each pipeline segment.
- Calculate the inventory estimate of each storage pool, using known storage pool configurations and measured pressure/temperature and metered injection/withdrawal rates.
- Maintain the estimated inventory volume of each storage pool.
- Maintain gas volumes by adding multiple segment and storage pool volumes for sub-systems and the total system.

Large complex pipeline systems are divided into multiple sub-systems based on operating regions, where inventory and storage pool volumes are totaled and monitored. Each sub-system is further divided into pipe segments and storage pool levels, where inventory volume and its changes are calculated.

The inventory monitoring function starts with the calculation of gas volumes in pipe segments and storage pools and then totals the calculated volumes for the sub-system level for inventory monitoring. The monitoring function also compares sub-system and/or system inventory levels and their changes with alarm limits.

Line pack for the pipe segment can be estimated in real time using gas supercompressibility and segment volumes by applying the average segment pressure and temperature. Accurate line pack calculation is only possible with a real-time transient model. Refer to Chapter 6.4.1.3 for this method of line pack calculation. The line pack management function maintains the following data:

- Start of gas day line pack for the pipe segments, sub-systems and system
- Current line pack and line packing/drafting rates in each segment and sub-system
- Daily line packing/drafting rates on an hourly basis
- Monthly line packing/drafting rates on a daily basis

Gas storage pool management allows the operator to monitor gas inventory in each storage pool and make injection/withdrawal operations in the course of managing gas transportation. Gas volume in a storage pool is estimated using storage pool volume, pressure, temperature, and supercompressibility, while volume changes are calculated using the metered injection and withdrawal rates.

See Reference (7) for an accurate volume calculation of gas storage pools.

The gas storage inventory and injection/withdrawal volume data is used for gas scheduling. A storage pool management function maintains the following data:

- Start of month gas storage inventory
- Current gas storage inventory volume and injection/withdrawal rates, measured at storage pool facility
- Daily injection and withdrawal rates on an hourly basis
- Monthly injection and withdrawal rates on a daily basis

These line pack and storage pool volumes are combined to give the inventory volumes belonging to the same sub-system and system. The following inventory data for sub-systems and system are maintained:

- Start of month total inventory volumes
- Current total inventory volumes and their changes
- Daily total inventory changes on an hourly basis
- Monthly total inventory changes on a daily basis

The user interface for the inventory monitoring system has to provide easy access to detailed inventory data including the line pack and storage pool volume. The user interfaces may include the following:

- Line pack data summary with trends
- Storage pool volume summary with trends
- Inventory sub-total summary with trends
- Inventory total summary with trends

4.5.2 Gas Load Forecasting

It is important to provide accurate forecasts of gas demand for short and long periods to operate the pipeline system efficiently and to make optimum use of the pipeline facilities. Short-term forecasts from day to day or from week to week are important for operations, particularly for local distribution companies (LDC's), while long-term forecasts are useful for planning and designing pipeline systems and their facilities. This section describes only the functionality of the short-term load forecasts.

The short-term load forecasting system allows gas companies to predict short-term pipeline system load. The main function of the system is to identify which weather forecast district a gas load area is located in and generate hourly gas loads so the effects of current and predicted weather conditions can be anticipated. Such a weather dependent load is often called send-out, or sometimes "firm send-out." The adjective "firm" is chosen because an LDC is legally bound to provide gas to firm customers, short of catastrophic circumstances such as pipeline accidents.

For example, residential and small commercial customers do not nominate for their gas; it is provided to them under a monthly service contract. Accurate send-out estimates are important to ensure that a LDC's supply nominations are done as economically as possible. The system can produce load forecasts of the current gas day and several days in the future. However, the results for the future are less accurate than the load forecast for the current gas day, mainly because the predicted weather conditions may not be reliable.

Since gas demand is largely influenced by temperature and other weather related parameters, reliable short-term load forecasts depend largely on an accurate expression of the relationship between gas demand and temperature with weather parameters. The relationship is statistical in nature and thus analyzed using a traditional statistical method such as a linear regression analysis or other more sophisticated statistical techniques. It uses multi-year historical data of gas load and weather in order to develop a proper equation and to determine similar gas day demand patterns.

The gas load consists of two main components: the fixed load and the predicted load. Fixed load customers are mostly industrial users such as power and fertilizer plants. Their demand requirements are relatively well defined, so load forecasting is not required. The predicted loads can be determined by the gas load forecasting system in a similar day method or statistical method. These two methods may be used independently or together. When they are used together, a relative weight is assigned to each method depending on its reliability and then applied to the predicted load calculated from each method to determine an overall load forecast.

The similar day method searches for the gas day load forecast from a historical database by matching a set of selection criteria. The historical database contains several parameters such as seasonal factor, weather, temperature averaging and volume lag, the time of day and day of the week, etc. The load forecast that matches the criteria is selected as the predicted load for the gas day being forecast. Since the current demand may have increased from the gas consumptions of the previous years, the selected load may need to be adjusted by a growth adjustment factor.

A linear regression model has been successfully implemented (8) and other statistical methods used to predict gas load (9, 10). Recently, several companies successfully implemented neural network techniques (11). The gas load forecasting system using a statistical method uses the same data as the similar day method. More specifically, the statistical forecasting system estimates the load forecast using a statistical method with the following data sets stored in an historical load database:

- Seasonal factor, because the uncertainty in consumption patterns is different for each season
- Weather related parameters, including ambient temperature, wind speed, sun light or cloud cover, humidity, etc., among which ambient

temperature is most important

- Temperature averaging and volume lag due to consecutive cold days
- Time of day and day types such as week day, weekend or holiday

The historical database contains multi-year hourly historical records of gas load corresponding to the above data for each gas load forecast area. These methods do not necessarily use all the above parameters. The choice depends on the load forecast area and the availability of data.

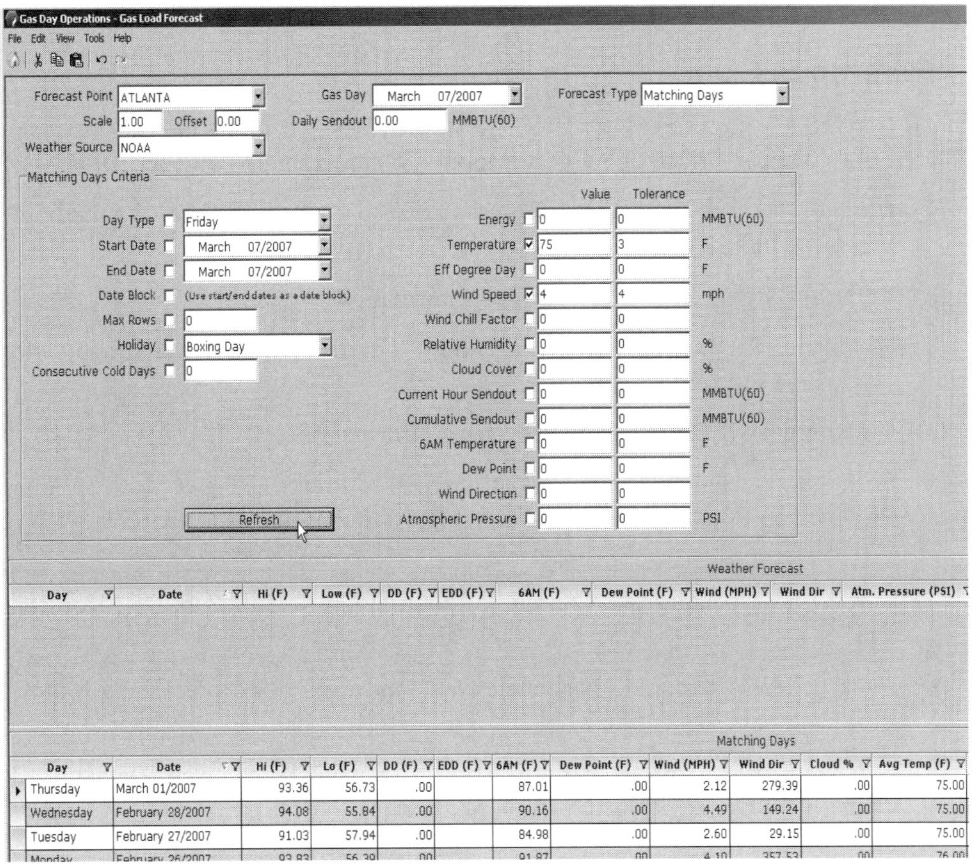

Figure 13 Load Forecasting Display (Courtesy of Telvent)

Specifically, the short-term gas load forecasting system needs to perform the following functions to support its main load forecasting capability:

- Enter the actual weather conditions of the current gas day and the predicted weather conditions of the future gas day for each gas load forecast area. If the number of gas load forecast areas is large, the data entering process needs to be automated.

- Build an historical load database by collecting and editing the hourly loads and associated parameters for each gas load forecast area.
- Enter fixed loads and include the loads in the final loads for their corresponding gas load forecast area.
- Search the database for gas days with matching similar day load criteria.
- Display the load forecast and actual load with weather conditions for each gas load forecast area as the gas day progresses as well as the load forecasts and predicted weather conditions for the future gas days.

References

(1) Bergen, H. and Treat, R. W., "EFM Provides Accuracy In Measurement Data", The American Oil & Gas Reporter, October and November, 1994

(2) Kimball, D. L., "Unbundling Prompts Shift to Daily Balancing and Billing", Pipe Line Industry, Oct., 1994

(3) Refer to www.GISB.org for detailed information

(4) McQuade, R., "GISB Standards Help Promote Seamless Marketplace for Gas", Pipe Line & Gas Industry, Apr., 2001

(5) Refer to www.EDIGAS.org for detailed information

(6) Seelinger, J. and Wagner, G., "Thermal Billing Using Caloric Values Provided by Pipeline Simulation", Pipeline Simulation Interest Group (PSIG), 2001

(7) Tek, M. R., "Underground Sorage of Natural Gas," Gulf Pub. Co., Houston, 1987

(8) Banks, C. W., "Colorado Insterstate Develops a gas sales forecast algorithm", Pipe Line Ind., Sep., 1986

(9) Lyness, F. K., "Consistent Forecasting of Severe Winter Gas Demand", J. Operational Research Society, Vol. 32, 1981

(10) Taylor, P. F. and Thomas, M. E., "Short Term Forecasting: Horses for Courses" J. Operational Research Society, Vol. 33, 1982

(11) Miura, K. and Sato, R., "Gas Demand Forecasting Based on Artificial Neural Network", Proc. of International Pipeline Conference, ASME, 1998

5 Liquid Pipeline Management System

5.1 Introduction

This chapter discusses several applications unique to common carriers of liquid products – batch scheduling systems with nomination management, batch tracking and liquid volume accounting systems. The transportation of liquid petroleum products starts with a request for product movement, usually in the form of a nomination, by shippers to the pipeline company. The pipeline company schedules and allocates the nominated volumes and then monitors the product movements when they are injected into the pipelines. After the products are delivered to the nominated delivery locations, the volumes are measured and accounted for billing to the shippers.

Common carriers publish tariffs that are dictated by FERC 68 in North America. Tariffs cover the transportation rates and rules including nominations and minimum batch size requirements. For common carrier pipelines, the nomination is a way for a shipper to reserve space in the pipeline to transport petroleum products from an origin to delivery locations via the pipeline system. Shippers are obliged to submit their initial nominations and the subsequent changes according to a certain set of rules in order to ensure that the nominations are accepted and their changes can be properly facilitated.

The tariff requires that all shippers submit to the pipeline company their intended shipping volumes and other relevant information on a certain date before the cycle lifting date. This initial nomination data permits the pipeline company to develop a plan to handle all shippers' transportation requirements. After initial nominations have been made, shippers are allowed to change their nominations until a specified date without incurring additional charges.

After the final changes are made, the pipeline company develops a transportation schedule to accommodate the shippers' nominated volumes. This is normally called a batch schedule because common carriers transport petroleum products in multiple batches. When the total nominated volume for all qualified shippers is greater than the pipeline capacity, the volume is prorated to allocate space on the pipeline. This prorationing or apportionment reduces the total volume to be moved in a cycle according to pre-assigned prorationing rules defined in the agreements between shippers and the pipeline company. This capacity constraint results in the total nomination for a shipper being limited to a maximum volume for the batching cycle. Lifted batches are continuously tracked to ensure batch movements are handled efficiently from lifting to delivery.

It is important that shippers adhere to their nomination because the schedule is built on the basis of the nomination. If a shipper fails to deliver what they nominated, then the schedule must be revised on short notice impacting other shippers.

Common carriers should accurately measure all lifted and delivered volumes. The measured volumes are validated and corrected to base conditions before billing the transportation charges. The transported products must be accurately accounted, because the corrected volumes are the basis of transportation fees charged by the pipeline company and of the custody transfer between producers and customers. The process of product transportation service, as summarized in Figure 1, is similar to the process of gas transportation service.

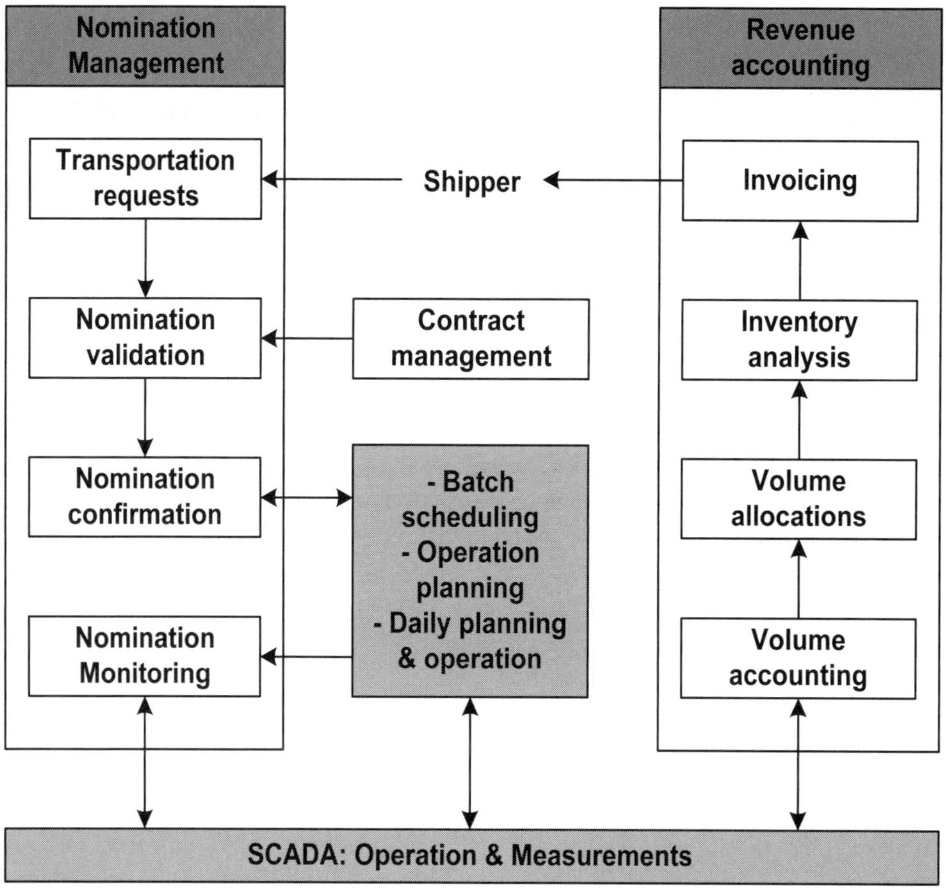

Figure 1 Process of Transportation Service

Normally, common carriers require the volumetric and revenue accounting on a monthly basis and calculation of tickets and inventories on a daily basis. A computer-based accounting system is based on the following transportation business process:

- The common carrier and shippers negotiate the contract and tariff. This task is performed by the marketing department of the common carrier, which maintains the contract and tariff database.
- The shippers enter and modify monthly nomination data, and the schedulers develop a monthly schedule using the inventory and nomination data, out of which a daily schedule is created.
- The pipeline dispatcher or field operation staff controls the batch injection and delivery according to the daily schedule, generating tickets of the injected batches for the shippers. Meter tickets are collected by the SCADA system through the metering station or manually entered by field operation staff if the ticketing process is not automated. All meter tickets are validated, daily scheduled volumes to actual tickets are verified, and then actual ticketed volumes are compared to scheduled volumes. All tickets that are accurately accounted for are closed. Tickets and inventories are reviewed by scheduled batch through a batch movement balancing process to allocate tickets in a daily schedule. Ticket allocation is required for revenue accounting. These functions are normally performed by a volume accounting system.
- The revenue accounting and invoice for each shipper is generated from the volume accounting and tariff. The revenue accounting system consolidates all billable transactions into a revenue database to generate the invoices for all shippers. It allows the system users to review all contract information and to calculate the prices for the transportation services based on tariffs. The system generates invoices and transportation service reports for shippers and internal customers such as marketing and management.

The complete transportation service system may be divided into the following systems:

- SCADA system, which is discussed in Chapter 1
- Operational applications such as pipeline leak detection
- Scheduling system with nomination management
- Shipper information system
- Volume accounting system
- Commercial system such as tariff management and revenue accounting

This chapter focuses on the nomination and scheduling system and volume accounting system in detail.

These systems can be integrated by means of data through a common database, interfacing with the scheduling and nomination, SCADA and other applications such as batch tracking, the volume accounting, and revenue accounting systems. As a minimum, the integrated system requires the following data:

- Pipeline configuration data including pipe segments, tanks, etc.
- Various static data such as location names, commodity names, shipper names, etc.
- Nomination data containing initial and modified nominations
- Scheduling data containing monthly and daily schedules
- Ticket data containing captured field data and reconciled tickets
- Pipeline inventory data containing current and historical line fill data
- Tank inventory data containing current and historical tank data
- Contract and tariff data containing the tariff and pricing data
- Revenue data containing invoicing and customer reporting information

The integrated system can improve the profitability and customer satisfaction level by rendering the following benefits:

- Streamline the nomination, scheduling, volume and revenue accounting, reporting and other business processes.
- Enhance the quality of data.
- Minimize redundant data entry.
- Reduce application interface errors.
- Provide data security by means of user grouping.
- Improve reporting capability by providing access to volumetric data on a timely basis.

5.2 Liquid Pipeline Operation

This section briefly describes liquid pipeline operations with emphasis on batch operations to provide a proper context to the liquid applications. Refer to Reference (1) for more detailed discussion of liquid transportation. Liquid pipeline operations are different from gas pipeline operations due to liquid being of high density, vapor pressure, low compressibility, and the fact that multiple liquid petroleum products are shipped in batch pipelines. High density causes large pressure changes for large changes in elevation, requiring strict control of vapor pressure and maximum allowable operating pressure. Since liquid compressibility is low and density is high, line pack change is negligible but surge pressure control is critical. Sequential but complex batch operations may be required to deliver multiple products in a single pipeline, particularly a long pipeline.

Liquid pipelines transport petroleum products in two different modes: single product transportation through a dedicated pipeline and multiple products in a sequence of batches. If the product specifications are very rigorous, dedicated pipelines can be used, usually at a higher cost. If multiple products need to be transported, it is more economical to build a single pipeline and operate it in batch mode instead of building multiple pipelines dedicated to single products. Since a single product pipeline operation is simple, this section focuses on the discussion of various subjects related to batch operations.

Batch operation allows the pipeline company to transport multiple products to multiple customers in a single pipeline. Each product is lifted from the originating station sequentially, and a fungible product may contain several batches owned by multiple customers. A batch is a specific quantity and type of product with uniform specifications, and may be delivered to multiple locations along the pipeline system. A typical batch operation diagram is shown below.

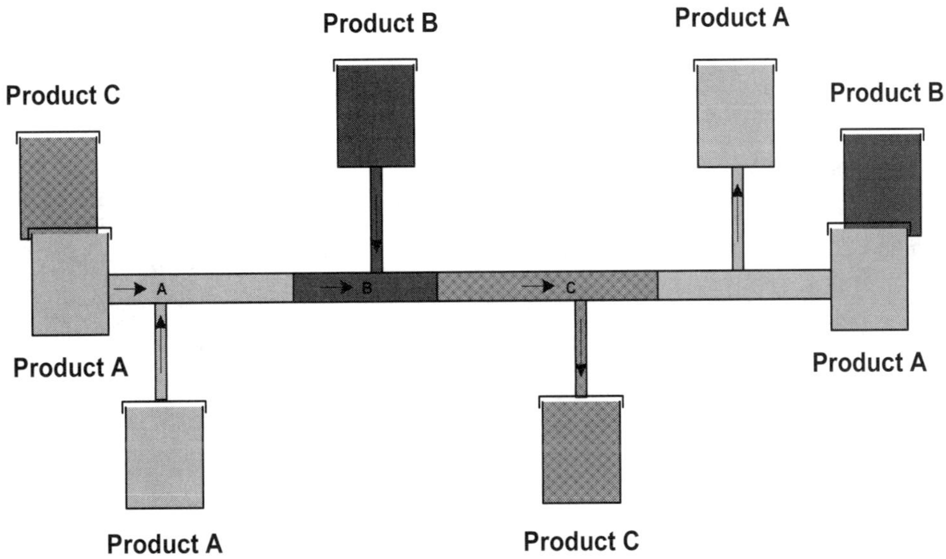

Figure 2 Batch Operation

When multiple products are transported in a batch mode, all products are pumped during a fixed period. This period is called a batching cycle. Usually there are multiple batching cycles in a single nomination period. The batching sequence is not always fixed, but practically it may be fixed for every cycle as long as the same products are lifted. The batching sequence is arranged in such a way that is likely to result in the least formation of batch contamination interfaces.

There are two types of batches: segregated batches and fungible batches. If two adjacent batches have different product specifications, the petroleum products should not be commingled during transportation through pipelines in order to maintain product quality and specifications. Segregation of the batches avoids commingling. Also, batches may be segregated if they are very large in volume or if the pipeline is short. Normally, a segregated batch is transported through the pipeline with a single batch identity and ownership. If batches from more than one shipper meet the minimum specifications, they may be put together into a single fungible batch still satisfying the same specifications. The specifications are established by the pipeline company, based on industry standards and regulatory requirements. The main reason for transporting products in fungible batches is to reduce the cost of transportation, providing flexible lifting and delivery operations. A typical batching cycle and sequences are shown Figure 3.

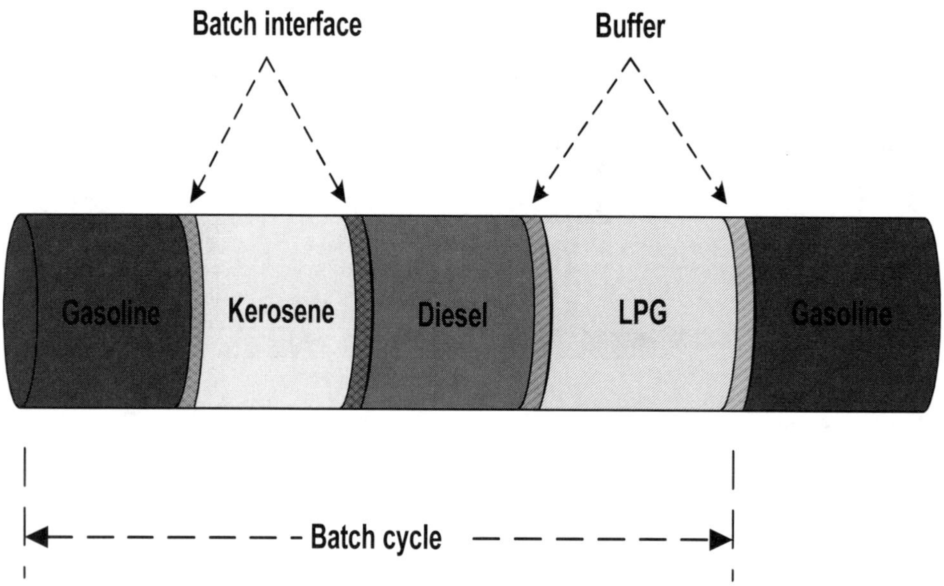

Figure 3 Batching Cycle and Sequence

Batches can be injected or delivered anywhere along the pipeline as long as injection or delivery facilities are available. If a batch is injected at an intermediate location along the main pipeline, the injection can be full stream or side stream injection. For full stream injection, the upstream section of the injection location shuts down, producing zero upstream flow and the downstream flow rate is the same as the full stream injection rate, while the downstream flow for a side stream injection is the sum of the upstream flow and side stream injection rate. Similarly, either full stream or strip (side stream) delivery can be made at some points along the pipeline. For full stream delivery, the upstream flow of the delivery location is the same as the delivery rate and the downstream flow is zero. Similarly, the upstream flow for the strip delivery is the sum of the downstream flow and the delivery rate. For optimum pipeline operation, it is desirable to schedule full stream deliveries to occur at the same time as full stream injections at the same location so that the pipeline does not have to be shutdown on either the upstream or downstream side.

As multiple products move along the pipeline, interfacial mixing takes place at the interface boundaries between two adjacent batches. The commingled petroleum product that does not meet the shipper's product specifications is called transmix or slop. This off-spec product is accumulated in a slop tank and then sold separately at a lower price, sent to a refinery for reprocessing, or blended with other tolerable product. The interface mixture may be cut into one or the other product, or divided between the two adjacent products at the mid-gravity point.

References (2) and (3) discuss the factors contributing to interface mixing and the method of estimating the length of a batch interface. In order to minimize interfacial mixing, batches are sized large and lifted in a pre-determined batching sequence. For this reason the tariff specifies the minimum batch size requirements. Normally, the sequencing of batches in the pipeline is such that products closely related are adjacent in descending or ascending order of quality or gravity in order to minimize batch interface sizes. If the product properties such as density and viscosity are significantly different between two adjacent batches, the interfacial mixing can grow large. To reduce the mixing of more expensive product, a buffer product may be inserted between the two adjacent batches. A separation pig or sphere can also be inserted in front of a new segregated batch to avoid any interfacial mixing, but this operation requires pigging facilities and extra operating cost. An example of an interfacial mixing profile is shown below.

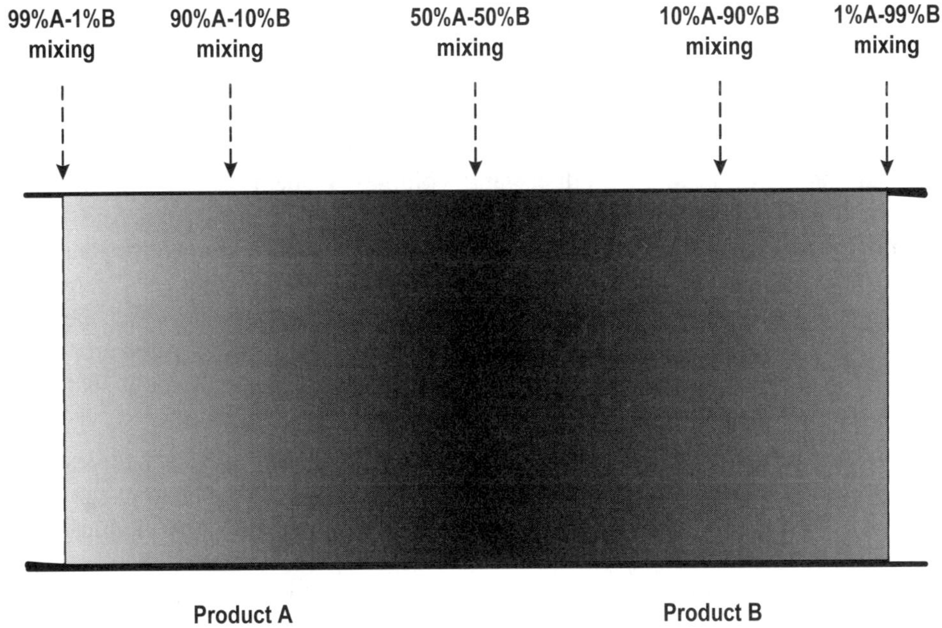

Figure 4 Interfacial Mixing Profile

When batches are lifted, they can be launched either manually or automatically. Automatic batch launchers are not only economical but also generate accurate and timely batch information. The information an automatic batch launcher generates includes accurate batch launch time and batch identification. The batch ID may identify the product of the lifted batch, batch number, size or quantity of the batch, and shipper of the batch.

Batches need to be delivered to the correct delivery locations and tanks. This need requires frequent tracking of batches and detection of batch interfaces. Periodic batch tracking can be performed manually or real-time tracking automatically using a real-time modeling technique. Batch interfaces are detected using interface detectors at the delivery locations. A densitometer is popular as an interface detector using the difference in densities between the adjacent products. If the density difference is too small to use a densitometer, dye may be injected between two batches and color change can then be used for interface detection.

5.3 Batch Scheduling System

Since common carrier pipelines deal with various products from many shippers, product shipping schedules are crucial to meeting the shipping requirements efficiently with available pipeline capacity and facilities. The batch scheduling function is to sequence the batches within the time windows for lifting from and delivering batches to nominated locations. Therefore, scheduling is a process of generating a workable schedule for the economical transportation of petroleum products along a pipeline, and a schedule provides detailed information about the locations, dates and times of product lifting and delivery along with volumes and flow rates.

The pipeline schedulers perform complex tasks of scheduling shippers' nominations. They arrange products and volumes sequentially at the injection locations, while determining injection and delivery dates and times so as to minimize pumping costs, arrange for pipeline maintenance and make the best use of the pipeline capacity. If the schedule doesn't meet certain shippers' requests, the scheduler informs the shippers of the shipping problems and asks them to modify their requests. Changes require the scheduler to develop a new schedule with the modified requests. The final schedule is sent to the pipeline dispatchers for operating the pipeline system.

In general, two types of schedules are produced: long-range schedules and operating schedules. A long-range schedule deals with monthly batching activities used for planning purposes. From the shippers' monthly nominations, the scheduler develops batch schedules including the sequence of batches to be lifted at the origins as well as approximate dates and times which may vary with future events. The main criteria for creating the long-range schedule are to optimize flow efficiency and provide evenly spread out deliveries throughout the scheduling period, while generating a minimum amount of transmix.

An operating schedule shows the batch schedule with the dates and times of short-term events. The operating schedules are normally updated daily largely due to changes in nominations and/or pipeline states. The schedule is used by the pipeline operators to provide the basis for the operating procedures. Since the schedule must be accurate, the scheduler uses the line fill data at the schedule starting time and then simulates batch movements with batching events to determine the operating schedule. The line fill data can be updated manually or automatically from the batch tracking data.

Common carrier pipeline companies create a batch schedule to accommodate the transportation requirements requested by shippers. The schedulers are responsible for the schedule creation, usually taking the following steps:

1. Nomination confirmation
    - Review the shippers' nominations and special requests and develop a preliminary schedule.
    - Publish the schedule to the shippers for their review.
    - Adjust the nominations if shippers request changes to the nominations or if the pipeline capacity is limited.
2. Create a batch schedule
    - Set up the line fill and tank inventory data at the time of the schedule creation.
    - Develop initial batch plans based either on an automated approach or scheduler's experiences.
    - Select a set of flow rates that will ensure the nominated volumes can be pumped in the nomination period, usually a month, and can be accommodated without incurring excessively high power or energy costs.
    - Simulate the product movement along the pipeline system using an initial batch plan. If the product movement is based on volume displacement, the scheduler may use a hydraulic model to examine schedules for hydraulic performance.
    - Determine feasible batch schedules.
3. Optimize the schedule
    - Determine the evenness and distribution of injections and deliveries over the nomination period.
    - Calculate the overall cost on power/energy, inventory and transmix for the feasible schedules.
    - Select an optimum schedule that minimizes the overall cost while balancing injections and deliveries throughout the nomination period.

5.3.1 Nomination Management

Shippers make a reservation to ship petroleum products through the pipeline by means of nominations. A nomination is a request by a shipper to transport a particular batch of product through the pipeline system, and identifies the products and their volumes, injection and delivery locations, etc. In addition, the nomination describes third parties that may supply the product, have the product consigned, or even provide tankage.

Shippers should submit their nominations in accordance with a set of rules and regulations to ensure that their initial nominations and subsequent changes will be accepted in time and without incurring additional fees. The tariff specifies the rules and regulations including the due dates for nominations and nomination changes. FERC 68 dictates that the pipeline company provides the scheduled lifting dates from the origin. According to the scheduled dates, shippers should submit nominations that they intend to ship in the coming month. This allows the pipeline company to analyze the shipping requirements to handle all shippers' products and if restricted by pipeline capacity allocate the volumes to all the shippers. Without this advanced nomination data, the pipeline company may not be able to plan the following month's shipping schedule and thus cannot accept the shipper's requests.

The pipeline company allows the shippers to change their nomination up to the final nomination change date and time. Nomination changes may be made by the shippers after the due date and time, but the pipeline company is not obliged to honor the changes. The nomination rules including the fee structure are designed to satisfy shippers' transportation requirements and maintain efficiency in the pipeline operations. Also, the pipeline company may be getting volumes through feeder pipeline in which case the shipper nominates to both the feeder pipeline as well as the common carrier pipeline. In this situation, the common carrier pipeline will verify the nominations are consistent between both the shipper and the feeder.

5.3.1.1 Shipper Information System

Many pipeline companies use a shipper information system at the core of their business. Broadly, a shipper information system may provide the electronic exchange of information needed to support the functions for managing tariffs, nominations, product injection and delivery status and schedule, product inventory, volume accounting, billing, pipeline operation announcements, and other important functions. Its functions can be automated via computer software and integrated not only to improve the pipeline company's business efficiency but also to provide the customers with fast and reliable service.

This section defines the shipper information system in a narrower sense, focusing on the automated process of nominating products by shippers and of reporting product scheduling and delivery status to shippers. Specifically, the system allows the shippers to:

- enter and change nomination data.
- monitor the progress of their nominations from scheduling and lifting to delivery.
- receive the reports pertaining to their nominations as well as transportation status and history.

The nomination data includes the shipper, origin locations and volumes, destination locations and volumes, products and fungibility, batching cycle, supplier, consignee, and possibly tankage and comments. A nomination is characterized and identified by batch codes. A batch code may include a shipper code, product code, cycle number, batch number within the cycle, point of origin, and possibly supplier and consignee codes. If a pipeline company operates multiple pipelines, the pipeline code may be included in the nomination code or the company may determine the pipeline or route that a batch is sent to its destination as part of building the schedule. The cycle number with a specified start date is available on the list of the annual cycle numbers posted by the pipeline company. The shipper is the company or legal entity requesting product shipment. The supplier code identifies the company or individual entity supplying the batch at an origin location, and the consignee is the party to whom the batch is delivered. They are intended for the pipeline company to provide the suppliers and consignees with the batch schedule and ticket information of their batches. In addition, the pipeline company may provide schedule and ticket information to the tankage provider related to the batch, if the tankage code is available.

In fungible systems, the pipeline company does not need to guarantee that the physical product lifted is the same product delivered, i.e. the pipeline company is permitted to exchange product batches of the same commodity meeting the same specifications. The actual product is exchanged between batches in the pipeline. For example, a shipper's nomination may be received in one batch but delivered from another batch. This process is called an exchange. This can lead to a further extension where a pipeline company can exchange batches from different pipelines. For example, a pipeline may receive product at location A and deliver it to location B, while a second pipeline receives product at location C and delivers it to location D. In a fully fungible system, it is possible to create a nomination in pipeline A – B and another nomination of the same commodity in pipeline C – D. The product can then be swapped such that the shipper that supplied product at A can take delivery at D while the shipper that supplied product at C can take delivery at B. In this case, the first nomination will be for product movement from location A to D and the second nomination will be for product movement from location C to B, even though there is not a physical pipeline connection from A to D or C to B. This is often called a 'virtual' nomination. In all exchanges, the product volumes must match such that all received volume is matched by delivered volumes.

The shipper information system accepts nominations and their changes from shippers manually or electronically. Manual data entry can be achieved by using faxes, telephones and other means of data entry, or by a computer-based system using a standard Electronic Data Interchange (EDI) or even internet-based approach. An internet-based shipper information system is further discussed in the next section.

The shipper information system provides the following nomination data management functions:

- Nomination data creation capability, showing the date and time stamp when new nominations are entered, and reading the nomination from other sources such as EDI or spreadsheet data
- Nomination editing functions with a version number to allow the user to view earlier versions of the request. Each time a change is made to a nomination, another version is created. All versions of a nomination can be viewed from the nomination display.

Version	Name	Prod	Fung	Loc	Nominated Vol	Prorated	Type	Scheduled	Status	Volume	Supplier	Consignee
0	LD-533-01	LSD	Yes		43800.00 m3	N		0.00 m3	Confirm	43800.00 m3		
				SN	43800.00 m3	N	Lift	0.00 m3		43800.00 m3	EX	SH
				NT	43800.00 m3	N	Del	0.00 m3		43800.00 m3		
0	OE-657-00	OEM	Yes		53000.00 m3	N		0.00 m3	Confirm	53000.00 m3		
				SN	53000.00 m3	N	Lift	0.00 m3		53000.00 m3	IO	PC
				SG	53000.00 m3	N	Del	0.00 m3		53000.00 m3		
0	HC-665-00	HCN	Yes		45800.00 m3	N		15300.00 m3	Confirm	45800.00 m3		
				SN	45800.00 m3	N	Lift	15300.00 m3		45800.00 m3	EX	SH
				NT	45800.00 m3	N	Del	15300.00 m3		45800.00 m3		
0	CL-486-00	CLHO	Yes		35000.00 m3	N		5000.00 m3	Confirm	35000.00 m3		
				SN	35000.00 m3	N	Lift	5000.00 m3		35000.00 m3		
				NNR	35000.00 m3	N	Del	5000.00 m3		35000.00 m3		
0	EX-566-01	EXP	Yes		43000.00 m3	N		0.00 m3	Confirm	43000.00 m3		
				SN	43000.00 m3	N	Lift	0.00 m3		43000.00 m3	CP	EX
				SG	43000.00 m3	N	Del	0.00 m3		43000.00 m3		
0	FR-477-00	FUR	Yes		67000.00 m3	N		0.00 m3	Confirm	67000.00 m3		
				SN	67000.00 m3	N	Lift	0.00 m3		67000.00 m3	IO	IO
				NT	67000.00 m3	N	Del	0.00 m3		67000.00 m3		

Figure 5 Example of Nomination Display (Courtesy of CriticalControl)

- Deleting a nomination, retaining the actual record of the nomination but zeroing the volume of the nomination
- Nomination list display that shows a full description of each shipper's nomination belonging to a given pipeline

An example of a nomination display is shown above. The nomination display lists the nomination detail received from shippers and tracks the amount of each nomination that has been scheduled. The display may contain a column to show the nomination volume and another column to show the volume that has been scheduled from each of the nominations. The fungibility column may indicate whether the product is fungible or segregated. The scheduled volumes can be highlighted in color to indicate whether the nomination is fully scheduled, partially scheduled, overscheduled or unscheduled. The nomination status shows the state of the nomination, and the status can be "scheduled", "pending", "deleted", or "rejected." Any change in the status needs to be updated to provide the shippers with an indication of the status of their nominations.

When a product is transferred from one pipeline system to another, the nomination is an inter-pipeline nomination that the shipper information system may be able to handle. Once an inter-pipeline nomination is created, it should appear in the nomination lists of both the originating and transfer pipelines. If the nomination is changed, the change has to be made on both sides. An example for an inter-pipeline nomination is given below. In this example, the batch has two lifting locations and two delivery locations, one for each of the pipelines through which the batch will travel.

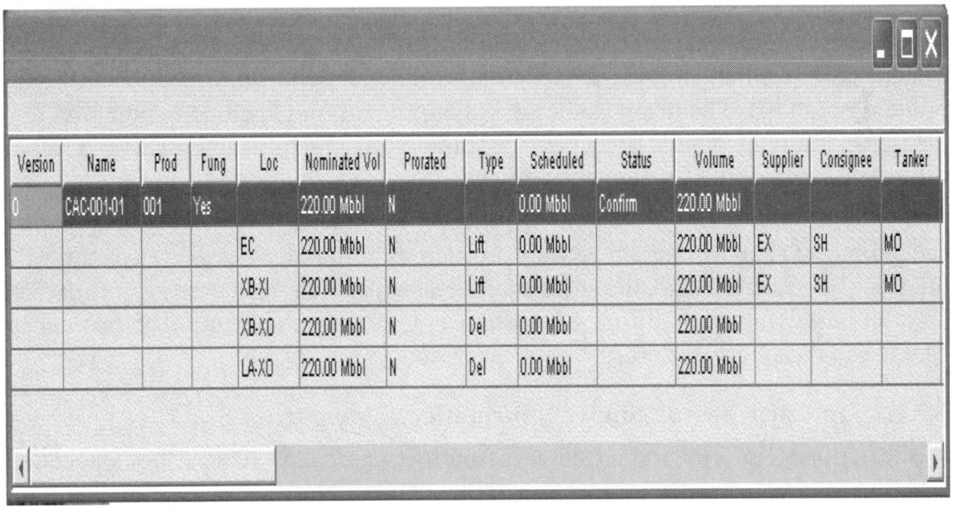

Version	Name	Prod	Fung	Loc	Nominated Vol	Prorated	Type	Scheduled	Status	Volume	Supplier	Consignee	Tanker
0	CAC-001-01	001	Yes		220.00 Mbbl	N		0.00 Mbbl	Confirm	220.00 Mbbl			
				EC	220.00 Mbbl	N	Lift	0.00 Mbbl		220.00 Mbbl	EX	SH	MO
				XB-XI	220.00 Mbbl	N	Lift	0.00 Mbbl		220.00 Mbbl	EX	SH	MO
				XB-XO	220.00 Mbbl	N	Del	0.00 Mbbl		220.00 Mbbl			
				LA-XO	220.00 Mbbl	N	Del	0.00 Mbbl		220.00 Mbbl			

Figure 6 Inter-Pipeline Nomination (Courtesy of CriticalControl)

The nominations should be validated to ensure the integrity of all data fields. The lifting or delivery location for a particular shipper is checked if the shipper is permitted to nominate to the location. If the nomination originator is different from the shipper, the originator should get an approval from the shipper and the pipeline company before the nomination is entered. Nomination volumes from all origin locations should be within the minimum and maximum volumes and match with the delivery volumes for every shipper, product and cycle.

The shipper information system maintains a nomination database, which contains not only the initial and modified nominations for all the shippers but also historical nominations even after the current batching cycle is completed. All current nomination data should be made available online. After a certain period of time the nominations need to be archived.

The shipper information system requires some type of security including the entry of a user name and password, which must be validated against the qualified user information. It provides different system access privileges depending on the users, who are restricted to their permitted pipeline or pipelines. Certain users such as shippers with permission for a pipeline are allowed to enter and change nomination data, while other parties may only be allowed to view the nomination data. Some users may have not only the nomination data entry and change access but also access for approving nominations entered or modified by another user. Only support personnel are permitted to access the database for maintenance purpose.

The shipper information system displays nominations, nomination status, batch movements, and schedule and tickets for the nominations whose shipper matches the particular system user. The system may display the pipeline configuration showing all origins and delivery locations and tank inventory. In addition, the system provides the shippers with the up-to-date shipping information through a bulletin board (4). The displays on the bulletin board may include nomination due dates, operational status of pipelines, notice of pipeline activities and other information possibly affecting shippers, etc.

The shipper information system should be interfaced with its corresponding scheduling system if the scheduling system is computer-based. Also, an EDI interface is required to enter nomination data in such instances. Recently, an internet-based shipper information system is more popular than other manual or EDI based data entry system.

5.3.1.2 Internet-Based Shipper Information System

If the shipper information system is internet-based (5, 6, 7), the associated functions can be accessed around the clock. This enables the pipeline company to exchange information electronically with shippers and other customers. The basic functions of an internet-based shipper information system are similar to a non-internet based system. Such systems allow shippers to nominate and manage their data in the following manner:

- Shippers enter and change their nominations on-line directly into the pipeline company's nomination database, so that data accuracy and transaction speed can be maintained. This eliminates the need for schedulers to manually type in all the nominations.
- Shippers can directly access nominations and schedules including their nomination status and history, injection and delivery schedules, inventories, operating information, and other news anywhere and anytime they wish.
- Shippers can approve nominations and manage inventories. This results in improved shippers' satisfaction and the improvement of the transportation business processes.
- Shippers can not only view their data including tickets, inventories and invoices, but also download data via email and then fax the data to associated parties.

Compared to a manual system, an internet-based shipper information system allows the users to save significant amounts of time and energy. The internet-based shipper information system offers shippers reliability and flexibility while satisfying their business requirements, which eventually result in higher profitability.

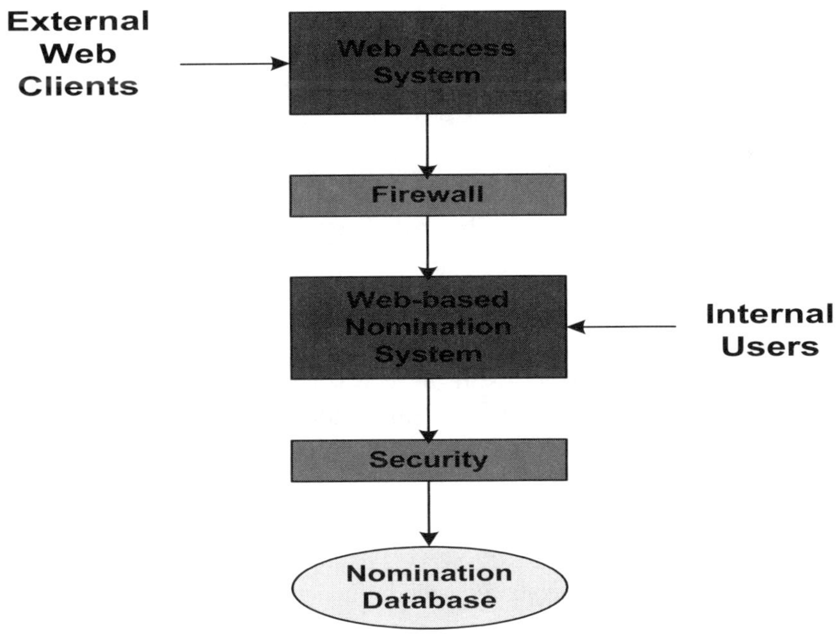

Figure 7 Data Security

Data security and integrity is most critical when shipper information is transmitted over the internet. Therefore, an internet-based shipper information system should provide authentication, authorization, and the encryption capability to secure the nomination and transportation transaction data, employ multiple firewalls to protect from any fraudulent access, and check data integrity to ensure that data is not damaged during transmission. Typically, data security can be achieved as shown in Figure 7.

Refer to the web site of Tranport4 (5) for the discussion of a hierarchical data security model, system structure diagram, and benefits to both shippers and pipeline companies.

As a minimum, the internet-based shipper information system should support the following functions entered through a main display screen:

- Select a pipeline on which products to be shipped are nominated, if more than one pipeline is operated by the carrier.
- Create new nominations or change existing nominations for the selected pipeline.
- View nomination status and shipping progress with the lifting and delivery date and time for scheduled batches.
- View nomination history with version numbers, showing historical data for all the nominations that have been created by the shipper.
- View shipper balance reports and shipper invoices.
- View transit times of a batch for its various routes.
- Manage data including shipper configurations and other system administration functions.
- Provide the capability to estimate tariff.

5.3.2 Computer-Based Batch Scheduling System

Many schedulers of simpler pipelines have traditionally used hand calculations to calculate batch movements and storage inventories. For more complex lines where multiple injection and delivery locations are present, many schedulers have used a manual method of scheduling using a spreadsheet, sometimes augmented with a graphical representation of the pipeline topography, popularly known as a railroad graph. A computer-based version of the graph is shown in later sections. However, these methods are laborious and time-consuming and often prone to error. They can also be slow in responding to changes in nominations or pipeline operations.

The key requirements of a computer-based scheduling system are:

- To quickly generate an optimum schedule that will meet shipper requests and that guarantees operation within pipeline constraints

- To ensure predictable, evenly spaced injections and deliveries that do not overtax the facilities of either the shippers, feeders or delivery facilities
- To ultimately optimize profit of the pipeline company.

A computer-based batch scheduling system can not only improve scheduling efficiency and accuracy but also offer quick response to shipper's changing requests and the ability to examine a variety of possible schedules. The schedulers can focus on their scheduling knowledge and capability to develop an optimum scheduling strategy, while being relieved of low level tasks and refining the schedule with speed. The speed and schedule accuracy allows the pipeline company to improve shipper service by offering the shippers the flexibility to quickly change their plans and by optimizing the utilization of their storage facilities.

At the present time, many schedulers use spreadsheets to do the calculations when manually creating batch schedules. This requires a certain level of experience on the part of the scheduler. This approach works well if the pipeline system and its batching operation is relatively simple or the number of products and shippers is small. Where this is not the case, the schedulers need an automated method with more sophisticated software tools that can help them improve scheduling efficiency in dealing with large numbers of products and shipper requests for complex batching operations with many injection and delivery locations.

Broadly, there are two types of batch scheduling system models: one based on a hydraulic model and the other on a volume displacement model. With many years of experience, the schedulers intuitively come up with reasonable initial batch plans using the model to test the viability of the plan. In the past, numerous attempts have been made to automate this process of developing optimum initial batch plans, using brute force methods, mathematical programming approaches, or even expert systems. Some approaches were successful for simple pipeline systems and operations, but as yet solutions for complex systems are still in the research phase.

A hydraulic model based system (8) uses an initial batch plan as an input to simulate batch movements along the pipeline system, calculating hydraulics for pressure profiles, flow velocity and estimated time of batch arrival. The pipeline system and operating constraints, including hydraulic and tank limits, should not be violated. If this simulation proves invalid due to a violation of the constraints, another batch movement simulation is tried with a different batch plan. A hydraulic model based system can produce a hydraulically accurate schedule. On the other hand, it may take a long time to run the hydraulic model with many long-term initial batch plans, particularly for a long pipeline with many batches.

A volumetric simulation method doesn't take into account hydraulics. This model may include hydraulic constraints indirectly with the specification of maximum and minimum flow rates. It requires the assumption that the fluid is incompressible, flow rate change occurs instantaneously, and fluid properties are independent of pressure and temperature. Even though this method may not produce schedules with the hydraulic accuracy of the hydraulic simulation model, the execution speed allows many initial batch plans to be tried in a short time. The performance in execution time is critical for a complex pipeline system with frequent changes in nominations.

A computer-based scheduling system requires initial batch plans and batch movement simulation model, in addition to various input data such as pipeline configuration, nomination, and inventory.

5.3.2.1 Input Data Requirements

The schedulers require two types of data: static data which does not change frequently and dynamic data which changes almost daily. The static data may include:

- Pipeline configuration – pipe size, pipe length or displacement, line fill adjustment factor, injection and delivery locations, junctions and lateral lines, transfer points, etc.
- Hydraulics and facility constraints – maximum and minimum flow rates and pressures, pumping capabilities, etc. The pressure limits are required only for hydraulic model based batch scheduling system.
- Product parameters – products permitted in the pipeline, Drag Reducing Agent (DRA) usage, maximum and minimum volumes, physical properties, etc.
- Tank data – tank capacities, maximum and minimum tank levels, product designation, ownership, etc.
- Batching rules and requirements – batch sequence rules, buffering rules, shipment rules for shippers, flow reversing operation, break-out operation, fungibility, cycle length, etc.
- Shipper data – information on shippers, consignees, facility operators, etc.
- Time parameters – time or date related constraints such as times of restricted flow rate or maintenance schedule times

In addition to the static data, the schedulers use the following data:

- Nomination data – products, volumes and other data nominated by shippers at origin and delivery locations
- Line fill data – products, volumes and locations in the pipeline at the time of batch scheduling
- Tank inventory data – product and volume with an estimated fill or depletion rate of each tank

5.3.2.2 Pipeline and Tank Inventory Data

The initial pipeline and tank inventory data is used as the initial pipeline state for simulating the movement of products through the pipeline system. The pipeline and inventory data is usually made available in real-time from the SCADA system or specific applications such as batch tracking and tank inventory systems, which provide batch and inventory tracking data. Refer to Chapter 6 for detailed discussion of the tracking functions.

The batch and inventory tracking data received from the SCADA or real-time tracking system may include the following pipeline state:

- Time stamp of data retrieval
- Line fill data, including the batch ID or name, product, location and segment volume in the pipeline, and estimated time of arrival (ETA)
- Valve status
- DRA concentrations if DRA is used
- Scraper locations
- Tank inventory data, including the product and tank level or volume
- Meter data at lifting and delivery points to indicate the lifted or delivered volume of batches that are lifting and delivering at the time the data was captured

Since pipeline dispatchers are not always able to meet a schedule, operations can become out of sync with the schedule. Therefore, the current operating data is required to create an accurate schedule by bringing the schedule back in line with real-time operations. This is called schedule reconciliation. Batch volumes are adjusted to the schedule start time based on the metered injection and delivery volumes. If a change of batch has occurred between the line fill collection time and the schedule start time, an indication from SCADA of the change is required to trigger the launch of a new batch at a lifting location or completion of a batch delivery at a delivery location.

The batch tracking data is made available to the scheduling system either manually or through a software interface. If the line fill data set is large, manual data entry is time consuming and prone to errors. A software method facilitates a fast and accurate transfer of the line fill data from the batch tracking system to the scheduling system. The link between the SCADA and scheduling system is through the interface software. This interface synchronizes or reconciles real-time operation with schedules. The real-time conditions include batch locations and tank volumes as determined by the SCADA system at specific times. This data is stored in a database that contains current and historical line fill and inventory data.

The interface software should provide the capability of capturing and editing the line fill and inventory data. It can be used to update schedule information in the following steps:

- Take a snap shot of the line fill and inventory data from the host SCADA and/or tracking system and then deposit data into a historical database.
- Select the real-time data from the historical database at the schedule starting time.
- Compare the actual pipeline state at a specified time, as determined by the real-time data, with the expected pipeline state as determined by the schedule.
- Edit the real-time or scheduled data to produce a match between the schedule and the real-time data that gives an accurate portrayal of the pipeline operation.

The interface software may display the location of batches in the pipeline predicted in the schedule at the selected time as compared to the location in the real-time data. The display includes the batch name, scheduled and actual product, scheduled and actual size, scheduled and actual DRA concentration, scheduled and actual lifting and delivery volumes, and scheduled and actual ETA. The interface software also provides a display to show the type and level of product in the tanks at the start of the schedule, at the scheduled time, and as determined from the real-time data. The tank display shows the tank and capacity, volume at the beginning of the schedule, volume at the selected time, and actual volume as determined by the SCADA system. Examples of these displays are given in Figure 8.

Selected Time: 09/15 07:06 Tracker Time: 09/15 07:06

Status	Batch	Sched Product	Track Product	Sched Size	Track Size	Sched Vol to Next	Sched Next Location	Track Vol to Next	Track Next Location	Sched ETA
M	SSW-235	LC		2854.85		12211.15	PINC			09/17 05:14
M	SHS-266	HSC	HSC	1799.97	1799.97	10411.18	PINC	10411.25	PINC	09/16 22:25
M	SCC-090	SCO	SCO	2006.40	2006.40	8404.78	PINC	8404.85	PINC	09/16 15:15
M	SHS-265	HSC	HSC	1010.50	1010.50	7394.28	PINC	7394.35	PINC	09/16 11:39
M	SSW-234	LC	LC	3493.09	3493.09	3901.19	PINC	3901.26	PINC	09/15 22:42
M	SHS-264	HSC	HSC	2491.70	2491.70	1409.49	PINC	1409.56	PINC	09/15 12:44
M	SSW-233	LC	LC	1409.56	1409.56	0.00	PINC	-0.00	PINC	09/15 07:06
M	SSW-237B	LC		2500.00		11555.07	HART			09/17 02:30
M	SSW-237	LC	LC	1498.59	1498.59	10056.48	HART	10056.55	HART	09/16 21:09
M	SHS-268	HSC	HSC	1994.00	1994.00	8062.48	HART	8062.55	HART	09/16 14:02

Tank	Capacity	Start Volume	Tracker Volume	Start Product	Selected Volume	Selected Product	Tracker Product	Tracker Status	Tracker Time
T22/23	2470.00	1198.70	1198.70	LC	1198.70	LC		good	09/15 07:06
T19	2773.00	2340.80	2340.80	BU	2340.80	BU		good	09/15 07:06
T18	7363.00	2834.40	2834.40	LC	2834.37	LC		good	09/15 07:06
T17	7305.00	2253.70	2253.70	CO	2253.70	CO		good	09/15 07:06
T16	7190.00	2629.20	2629.20	LC	2629.11	LC		good	09/15 07:06
T15	7184.00	4138.90	4138.90	HC	4138.97	HC		good	09/15 07:06
T14	3935.00	20.50	20.50	HSC	20.50	HSC		good	09/15 07:06
T34	7190.00	4366.70	4366.70	HSC	4366.70	HSC		good	09/15 07:06
T33	2562.00	874.30	874.30	HSC	874.30	HSC		good	09/15 07:06

Figure 8 Pipeline and Tank Inventory (Courtesy of CriticalControl)

5.3.2.3 Initial Batch Plan

An initial batch plan is a proposed batch line up that can be either generated manually or by means of a model. Normally, the schedulers can develop initial batch plans using not only the pipeline configuration and other facility data but also the nomination, tank inventory and line fill data. This plan is similar to a batch schedule except that some plans may not be feasible as schedules or require batch re-sequencing to meet some pipeline constraints. The initial batch plan includes batch lift start/end times, lifting and delivery volumes, locations, and/or flow rates. A plan can be developed as a schedule if it is shown to be feasible by satisfying all the required rules and constraints. The feasibility can be confirmed by simulating batch movements through the pipeline system.

There can be a large number of initial plans, because a large number of combinations are possible with various changes in the following variables:

- Batch size and time to be lifted to meet the delivery requirements
- Range of flow rates between the allowable limits
- Starting/stopping pumps
- Pumping configurations
- Power consumption at pump stations
- DRA injection for high flow rates, particularly those exceeding the maximum line rate
- Product sequence (if it is not fixed)

In addition, the following factors and constraints should be taken into account:

- Multiple injection and delivery locations and operations such as side-stream/ full-stream injection and strip/full delivery

- Fungible and segregated products
- Maintenance schedules
- Time period for each cycle, if multi-period scheduling is required
- Product sequence, if a fixed product sequence is established
- Volume exchange, if this operation is allowed
- Injection terminal constraints such as injection rate, storage capacity, injection time limit, etc.
- Delivery terminal constraints such as delivery rate, storage capacity, delivery time limit, etc.
- Pipeline capacity
- Power usage restrictions

The schedulers with many years of experiences in scheduling may be able to reduce the number of combinations. They normally start with a base plan to simulate the batch movements and then interactively change one of the above variables to create feasible schedules that satisfy established optimality criteria (sometimes called 'fitness criteria'). They may select an optimum schedule out of multiple feasible schedules.

An automated method can also be used to create feasible batch plans. Mathematical programming techniques were successfully applied to simple batch scheduling problems (9, 10, 11). These techniques may generate one or more feasible plans based on the same input data as well as batching rules and constraints on the pipeline system and operational rules a human scheduler applies. The plan includes all the information about a batch schedule such as schedule orders, batch names and sizes, injection start times and durations, flow rates, and volumes to be delivered to tanks. These feasible plans are used to simulate the batch operations and movements to confirm if they can satisfy all the batching requirements. An optimum schedule can be selected by applying fitness criteria. As pointed out earlier, this method is limited to simple scheduling problems.

5.3.2.4 Volumetric Simulation Model

A volumetric simulation model is based on the following assumptions:

- The fluid is not compressible.
- Flow rate changes occur instantaneously.
- Product movement is independent of pressure and temperature variables.

These assumptions allow the model to disregard factors such as pipeline topographical profiles, volume expansion and contraction during the movement of batches along the pipeline, and the secondary effects of pump start and stop actions. Fortunately, these factors do not adversely affect the batch scheduling accuracy in a significant manner.

To assure the schedule does not violate hydraulic constraints such as slack flow or maximum allowable operating pressure, these models must impose rate maximum and minimum limits. An enhancement to the volumetric model can be made to adjust rates based on product density, bringing the resulting calculated times very close to those supplied by hydraulic models. The resulting calculations are thus reduced to linear relationships. The schedules produced by such a model provide results that are as close to actual movements as are required for scheduling purposes, and the calculations are simplified enough that fast computer simulation times can be achieved. An additional advantage of the volumetric model is that the tuning and maintenance required by a hydraulic model are eliminated.

The simulation model ultimately creates the schedule or pumping orders, which provide instructions and estimated time of arrival (ETA's), by performing the following functions:

- Allow the addition, deletion, and sequence changes for pipeline events as well as modification of existing events.
- Provide the capability of viewing the results of changes in both tabular and graphical formats useful to schedulers.
- Allow the scheduler to set up the injection and delivery events for the pipeline system and to select from multiple operational product paths. A tender in a cycle can be set up as any number of lifts and any number of deliveries at multiple locations.
- Allow the scheduler to define operations such as batch splitting, batch blending, pipeline reversal operations, and strip delivery events.
- Generate event times for all the injection and delivery operations.
- Provide a pipeline rate schedule that can be used by the schedulers to determine pipeline operations. This information is used to construct the graphical view of the pipeline system as well as providing event start and end times for all pipeline events.
- Adjust rates across pipe segments to ensure optimum pipeline operation. For example, the flow rate of a delivery into a location may be slowed down to ensure that it finishes deliveries at the same time as an injection taking place at the same location in the downstream segment.

The volumetric simulation model requires an initial batch plan and the same static input data. With the initial batch plan, the schedulers can interactively use the simulation model to create a new schedule or update an existing schedule. They may take the following steps repeatedly to get an optimum schedule:

- Obtain the line fill and tank inventory data from the SCADA system or batch/inventory tracking.

- Create an initial batch plan or multiple plans, either manually or by using a computer-based model, which will serve as the starting point for creating batches and building a schedule.
- View the scheduling results with displays, which are discussed in the next section.
- Modify the schedule, changing variables such as the batch size and flow rate, and even replacing or deleting batches.
- Monitor the impact of scheduled activities over time along the pipeline and at the lifting and delivery locations.

A computer-based scheduling system should be able to provide the capability to perform the above functions. Also, the following control actions are required to simulate batch movement along the pipeline and batch operations at facility locations:

- Start/stop lifting a new batch at a lifting location such as tank farm or refinery.
- Side stream injection by adding a batch to a compatible product, making the host batch size larger. If the flow rate downstream of the injection point exceeds the maximum rate for that line segment, the upstream flow will be reduced to meet the maximum flow restriction.
- Blend a batch at a downstream lifting point into a passing batch of compatible product.
- Split the passing batch by putting the new batch into the line at a junction full stream and stopping the upstream flow until this batch is completely in the pipeline.
- Insert a batch into the pipeline while another batch is delivered upstream at the same time.
- Transfer a batch from the current pipeline to another pipeline at a transfer point.
- Start/stop delivering a batch at a delivery location.
- Strip delivery by stripping the flow at a junction. As a result, the flow rate downstream of the junction is reduced by the amount of the strip delivery rate.
- Automatically re-sequence a batch in a downstream pipeline if the batch's position is altered in an upstream pipeline.
- Generate transmix or interface contamination to simulate the intermix between batches.

Other actions such as pump start/stop and DRA injection start/stop are required to move and schedule the batches.

In the scheduling process, the schedulers apply the rules and constraints unique to the pipeline system and its operations. Therefore, the scheduling system should be able to provide the capability to enter and edit the rules. The following is a partial list of potential rules:

- Fungibility rules that guarantee segregated products are separated from fungible products.
- Blending rules that guaranttee incompatible products are not blended.
- Optimal inventory of products at terminals with just-in-time delivery
- Minimum transmix size and maximum number of interfaces
- Minimum power or energy cost
- Minimum and maximum batch sizes
- Acceptable upper and lower tank levels
- Minimum and maximum flow rates
- Buffer constraints
- Time period when pipeline maintenance needs to be carried out

Using the initial batch plans and simulation model, the above scheduling process can be somewhat automated to determine an optimum schedule.

5.3.2.5 Scheduling Optimization

Profitable operation of the pipeline system is one of the key objectives for a pipeline company. One way to increase profits is to minimize operating costs, and the pipeline schedulers and dispatchers play a critical role in saving costs. The schedulers need to create an accurate schedule with minimum batch interface losses and inventory cost, while minimizing energy consumption and maximizing the use of the existing facilities including volume exchanges. The dispatchers are then responsible for carrying out the orders defined by the schedule without incurring penalty.

Whether a batch schedule is developed manually or automatically, the objective of the scheduling process is to obtain an optimum batch schedule. One or more of the following optimization criteria may be applied, while ensuring all nominations are pumped within the nomination period:

- Pipeline throughput maximization
- Energy consumption minimization
- Contamination or transmix minimization
- Minimum use of tankage
- Even distribution of delivering batches to shippers

A simple approach to measuring schedule optimization is to use some type of measurement to determine how well the schedule performs. For this discussion, we will refer to the set of criteria that make up this measurement as the fitness criteria. Once a schedule simulation is complete, the effectiveness of the schedule is measured by applying a set of fitness criteria. A number of cost parameters can be measured such as product interface mixing, product inventory levels, flow rates, energy consumption, and so on. Weighting factors are applied to the various parameters to give an indication of which parameters are the major costs in a given schedule. This allows the scheduler to adjust the schedule to optimize the overall fitness of the schedule.

A more sophisticated approach would be to use a mathematical programming technique and scheduling rules to create an optimized schedule. In this approach, the objective function, consisting of the weighted fitness criteria or optimum criteria, would be minimized by an optimization technique. The optimized schedule should meet the optimum criteria without violating any pipeline system constraints.

So far, no fully automated scheduling optimization system has been reported. Both an expert systems and artificial intelligence methods have been attempted (12). An expert system could solve even a complex scheduling problem. However, maintenance of the rules and their changes is labor intensive, with the result that these systems were not practical for anything beyond the simplest pipelines. An artificial intelligence technique using a genetic algorithm may be used to generate an optimum schedule if the search space can be narrowed within an applicable range. However, no tangible progress has been reported and it is still too early to automate the process of solving general scheduling problems using an artificial intelligence technique.

Another approach to scheduling automation is to use a constrained combinatorial optimization technique. This technique achieves an optimum plan by enumerating possible combinations of batches while applying required rules and constraints. This technique has been successfully applied to some practical scheduling problems.

5.3.2.6 Scheduling Displays

The scheduling displays help the schedulers to perform the scheduling tasks efficiently. Since the shippers change their nominations frequently, the schedulers need to respond to their changes and modify the schedule accordingly. Described below are examples of various displays. These displays are not necessarily required or provided by all scheduling systems.

1. **Batch List**

   The batch list provides a tabular view of a schedule, listing each batch along with specific details regarding the batches. Batch information includes, at a minimum, the batch identification, the product being moved, a route (lifting and delivery locations), the batch size and the time at which the batch lifts and delivers. Additionally, it may include other data such as the flow rate at which the batch will move, the nomination from which the batch was taken and some information about the trigger that causes the batch to lift and/or deliver. Editing capability is provided from the batch list to edit the data within some of the data cells. An example of the batch list is shown below.

File Edit Options Help

Name	Product	Lifting	Delivery	In Rate m^3/hr	Out Rate m^3/hr	Size m3	Start Lift	End Lift	Start Deliv	End Deliv	InTrig Type	InTrig Batch
BR-426	BR	EP	CR	2200.0	2200.0	7389.0	01/04 23:35	01/05 06:58	01/14 20:41	01/15 04:05	Follow	25983 : AHS-422
BR-426	BR	EP	CR	2200.0	2200.0	6211.0	01/05 06:58	01/05 13:11	01/17 17:23	01/17 23:35	Follow	25984 : BR-426
LLG-428	LLG	EP	CR	2200.0	2200.0	10800.0	01/05 13:11	01/05 23:59	01/17 23:35	01/18 10:23	Follow	25988 : BR-426
M-318	M	EP	CR	2200.0	2200.0	16700.0	01/05 23:59	01/06 16:41	01/18 10:23	01/19 03:05	Follow	25985 : LLG-428
OSH-336	OSH	YP	CR	2500.0	2500.0	18700.0	01/06 16:41	01/07 11:23	01/15 04:05	01/15 22:47	Follow New Site	25986 : M-318
PBS-336	PBS	YP	CR	2500.0	2500.0	25700.0	01/07 11:23	01/08 13:05	01/15 22:47	01/17 00:29	Follow	25987 : OSH-336
SO-321	SO	YP	CR	2500.0	2500.0	16900.0	01/08 13:05	01/09 05:59	01/17 00:29	01/17 17:23	Follow	25989 : PBS-336
LLK-426	LLK	EP	CR	2500.0	2500.0	14900.0	01/09 05:59	01/09 20:53	01/19 03:05	01/19 17:59	Follow New Site	25990 : SO-321
LSB-427	LSB	EP	CR	2500.0	2500.0	17800.0	01/09 20:53	01/10 14:41	01/19 17:59	01/20 11:47	Follow	25991 : LLK-426
BR-453	BR	EP	CR	2300.0	2300.0	19000.0	01/10 14:41	01/11 09:41	01/20 11:47	01/21 06:47	Follow	25992 : LSB-427
LLB-455	LLB	EP	CR	2300.0	2300.0	17600.0	01/11 09:41	01/12 03:17	01/21 06:47	01/22 00:23	Follow	25993 : BR-453
LSB-459	LSB	EP	CR	2300.0	2300.0	22000.0	01/12 03:17	01/13 01:17	01/22 00:23	01/22 22:23	Follow	25994 : LLB-455
OSS-462	OSS	EP	CR	2300.0	2300.0	17800.0	01/13 01:17	01/13 19:05	01/22 22:23	01/23 16:11	Follow	25995 : LSB-459
SO-439	SO	EP	CR	2300.0	2300.0	19200.0	01/13 19:05	01/14 14:17	01/23 16:11	01/24 11:23	Follow	25996 : OSS-462
AHS-423	AHS	EP	CR	2200.0	2200.0	12700.0	01/14 14:17	01/15 02:59	01/24 11:23	01/25 00:05	Follow	25997 : SO-439
BR-428	BR	EP	CR	2200.0	2200.0	7389.0	01/15 02:59	01/15 10:22	01/25 00:05	01/25 07:29	Follow	25998 : AHS-423
BR-428	BR	EP	CR	2200.0	2200.0	6211.0	01/15 10:22	01/15 16:35	01/27 20:47	01/28 02:59	Follow	25999 : BR-428
LLG-430	LLG	EP	CR	2200.0	2200.0	10800.0	01/15 16:35	01/16 03:23	01/28 02:59	01/28 13:47	Follow	26000 : BR-428
M-319	M	EP	CR	2200.0	2200.0	16700.0	01/16 03:23	01/16 20:05	01/28 13:47	01/29 06:29	Follow	26001 : LLG-430
OSH-335	OSH	YP	CR	2500.0	2500.0	18700.0	01/16 20:05	01/17 14:47	01/25 07:29	01/26 02:11	Follow New Site	26002 : M-319
PBS-337	PBS	YP	CR	2500.0	2500.0	25700.0	01/17 14:47	01/18 16:29	01/26 02:11	01/27 03:53	Follow	26003 : OSH-335
SO-323	SO	YP	CR	2500.0	2500.0	16900.0	01/18 16:29	01/19 09:23	01/27 03:53	01/27 20:47	Follow	26004 : PBS-337
LLK-431	LLK	EP	CR	2500.0	2500.0	14900.0	01/19 09:23	01/20 00:17	01/29 06:29	Not Finished	Follow New Site	26005 : SO-323
LSB-433	LSB	EP	CR	2500.0	2500.0	17800.0	01/20 00:17	01/20 18:05	Not Started	Not Finished	Follow	26006 : LLK-431
BR-451	BR	EP	CR	2300.0	2300.0	19000.0	01/20 18:05	01/21 13:05	Not Started	Not Finished	Follow	26007 : LSB-433
LLB-456	LLB	EP	CR	2300.0	2300.0	17600.0	01/21 13:05	01/22 06:41	Not Started	Not Finished	Follow	26008 : BR-451
LSB-463	LSB	EP	CR	2300.0	2300.0	22000.0	01/22 06:41	01/23 04:41	Not Started	Not Finished	Follow	26009 : LLB-456
AHS-470	AHS	EP	CR	2300.0	2300.0	18000.0	01/23 04:41	01/23 22:41	Not Started	Not Finished	Follow	26010 : LSB-463
BR-474	BR	EP	CR	2300.0	2300.0	21000.0	01/23 22:41	01/24 19:41	Not Started	Not Finished	Follow	26011 : AHS-470

Figure 9 Batch List Display (Courtesy of CriticalControl)

2. Batch Graph

The batch graph is often called a "rail road chart." This graph displays product movement in a distance vs. time relationship. The vertical axis is distance (measured in volume units) and the horizontal axis represents time. Each batch is represented as a contiguous polygon, moving down the pipe in time. The batch interfaces are represented by oblique lines. The slope of these lines represents rate (distance over time). The colors can be used to distinguish the product moved. The batch graph contains the information about batch movements including the products, batch sizes, volume flow rates through specific locations over time, and route. The following figure shows a batch graph that gives movement across a section of pipeline.

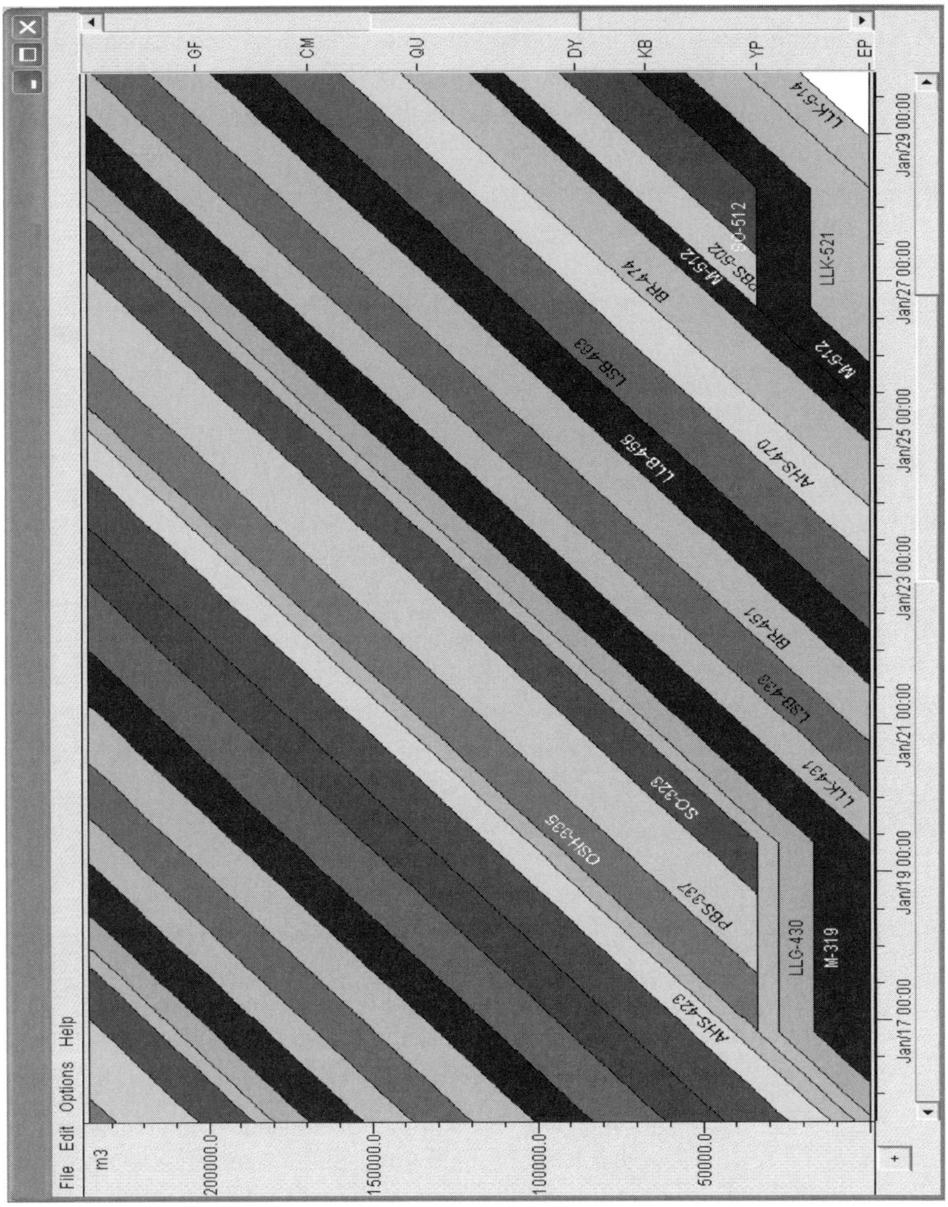

Figure 10 Railroad Graph (Courtesy of CriticalControl)

3. Batch Flow Chart

The batch flow chart provides a view of the batch flow through various locations. It is useful for determining activities at a location along the pipeline. This is beneficial when scheduling the batches that transfer from one pipeline to another. Different colours can be used to indicate different products.

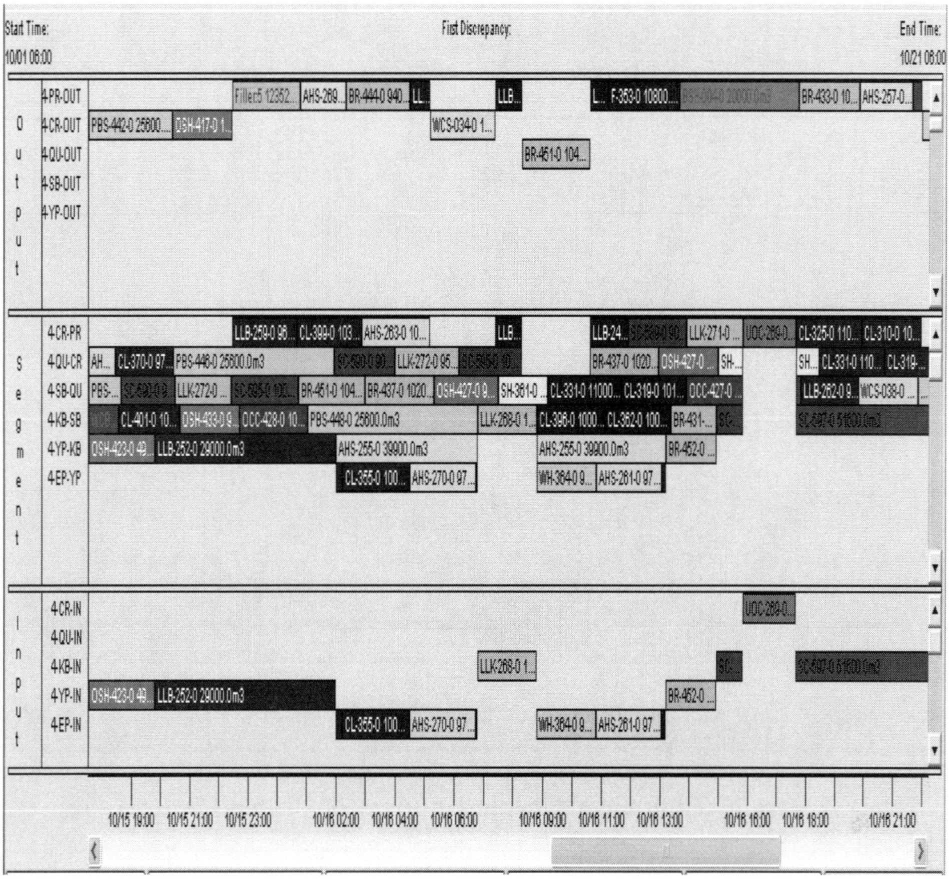

Figure 11 Batch Flow Chart (Courtesy of CriticalControl)

The batch flow chart provides the batch flow monitoring capability. Product flow appears as a single row of activity blocks at each location including injection and delivery points. The chart shows the progression of the batches as they move through the pipeline system. The individual blocks in the product flow represent batches. Any gap between the blocks indicates the product flow has stopped until the appearance of the next product block.

4. Tank Trend Graph

The tank trend display provides a graphical representation of the movement of product through a tank. It serves as a guide to ensure tank levels are operated within operational limits. An example of a tank trend graph is shown below.

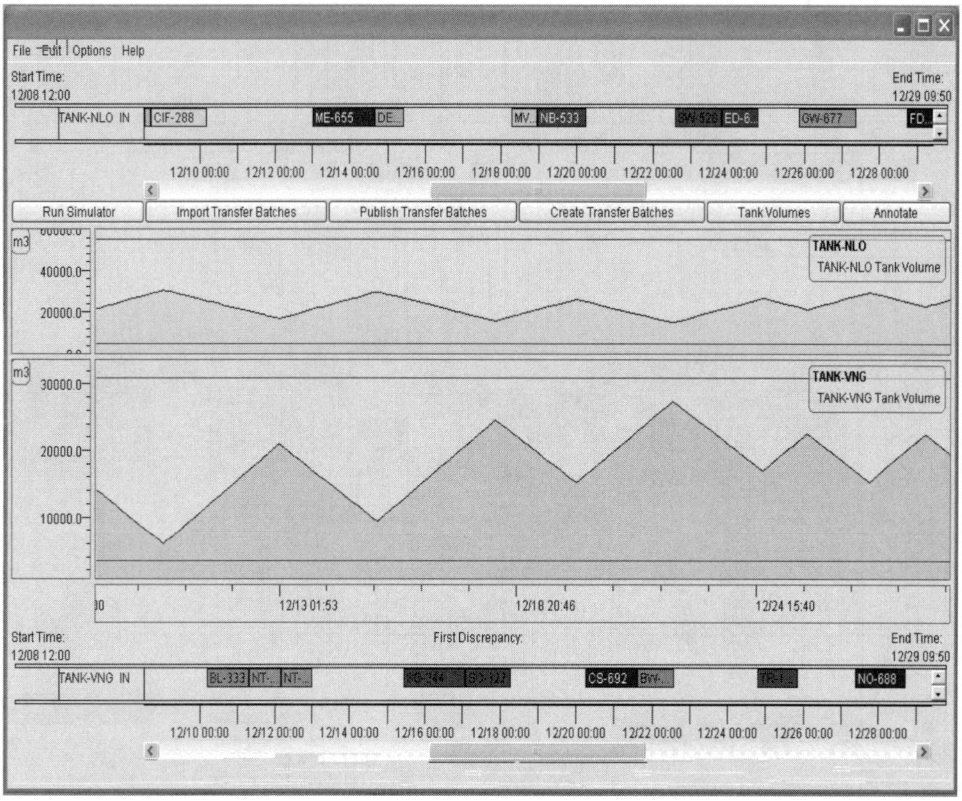

Figure 12 Tank Level Graph (Courtesy of CriticalControl)

5. Schedule Orders and Dispatcher Schedule

The scheduling system ultimately provides a schedule including a set of orders that describe the actions required to execute the schedule during the scheduling period. These orders dictate the actions required by the dispatcher to achieve the product movement predicted by the scheduler. Orders from different schedules (representing different pieces of the pipeline network) may be combined to form a full set of dispatcher orders.

The actions included in the schedule orders need to be complete enough to permit the pipeline dispatcher to achieve the objectives set by the scheduler. These action may include start/stop injection of batches, start/stop delivery of batches, start/stop pumps, change lifting/delivery rates, start/stop DRA injection, open/close valve, start/stop tightline, expect an interface change, launch/receive a scraper etc. The Schedule

Orders Display below shows an example of these actions.

Action	Time	Location	Batch ID	Product	Batch Volume	Flow Rate	Metered Volume	Pump Name	Trigger Reason
start lifting	01/08 20:42	EP	LLK-318	LLK	12289.2	1000.0	8711.0		Quota
stop lifting	01/09 08:59	EP	LLK-318	LLK	12289.2		21000.0		Quota
start pump	01/09 08:59	EP_PS						EP_P1:EP_PS	Batch Head Arrival Event
start lifting	01/09 08:59	EP	LSB-422	LSB	21000.0	1000.0	0.0		Quota
start lifting	01/10 05:59	EP	OSS-431	OSS	18900.0	1000.0	0.0		Quota
stop lifting	01/10 05:59	EP	LSB-422	LSB	21000.0		21000.0		Quota
start lifting	01/11 00:53	EP	PH-451	PH	21000.0	1000.0	0.0		Quota
stop lifting	01/11 00:53	EP	OSS-431	OSS	18900.0		18900.0		Quota
stop pump	01/11 21:53	EP_PS						EP_P1:EP_PS	Batch Head Arrival Event
stop lifting	01/11 21:53	EP	PH-451	PH	21000.0		21000.0		Quota
start lifting	01/11 21:53	EP	SC-452	SC	5600.0	1000.0	0.0		Quota
start lifting	01/12 03:29	EP	PH-455	PH	22000.0	1000.0	0.0		Quota
stop lifting	01/12 03:29	EP	SC-452	SC	5600.0		5600.0		Quota
stop lifting	01/13 01:29	EP	PH-455	PH	22000.0		22000.0		Quota
start lifting	01/13 01:29	EP	SC-453	SC	7100.0	1000.0	0.0		Quota
start lifting	01/13 08:35	EP	LLK-461	LLK	22300.0	1000.0	0.0		Quota
stop lifting	01/13 08:35	EP	SC-453	SC	7100.0		7100.0		Quota
start lifting	01/14 06:53	EP	LSB-463	LSB	21000.0	1000.0	0.0		Quota
stop lifting	01/14 06:53	EP	LLK-461	LLK	22300.0		22300.0		Quota
start delivery	01/14 10:05	CR	AHS-405	AHS	15000.0	1000.0	0.0		B/C
expect scraper	01/14 14:42	CM_STN	Inpsection						Batch Head Arrival Event
stop delivery	01/15 01:05	CR	AHS-405	AHS	15000.0		15000.0		B/C
start delivery	01/15 01:05	CR	BCH-407	BCH	19000.0	1000.0	0.0		B/C
stop lifting	01/15 03:53	EP	LSB-463	LSB	21000.0		21000.0		Quota
start lifting	01/15 03:53	EP	OCC-464	OCC	12710.8	1000.0	0.0		Quota
start lifting	01/15 16:36	YP	PBS-470	PBS	16700.0	1000.0	0.0		Quota

Figure 13 Schedule Orders Display (Courtesy of CriticalControl)

The schedule orders can be exported to make them available for other users. They may be exported in various formats to satisfy their requirements. Popular formats are HTML, spreadsheet, and CSV.

5.3.2.7 Schedule Publication and Reports

When the schedule has been finalized, the results must be published to the user community. This community includes other schedulers, pipeline dispatchers, shippers and terminal operators. The data published must include, at a minimum, the time of batch movements at each significant location along the pipeline. In addition, other actions may be included, such as tightlining operations, scraper movements, pump start/stop actions and valve movements. This data must be parsed so that only data required is provided to each recipient. For example, one shipper should not receive notice of another shipper's product movements.

The publication of this data can be manually transmitted. However, with a computerized scheduling system, this can be done electronically via FTP file transmission, database interfaces, e-mails or other such transmissions. In many cases, the schedule results are sent to the shipper information system so that the shipper community can view their nominations and the resulting schedule from the same computer interface.

When the schedule orders are exported they are integrated into the pipeline company's databases for use by the various users. This data can be distributed manually via FAX or hard copy, but with a computerized scheduling system this data can be distributed automatically and updated on a regular basis. The data presentation format can vary according to the company needs. Provision can be made in these reports to show if the pipeline is operating ahead or behind schedule to help field operators to know when to expect batch arrivals. An example of such a report is given in the following figure.

Terminal Delivery schedule

Mod	PL	CalcTime	Stale	SchedTime	Hrs_Off	Loc	Action	BtchID	Prod	TotBtchVol	BtchVol	DelTanks	Notes
	sppl	06/28/2005 08:47		06/28/2005 04:56	-3.9	NT	start delivery	LSL-417	LSL	2096	585	TK251	
	sppl	06/28/2005 09:57		06/28/2005 07:11	-2.8	NT	stop delivery	LSL-417	LSL	2096	585	TK251	
	sppl	06/28/2005 09:57		06/28/2005 07:11	-2.8	NT	start delivery	RUL-500	RUL	424	424	TK150/151	
	sppl	06/28/2005 10:48		06/28/2005 14:23	3.6	NT	stop delivery	RUL-500	RUL	424	424	TK150/151	
	sppl	06/28/2005 16:10		06/28/2005 14:23	-1.8	NT	start delivery	RUL-507	RUL	7033	7033	TK150/151	
	sppl	06/29/2005 15:17		06/29/2005 13:26	-1.9	NT	stop delivery	RUL-507	RUL	7033	7033	TK150/151	
	sppl	06/29/2005 15:17		06/29/2005 13:26	-1.9	NT	start delivery	LSD-510	LSD	5934	5934	TK677/678	
	sppl	06/30/2005 04:38		06/30/2005 02:56	-1.7	NT	stop delivery	LSD-510	LSD	5934	5934	TK677/678	
	sppl	06/30/2005 04:38		06/30/2005 02:56	-1.7	NT	start delivery	FUR-516	FUR	7951	3229	TK581	
	sppl	06/30/2005 11:52		06/30/2005 10:16	-1.6	NT	stop delivery	FUR-516	FUR	7951	3229	TK581	
	sppl	06/30/2005 16:34		06/30/2005 14:59	-1.6	NT	start delivery	FUR-516	FUR	7951	921	TK581	
	sppl	06/30/2005 18:26		06/30/2005 16:49	-1.6	NT	stop delivery	FUR-516	FUR	7951	921	TK581	
	sppl	07/02/2005 16:59		07/02/2005 15:27	-1.5	NT	start delivery	FUR-516	FUR	7951	3800	TK581	
	sppl	07/03/2005 00:35		07/02/2005 23:06	-1.5	NT	stop delivery	FUR-516	FUR	7951	3800	TK581	
	sppl	07/03/2005 00:35		07/02/2005 23:06	-1.5	NT	start delivery	RUL-600	RUL	921	921	TK150/151	
	sppl	07/03/2005 02:26		07/03/2005 00:54	-1.5	NT	stop delivery	RUL-600	RUL	921	921	TK150/151	
	sppl	07/03/2005 02:26		07/03/2005 00:54	-1.5	NT	start delivery	RUL-607	RUL	11500	7279	TK150/151	
	sppl	07/03/2005 20:38		07/03/2005 19:06	-1.5	NT	stop delivery	RUL-607	RUL	11500	7279	TK150/151	
	sppl	07/04/2005 06:21		07/04/2005 04:49	-1.5	NT	start delivery	RUL-607	RUL	11500	4221	TK150/151	
	sppl	07/04/2005 16:54		07/04/2005 15:22	-1.5	NT	stop delivery	RUL-607	RUL	11500	4221	TK150/151	
	sppl	07/04/2005 16:54		07/04/2005 15:22	-1.5	NT	start delivery	LSD-610	LSD	4000	4000	TK677/678	
	sppl	07/05/2005 04:01		07/05/2005 02:29	-1.5	NT	stop delivery	LSD-610	LSD	4000	4000	TK677/678	
	sppl	07/05/2005 04:01		07/05/2005 02:29	-1.5	NT	start delivery	FUR-616	FUR	6500	2279	TK581	
	sppl	07/05/2005 10:21		07/05/2005 08:49	-1.5	NT	stop delivery	FUR-616	FUR	6500	2279	TK581	
	sppl	07/07/2005 17:18		07/07/2005 15:45	-1.6	NT	start delivery	FUR-616	FUR	6500	2500	TK581	
	sppl	07/08/2005 00:15		07/07/2005 22:42	-1.6	NT	stop delivery	FUR-616	FUR	6500	2500	TK581	
	sppl	07/08/2005 09:58		07/08/2005 08:25	-1.6	NT	start delivery	FUR-616	FUR	6500	1722	TK581	
	sppl	07/08/2005 14:45		07/08/2005 13:12	-1.6	NT	stop delivery	FUR-616	FUR	6500	1722	TK581	
M	sppl	07/08/2005 14:45		07/08/2005 13:12	-1.6	NT	start delivery	RUL-120	RUL	500	500	TK150/151	
M	sppl	07/08/2005 16:08		07/08/2005 14:35	-1.6	NT	stop delivery	RUL-120	RUL	500	500	TK150/151	
	sppl	07/08/2005 16:08		07/08/2005 14:35	-1.6	NT	start delivery	RUL-127	RUL	2000	2000	TK150/151	
	sppl	07/08/2005 21:42		07/08/2005 20:09	-1.6	NT	stop delivery	RUL-127	RUL	2000	2000	TK150/151	
	sppl	07/08/2005 21:42		07/08/2005 20:09	-1.6	NT	start delivery	LSD-130	LSD	4000	4000	TK677/678	
	sppl	07/09/2005 08:48		07/09/2005 07:15	-1.6	NT	stop delivery	LSD-130	LSD	4000	4000	TK677/678	
	sppl	07/09/2005 08:48		07/09/2005 07:15	-1.6	NT	start delivery	FUR-136	FUR	5000	1779	TK581	
	sppl	07/09/2005 13:45		07/09/2005 12:12	-1.6	NT	stop delivery	FUR-136	FUR	5000	1779	TK581	
	sppl	07/09/2005 17:55		07/09/2005 16:22	-1.6	NT	start delivery	FUR-136	FUR	5000	1000	TK581	
	sppl	07/09/2005 20:41		07/09/2005 19:09	-1.5	NT	stop delivery	FUR-136	FUR	5000	1000	TK581	
	sppl	07/12/2005 07:55		07/12/2005 06:22	-1.6	NT	start delivery	FUR-136	FUR	5000	2000	TK581	

Page 1 of 1

Figure 14 Example of Schedule Report (Courtesy of CriticalControl)

Schedule orders are normally kept for a period of time to allow the pipeline company to match nomination and batch data with ticket data from operations. This helps to resolve any issues that may arise when adjustments are required.

5.3.2.8 Integrated Scheduling System and Benefits

An integrated scheduling system should be easily configurable and data driven to handle new shippers, new product types, new facilities and even new pipelines as the system grows over time. It may consist of the following components:

- Shipper information system
- Interface to retrieve and store real time pipeline conditions including line fill and inventory data
- Initial batch plan generation
- Scheduling simulation, including software to view and edit schedules
- Software to determine optimal operations
- Optionally, routines that can automatically update schedules to match operational conditions

An internet-based shipper information system improves the profitability not only for the pipeline companies but also for shippers by simplifying the transportation service business process. Real-time data transfer from SCADA and batch/inventory tracking systems ensures the accuracy of line fill and inventory data while reducing tedious manual efforts. A batch plan generation program can produce feasible batch plans automatically, so that experienced manpower requirements can be reduced. A computer-based simulation model helps the schedulers to create accurate schedules and respond to changing conditions expediently. A schedule optimization model may be able to generate an optimum schedule that minimizes the energy, interface mix and inventory costs. In addition, software may be provided to adjust future scheduled events and ETA's based on current operating conditions. Such a system could also adjust the schedule if actual volumes do not match scheduled volumes.

The database is at the core of an integrated scheduling system. The database contains not only the scheduling related data such as pipeline and tank description, rules and constraints, nominations, line fill and inventory, schedules, and pumping orders but also various reports including tickets, volume measurements and invoices. The access to this integrated database can be made available to both internal and external stakeholders. Via the database, schedulers have access to information from shippers, shippers have access to scheduling information, dispatchers have access to schedules, field personnel can be informed of schedule actions and so on.

The integrated scheduling system is beneficial to all stakeholders but particularly to the schedulers and dispatchers. It can render great benefits to the schedulers in their efforts to maximize pipeline throughput while minimizing energy consumption and transmix volumes. The benefits can be even greater for a complex network that may be divided into multiple sections, because several schedulers may be assigned to the network and can minimize communication problems between schedulers. For the operators, an optimized schedule provides them with a plan that optimizes the pipeline movements. They do not need to 'guess' the next action to provide the optimal product movement.

5.3.2.9 Scheduling System Implementation

The implementation of scheduling systems can include a range of automation, depending on the pipeline configuration and the business processes involved. Pipeline companies with scheduling systems will implement those pieces of a complete system that are economically feasible for their application.

Every batch pipeline system must have some method or tool for generating the initial schedules that pipeline dispatchers use to operate the pipeline. In the simplest cases this may involve a hand generated schedule or a spread sheet schedule package. As more sophistication is added, the schedule generation tool may include more functionality for adding more operational options and/or more accurate batch movement predictions. For example, the tool may provide batch delivery options, rate adjustment options and hydraulic limit checking. The most sophisticated package would include an automated schedule generator that can produce schedules that are optimized to meet some performance objectives. Such a system would not only improve productivity, but would ensure consistent schedules that meet predetermined optimization criteria.

The basic scheduling system can be enhanced to include some type of shipper information system that allows the scheduler to review nominations from the shippers and electronically retrieve the nominations into the schedule. An example of a display from such a tool is shown below. The integration of nomination input data into a scheduling system is a common practice in the industry. Such systems place the onus on the shippers to enter the data required to define the batch movements they require to meet the schedule objectives. This data is then automatically imported into the scheduling system and made available to the schedulers to use in the generation of the schedule.

The shipper information system may also feed back the current status of the nominations and may provide the shipper with the current location for the batches created for the nominated volumes.

Once schedules have been created, they must be distributed. Various methods can be employed for this distribution – from manual delivery to automated electronic delivery. The automated delivery may include both human and electronic recipients. In more sophisticated applications, the schedules are provided to station automation devices which can use the information provided to produce tickets when batch change indications are received.

After an initial schedule has been produced and issued, a method of updating the schedule must be provided to allow the schedule to reflect operational changes that arise through the scheduling period. These changes may include changes to nominations by the shippers, or changes to product movement made by pipeline dispatchers. Nomination revision management is normally provided by the shipper information system that retrieves and stores the nominations. The changes received can be electronically integrated into the existing schedule and the schedule recalculated to include these changes.

Version	Name	Prod	Fung	Loc	Nominated Vol	Prorated	Type	Scheduled	Status	Volume	Supplier	Consignee
0	LD-533-01	LSD	Yes		43800.00 m3	N		0.00 m3	Confirm	43800.00 m3		
				SN	43800.00 m3	N	Lift	0.00 m3		43800.00 m3	EX	SH
				NT	43800.00 m3	N	Del	0.00 m3		43800.00 m3		
0	OE-657-00	OEM	Yes		53000.00 m3	N		0.00 m3	Confirm	53000.00 m3		
				SN	53000.00 m3	N	Lift	0.00 m3		53000.00 m3	IO	PC
				SG	53000.00 m3	N	Del	0.00 m3		53000.00 m3		
0	HC-665-00	HCN	Yes		45800.00 m3	N		15300.00 m3	Confirm	45800.00 m3		
				SN	45800.00 m3	N	Lift	15300.00 m3		45800.00 m3	EX	SH
				NT	45800.00 m3	N	Del	15300.00 m3		45800.00 m3		
0	CL-486-00	CLHO	Yes		35000.00 m3	N		5000.00 m3	Confirm	35000.00 m3		
				SN	35000.00 m3	N	Lift	5000.00 m3		35000.00 m3		
				NNR	35000.00 m3	N	Del	5000.00 m3		35000.00 m3		
0	EX-566-01	EXP	Yes		43000.00 m3	N		0.00 m3	Confirm	43000.00 m3		
				SN	43000.00 m3	N	Lift	0.00 m3		43000.00 m3	CP	EX
				SG	43000.00 m3	N	Del	0.00 m3		43000.00 m3		
0	FR-477-00	FUR	Yes		67000.00 m3	N		0.00 m3	Confirm	67000.00 m3		
				SN	67000.00 m3	N	Lift	0.00 m3		67000.00 m3	IO	IO
				NT	67000.00 m3	N	Del	0.00 m3		67000.00 m3		

Figure 15 Schedule Information Display (Courtesy of CriticalControl)

To integrate operational changes, some method of data retrieval from the SCADA and/or batch tracking system is required. In the simplest case, this can be a manual data retrieval and integration. In more sophisticated systems the data is retrieved electronically. The data will include batch locations for batches in the pipeline as well as inventory for all storage locations included in the schedule.

In the most sophisticated systems, the data retrieved from the real-time systems can be automatically included in the schedule so that the schedule is up to date even when no human intervention is available.

5.4 Volume Accounting System

Chapter 2 describes the flow and other measurement issues of each instrument. The data supplied by each instrument may not be sufficient to provide the information useful for transportation services. A volume accounting system addresses various measurement issues, which make the measured values useful for the transportation services and to provide the required information and reports on tickets.

The volumetric accounting system performs various functions such as ticketing and volume balancing functions. It also determines net metered volumes for each meter in the pipeline system. Since the accounting system collects and maintains a large amount of data, it should be able to manage the large amount of data efficiently and reliably. This requires not only efficient data storage and retrieval management capability but also data entry and editing functions with auditing capability. In addition, the system should have the reporting functions for shippers and internal customers such as marketing and management.

Depending on pipeline operations, flow meters are classified as batch meter or continuous meter. The batch meter is applicable to batch operation and each batch is measured and represented by its corresponding ticket for the batch meter, while the continuous meter is applicable to a single product operation and a ticket is issued on a daily basis. A ticket is a record of metered batch volume for batch operation issued in the event of batch receipt/delivery or of metered receipt/delivery volume for single product issued daily or at a specified interval.

5.4.1 Ticketing Functions

The first step in volume accounting is to perform the ticketing functions. The host SCADA system collects the data from the measurement points throughout the pipeline system including tank farms, and the volume accounting system consolidates all the required data for further processing. The ticketing functions include the following:

- Capture tickets from field instruments either automatically and/or manually. Ticket types are meter and tank tickets.
- Validate meter tickets by applying validation rules to remove errors, verify scheduled volumes to actual tickets, and convert gross to net volumes for the customers at all measurement points on the pipeline system.
- Collect pipeline and tank inventory data.

Normally, the following input data are required to perform the ticketing functions:

- Meter configuration for each meter includes the name, size, location, type of meter, meter factors, applicable standard, etc. The meter configuration data is not frequently changed, unless a new meter is added, an existing meter is replaced or modified, or a new facility such as storage tank is added.

- Product definition includes the density and/or composition, volume correction factors for temperature and pressure, etc. The product data is not frequently changed, unless a new product is included in the transportation service or an existing product is removed from the service.
- Tickets are produced at the beginning and end of the batch for batching meters, while tickets are produced on a daily basis for continuous meters.
- Differential pressures for differential pressure meters are collected from the associated RTU or flow computer to calculate flow rate and net volume using a standard such as API 14.3.
- Accumulator volumes for linear flow meters – A gross volume is corrected to net volume either by the flow computer or the host SCADA. An incrementing volume of a rolling accumulator rolls over to zero at a specified maximum value, and that of a batching accumulator resets to zero at the onset of batch change flowing through the meter.
- Tank gauge input requires the tank level change and direction of flow, i.e., in or out of the tank to accumulate tank volume and calculate its net volume.
- Pressure, temperature, and density measurements are required to correct volumes to base conditions.

In addition, pipeline inventory and tank inventory data are required for daily scheduling, volume balancing and gain/loss analysis. These data are available through the SCADA and batch tracking application.

- Pipeline inventory data includes dates and times, line fill volumes, products and shippers for the pipeline.
- Tank inventory data includes dates and times, tank volumes, products, and shippers for each tank.

All these data items need to be integrated into a single ticket database, because a single integrated database allows the system users to easily access the required data, improve data quality, and reduce redundant efforts.

5.4.1.1 Meter Ticket

The meter and tank ticketing function works according to the daily schedule, from which a pending queue is created, listing pending tickets. A new ticket is started when a ticket is cut for the current batch and the status of the next batch in the pending queue turns to active ticket. If a single product is transported, a new ticket is cut at the designated time of the day. When a ticket is started, the volume of the ticket is initialized, its product is identified, and start time and ticket number with the corresponding schedule are created.

As the ticket meter is accumulating the ticket volume, the scheduled volume is compared to the ticket in order to ensure that the ticket volume does not exceed the scheduled volume beyond the specified tolerance. When the receipt or delivery of the batch is completed, a ticket for the batch is cut and the ticket status changes. The ticket for the current batch may be split, if the meter factor is changed or if the contract day ends for continuous flow metering. If a ticket for an unscheduled batch is created, it has to be reconciled. When an active ticket is completed, the ticket becomes a completed ticket. The completed ticket needs to be reviewed and edited, if necessary, for validation and verification before it is stored in a completed ticket database. Figure 16 exhibits an example of a meter ticket display.

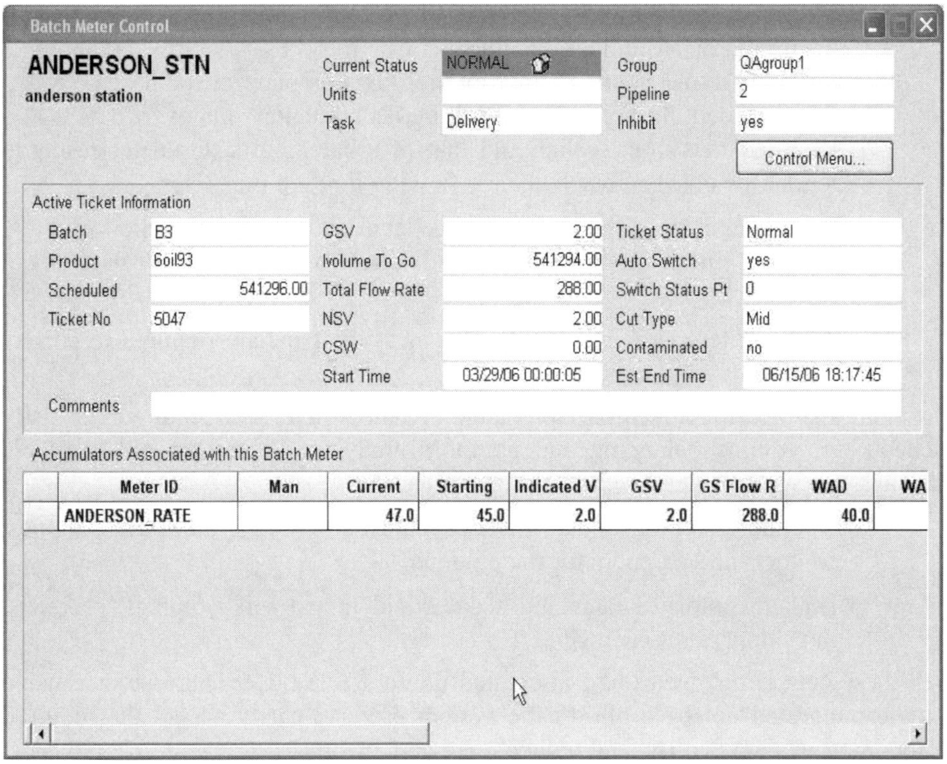

Figure 16 Meter Ticket Display (Courtesy of Telvent)

To sum up, the ticket status can be one of the following:

- The pending ticket is a ticket that is not yet started.
- The active ticket is a ticket whose product is flowing and thus volume is collected.

- A ticket becomes the completed ticket when receipt or delivery is completed.

Depending on the ticket status, different data are allowed to be modified, entered or deleted. For example, the pending ticket allows the batch ID and scheduled data to be edited, while the active ticket allows the batch ID and current measured quantities such as product and volume to be edited. After the active ticket is changed to the completed ticket, the batch ID, shipper, measured or manually entered quantities such as meter factor and product gravity, and product quality such as Basic Sediment and Water (BS&W) content can be edited. Any manual data editing requires the editor to record the reasons for future auditing.

A meter ticket may include the ticket number, shipper, ticket start and stop times and location, ticket status, meter ID and meter factor, average temperature and pressure, scheduled and ticket gross and net volumes, batch ID, product properties such as density, volume correction factor, BS&W content, etc. When the ticket is completed, the ticket data are stored in the ticket database after the data are validated and verified. They can be validated automatically with specified rules or manually verified before they are stored.

5.4.1.2 Tank Ticket

The tank inventories are mostly monitored and their operations are controlled by the common carrier. However, some tanks belong to its customers so that their inventories cannot be used for scheduling and not accounted for. The common carrier is still responsible for the lifting from and delivery to these tanks and thus responsible for producing tickets.

The data for a tank ticket are similar to the meter ticket data, except that the tank ticket includes the tank ID, flow direction, tank type with roof adjustment, tank gauges associated with this ticket, and tank strapping table to calculate tank volume. A strapping table is a set of calibrations that mathematically relate tank level to tank volume. The flow rate or volume in and out of a tank is measured using a tank gauge or flow meter connected to the tank. The flow rate, determined from the tank gauge, is less accurate than the flow rate measured by a flow meter. A tank ticket can be started or stopped automatically when receipt or delivery is completed or manually on dispatcher's or field operator's command. A floating roof tank with heavy snow or rain can effect the volume measurement reading significantly, so it must be taken into account when taking tickets based on tank levels.

A tank ticket can be generated either automatically or manually. Automatic ticket generation is done based on a sequence of the valve status changes. For manual operation, either the dispatcher or field operator issues the ticket with a unique tank ticket number and sends it to the SCADA system. To operate a tank safely, the dispatcher or operator requires the following data:

- Percentage of full value of the tank

- Volume to fill the tank to the maximum tank level
- Time to fill the tank to the maximum tank level (at a given flow rate)
- Volume to pump out from the tank to the minimum tank level
- Time to empty the tank to the minimum tank level (at a given flow rate)

Tank alarms are generated if certain conditions are violated. Some of the tank alarms are:

- Tank level alarms for the violation of the maximum or minimum level
- Time to fill or empty alarms if the calculated time-to-fill/empty passes a specified time limit
- Rate of change alarms for exceeding the highest rate of change

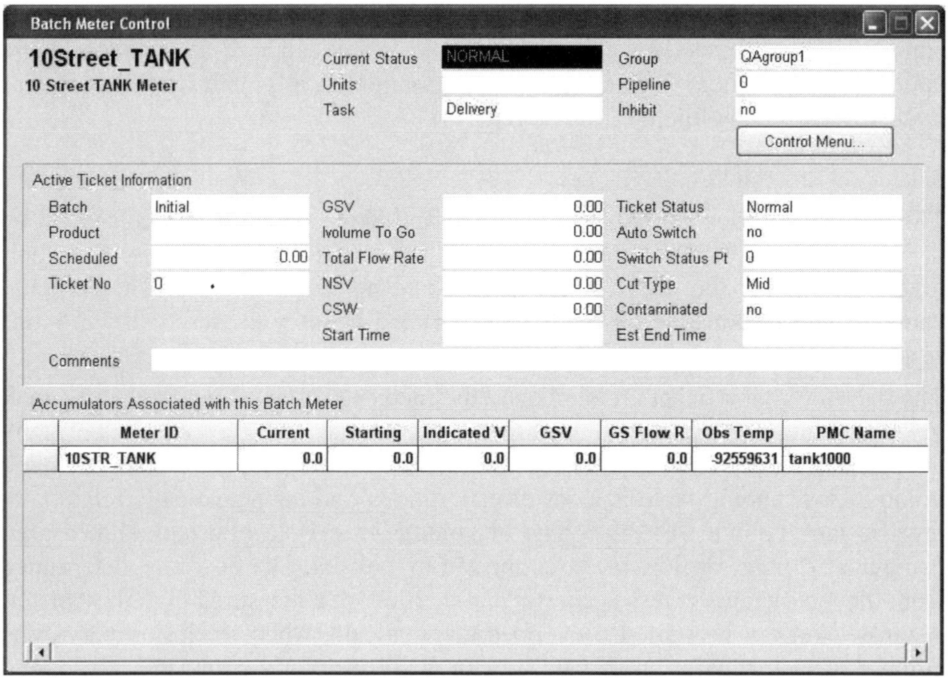

Figure 17 Tank Ticket Display (Courtesy of Telvent)

5.4.1.3 Volume Tracking

The dispatcher often requires the volume remaining in a batch to control the batch volume to be lifted or delivered. The volume tracking can be performed either automatically or initiated by the dispatcher. The warning message is generated when the volume remaining is less than a preset warning volume or when the time remaining for completion is less than a preset limit.

The volume to be tracked is taken from the batch schedule and then reset with each batch. The remaining batch volume is calculated by subtracting the lifted or delivered volume from the scheduled volume. In addition, the expected completion time of the receipt or delivery is determined using the current flow rate. The volume tracking allows the dispatcher to see the remaining volumes and the expected time to complete the scheduled volume.

5.4.2 Pipeline System Inventories

The pipeline system inventories include both the tank inventory and pipeline inventory. The common carrier is only responsible for monitoring and accounting for its own tanks, while it is responsible for tracking batches and managing inventories in the pipeline.

5.4.2.1 Tank Inventory

The main tank inventory functions are tank inventory data collection and storage, volume validation and correction, and inventory data update. The host SCADA system collects the tank data from each tank and sends them to the tank inventory database where the data are stored, and also where tank tickets can update the database. The tank data received from the SCADA system are the tank ID, inventory date and time, shipper, product, temperature, tank gauge level, and roof loading value. The tank inventory and ticket data are used for daily scheduling, volume balancing, and gain/loss analysis.

The tank inventory data needs to be validated automatically and/or manually and verified against the daily schedule. If the tank is connected to an RTU, it collects the gauge level, which is converted into the gross volume of the tank using an increment or level strapping table together with a floating roof correction. The RTU may be capable of converting the gross volume into the net volume and uploading all the measured and calculated values to the SCADA.

Tank inventory volume is determined from the measured gauge level through a multi-step process. First, the gauge level is converted to a gross volume using a volume conversion table. The table is built by means of increment or level strapping table. A level strapping table builds a relationship between the gauge levels and corresponding volumes, while an increment strapping table defines incremental volume for each level increment. One of these tables is used to calculate the gross volume of a tank. If the tank has a floating roof, the roof has the effect on the volume of the tank and thus the volume has to be adjusted to obtain the true gross volume. Assuming that free water is present on the bottom of the tank, its volume is also calculated from a strapping table to adjust the gross volume. Lastly, this gross volume is converted into the net volume by multiplying the volume correction factor for temperature or by using the appropriate API volume correction tables. The product density or gravity is also needed, because the correction factor is a function of the gravity. The volume correction for pressure is not required because the tank pressure is low and thus the correction is negligible. If the BS&W content is measured, the final net volume on the tank ticket is obtained by subtracting the BS&W content.

When the tank ticket and inventory data are processed, the tank inventory database may include the tank facility name or location, shipper, tank ID, ticket status, tank type with roof adjustment, free water gauge and volume, tank gauge level, tank gross and net volumes, temperature correction factor, etc. If the tank level is measured manually, the person's name and time need be recorded with comments. The database is normally updated daily or more frequently if required.

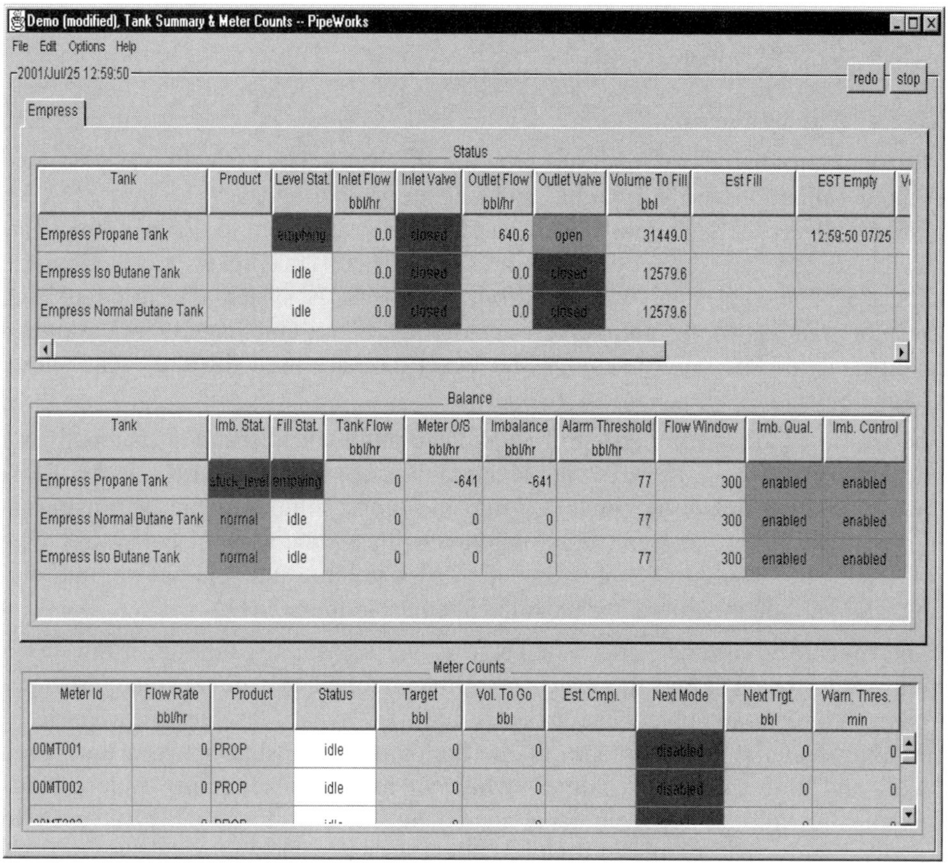

Figure 18 Tank Inventory (Courtesy of CriticalControl)

In addition to individual tank inventory, a tank farm inventory needs to be taken. Tank farm inventory is a balancing process, typically performed on a regular hourly or daily time period. All transactions at a tank farm are analyzed, receipts to tanks, tanks to pipeline, pipeline to tanks, tank to tank transfers, etc., to ensure that the accumulated transaction volumes from all inputs and outputs match the actual inventory in the tanks.

5.4.2.2 Pipeline Inventory

As discussed earlier, the pipeline inventory data is received from the host SCADA or batch tracking application. The data received from these sources are the pipeline segment, inventory date and time, batch code with shipper and product information, lifting time, and line fill volumes. The pipeline inventory data should be verified against batch code and volume. Since the batches in the pipeline are constantly moving, the line fill data are usually updated more frequently than daily. The pipeline inventories are used for daily scheduling, line balancing and batch movement analysis, and gain/loss analysis.

The pipeline inventory function may be performed automatically or manually. The pipeline inventory, if performed automatically, can be determined in real-time, and its calculation process and functions are described in Chapter 6. A manual process can be laborious if the number of batches is large. However, it may be simplified if only the changes from the previous time to the current time are incorporated to the previous inventory.

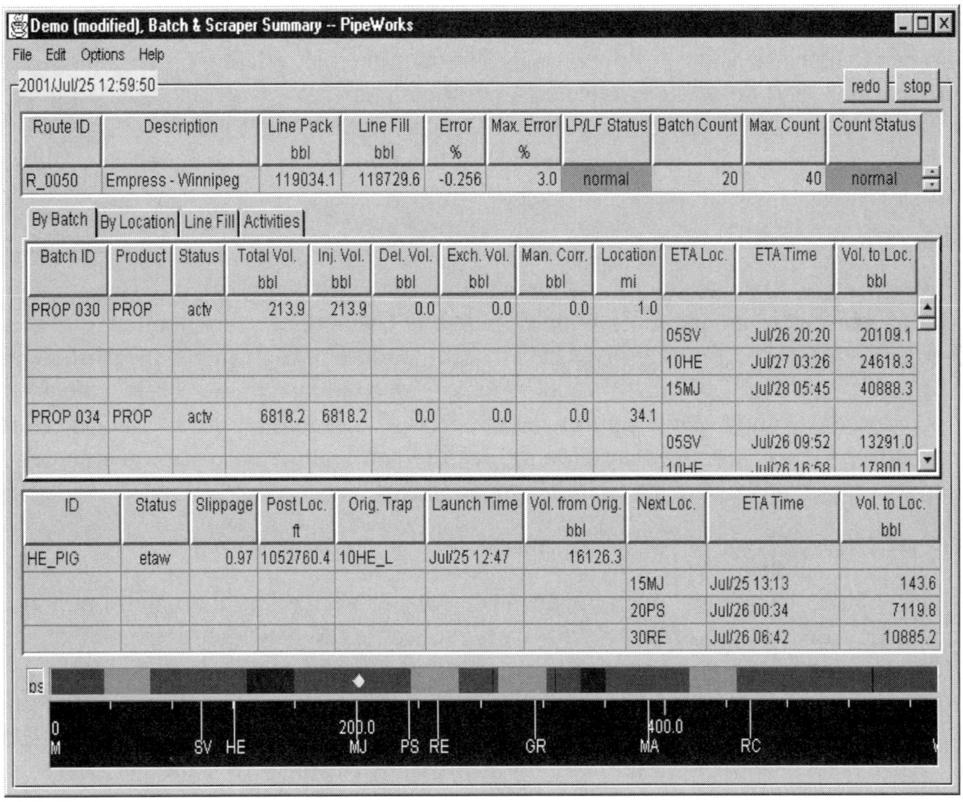

Figure 19 Pipeline Inventory (Courtesy of CriticalControl)

5.4.3 Volume Calculation and Balancing

5.4.3.1 Volume Calculation

Transportation services are charged based on net volumes or mass. Since mass is independent of flowing pressure and temperature, volume correction due to the pressure and temperature is not necessary. As described in Chapter 2, most of the metering devices measure raw flow or volume and also require regular calibration of meter and its associated measurements. Therefore, several steps with their parameters are needed to determine the net volume.

1. Meter Factor and Calibration

The calibration is required to maintain the meter accuracy. The meter factor and calibration is discussed in Chapter 3. It is determined as a result of meter proving. The meter factor is a meter correction factor, varying about 1.0. It accounts for fluid and mechanical effects on measurement, depending on product gravity and flow rate. Since it varies with gravity and flow rate, the factors are placed in a two-dimensional table; column for gravity and row for flow rate range. Meter factors are tracked over time to identify meters going off and thus being in need of maintenance.

When a meter measures a flow rate of a product, the meter factor corresponding to the product and flow rate is selected from the meter factor table and applied to the flow rate in order to adjust the value. The selected meter factor is checked against the flow range, and if the flow limit is violated an alarm is generated to indicate invalid meter factor.

2. Volume Correction Factor

The volume of a product depends on pressure and temperature. The volume correction factor is the ratio of a unit volume at base pressure and temperature to a unit volume at the flowing pressure and temperature at the time of measurement. There are several volume correction factors depending on the product density, because the volume correction factor varies with the product density and there is no universal equation to determine the factor. The volume correction method and applicable standards are given in the appendix.

5.4.3.2 Volume Balancing

The volume balancing functions are necessary to verify that the batch operation and ticketing is consistent with daily schedule. When the active ticket is completed and the volume is corrected to base conditions, the ticketed volumes should be reconciled to scheduled volumes and the tickets are adjusted for the product by shipper. The volume balancing functions are performed on a regular basis, hourly, daily and/or monthly, to identify any overages or shortages.

The ticketed volumes are reconciled to scheduled volumes based on daily scheduling information, determining the balance between scheduled and actual volumes. When a ticket is completed for the scheduled batch, the actual ticket is compared to the scheduled volume. It is intended to catch errors up-front by reviewing the ticket earlier in the accounting process and to verify that the scheduled batch is entered and the actual ticket is correct with respect to the schedule of the batch.

In addition, batch movements should be balanced as part of the verification process. Balancing of batch movements is required to check the correctness of both inventories and batch movements with respect to the daily schedule. It results in the comparison of all batch movements of actual volumes to scheduled volumes and eventually gain/loss for the specified periods such as daily and monthly.

Gain/loss can be calculated for shipper, product or their groupings, adding receipts to the beginning inventory of the specified period and then subtracting deliveries and ending inventory. The percentage of gain/loss is the percentage of gain/loss volume with respect to the delivered volume. As shown in the calculation method, the data sources of the gain/loss calculation include both the ticket and inventory data.

The completed ticket is closed after the verification process, confirming that the ticket is accounted for and that the scheduled volume vs. ticket volume and the gain/loss percentage are within the respective specified tolerances. This verification can be performed automatically or manually. If violations are detected during an automatic verification process, manual review and correction of the violations is required before it is closed. Otherwise, the ticket may be closed automatically.

As a result of the volume balancing including gain/loss, ticket volumes are adjusted and allocated to shippers. When ticket volumes are allocated at the end of the month, the actual volumes delivered to a shipper or customer is known. The allocated delivered volume is the basis for revenue accounting and used for transportation charges.

5.4.4 Product Quality

All products should satisfy tariff and product quality requirements. The product quality can be specified in terms of the following;

- Content of water and other impurities or basic sediment and water (BS&W) must be free or less than a specified percentage in order to avoid various measurement and operation problems including meter accuracy and pipe erosion.
- Range of product density or gravity should be allowable limites.
- Vapor pressure should be greater than a specified pressure at a specified temperature.
- Pour point should be greater than a specified temperature.

- Sulfur contents should be within specified tolerances.
- Viscosity should be within specified tolerances.
- Air content – Air has to be removed to avoid cavitation problems.
- Transmix – A transmix occurs as a result of the mixing of two adjacent products in a batch operation. Transmixes have to be handled as off-spec products and may be collected in a slop tank or refined again to meet the required specifications.

Sometimes, a color or injection temperature is specified as a quality measure. Sequencing and interface cutting procedures are used to maintain the product quality.

5.4.5 Volume Accounting System Interfaces

A volume accounting system requires several interfaces to receive input data and send the volume accounting results for invoicing. Listed below are the interfaces required by a volume accounting system:

5.4.6.1 SCADA Interface

As discussed in Chapter 1, the host SCADA system collects the real-time measurements from the pipeline system. The modern SCADA system usually validates the measured data to some degree. Since the volume accounting system uses the real-time measurements for its processing, the required measurement data are sent to the accounting system via software and hardware interfaces. Listed below are the key data sent to the volume accounting system:

1. Meter ticket: The SCADA data for meter tickets include the unique ticket number, ticketing date and time, schedule ID and batch code associated with the ticket, meter number of the ticket with meter factor and meter location, flow direction with receipt or delivery indicator, gross or net volume with the meter indicator, product with density or gravity and BS&W content, and shipper or customer of the ticket. If these data are not available through the SCADA, they should be entered manually.
2. Tank ticket: The SCADA system usually collects tank tickets. The data included in the tank tickets are similar to the meter ticket, except that the tank specific data such as the type of tank gauge level measurement and possibly roof correction are required. If the field measurements are not made automatically, they should be entered in the SCADA manually.
3. Pipeline inventories: The batch tracking system through the SCADA may provide the pipeline inventory data. Usually, the inventory data are collected at the beginning of each batch receipt and also in regular intervals like hourly and/or daily. The data include inventory date and time, batch codes, products and line fill volumes. The manual updating process is tedious and prone to

error, especially for large number of batches.

4. Tank inventories: The tank inventory data are made available to the SCADA either automatically or manually. The inventory data are updated in regular intervals, which include inventory date and time, tank ID, product, gross and/or net volumes, and shipper or customer.

5.4.6.2 Scheduling Interface

The scheduler creates initial schedule, which can be interfaced directly with the volume accounting system or indirectly through the SCADA system via the scheduling interface. The volume accounting system, subsequently revenue accounting system, uses the schedule data for ticket validation and batch movement balancing.

If the schedule is interfaced with the SCADA system, the system may have to provide the scheduler or dispatcher with the capability to edit the schedule. Then, the dispatcher processes the initial schedule for the batch operation as it is, or the scheduler modifies scheduling changes or even delete the schedule and to create a new schedule in order to accommodate current operating conditions and requirements.

In either way of interfacing, the volume accounting system may require the following scheduling data from the daily schedule:

- Schedule ID with revision number associated with a ticket
- Scheduled date and time
- Pipeline system or segment the batch movement is scheduled on
- Product scheduled to be transported
- Scheduled volume, receipt location and delivery location
- Meter number and/or tank number
- Shipper or customer

Additional data such as scheduler name and comments may be required.

5.4.6.3 Revenue Accounting System Interface

In general, the revenue accounting system requires not only the scheduling, ticketing and inventory data but also the net volume and batch movement balance with gain/loss volume.

5.4.6 User Interfaces

As discussed above, the number of functions and associated data is large, and several different types of users use the volume accounting system. This system needs to allow this diverse set of users to support these functions and to maintain large amount of data effectively. Therefore, an effective display tool, providing the users with easy navigation of displays and comprehensive capability of supporting these functions and editing data, is required. The display tool should provide the following capabilities to:

- enter the volume accounting data such as tickets and inventories manually, if the data are not made available from the SCADA automatically.
- view, validate, and edit the data stored in multiple databases.
- select a set of data applying selection criteria. The selection criteria may include pipeline system if there are more than one pipeline, schedule ID, tank ID, shipper or customer, etc.
- support industry standards and various calculations required for volume accounting. A few examples of standards are API and ISO, and those of calculations are net volume conversion from gross volume, strapping table conversion from gauge to gross volume and floating roof correction.

Discussed below are several user interfaces that are of high priority to end users:

5.4.6.1 Configuration Interfaces

The configuration interfaces are used to configure metering and tanking databases. The metering database includes the meter related data such as meter ID and location, meter type, meter factor, RTUs, analog points, status, deadband, alarm limit, products, volume correction factor, etc., while the tanking database includes the tank ID and location, tank type, tank target volumes, strapping table, product, etc.

5.4.6.2 Ticket Data Management

The ticket data management requires meter ticket and tank ticket management displays. The measurement staff may need the following displays:

- Meter ticket management displays: A series of meter ticket management displays allow the measurement staff to manage the meter ticket data efficiently. The data directly related to the meter ticket include meter ticket and schedule.
- Tank ticket management displays: The requirements of the tank ticket management displays are similar to those of the meter ticket management displays, except that the former requires gauge level entry and product sample displays.

5.4.6.3 Inventory Data Management

To manage inventory data, the required displays include tank inventory and pipeline inventory displays. The tank inventory displays are selected by tank ID and pipeline inventory displays by pipeline system. The inventory displays allow on-line listing and manual entries of inventories. Graphic displays of batch tracking help the users to view batch movements including batch interfaces and manage the pipeline inventory data.

5.4.6.4 Operational Interfaces

These interfaces are used by the dispatchers for operational purposes. The interfaces include meter ticketing, tank ticketing and volume tracking data.

5.4.6.5 Miscellaneous Displays

In addition to the ticket and inventory management displays, the following displays are also required:

- The scheduling system interface displays allow the dispatcher to view and process the schedule and the scheduler to update the schedule.
- The meter proving displays allow the users to control meter proving and manage the meter factor table.

5.4.7 Volume Accounting Reports

The volume accounting reports are very important, because they are the basis of revenue accounting and support the transportation charges. The reports include both detailed and summary. The reports are generated mainly for shippers to officially communicate the carrier's transportation services and for internal management to review its marketing and operation. Listed below are examples of the essential accounting reports:

- The ticket reports contain the information on meter and tank tickets entered through the SCADA or through manual entry as well as on ticket allocation after the tickets are verified. The ticket reports may be produced on the basis of selection criteria such as shipper, location, period (daily or monthly), etc.
- The schedule and batch movement reports contain schedule events, scheduled volumes vs. actual tickets, batch movement balance including the information on tickets, inventories and gain/loss analysis. A batch report may include batch interface or transmix profiles containing the volumes gained from interfaces, volume lost to interfaces, slop volumes for recycling, etc. The reports may be produced for shippers daily or monthly.
- A monthly volume balance report is required on a per product basis. The report may include the information such as line fill at the start of the month, total volume of each product supplied at each lifting location, total volume of each product delivered to each destination, line fill at the end of the month, volume error per product, etc.

- The revenue accounting reports describe all billable transportation services, including recurring charges, gain/loss settlements, and invoices.

References

(1) Mohitpour, M., Szabo, J., and Van Hardeveld, T., "Pipeline Operation and Maintenance" ASME, New York, 2004

(2) Levenspiel, O., " ", The American Oil & Gas Reporter, October and November, 1994

(3) Austin, J. E. and Palfry, J. R., "Mixing of Miscible but Dissimilar Liquids in Serial Flow in a Pipeline", Proc. Inst. Mech. Engineers, Vol. 178, Part 1, No. 15, 1964

(4) Holbrook, D.L., "Colonial Pipeline's info system speeds shipper nominations," Oil & Gas J., Dec. 1, 1986

(5) Refer to www.Transport4.com for further information. Four pipelines operating companies, Colonial, Buckeye, TEPPCO and Explorer, jointly developed the internet-based shipper information system and used by several pipeline companies and their shippers.

(6) Refer to www.atlas-view.com for further information. This system, called ATLAS, is used by Magellan Midstream Partners and its shippers.

(7) Refer to www.enbridge.com for further information. This system, called OM2, is used by Enbridge Pipeline and its shippers.

(8) Krishnan, T. R. V., et al, "Crude Scheduling Package for an Indian Cross Country Crude Pipeline", PSIG, 2003

(9) Sparrow, D. J., "Computer Aided Scheduling of Multi-Product Oil Pipe Lines," in Computer Assisted Decision Making edited by G. Mitra, pp 243 – 251, Elsevier Science Publishers B.V., 1986

(10) Neiro, S. M. S. and Pinto J. M., "A general modeling framework for the operational planning of petroleum supply chains", pp 871 – 896, Computers and Chemical Engineering 28, Elsevier Science Publishers B.V., 2004

(11) Cafaro, D. C. and Cerda, J., "Multiperiod Planning of Multiproduct Pipelines", pp 871 – 896, Computers and Chemical Engineering 28, Elsevier Science Publishers B.V., 2004

(12) Private communication

6 Applications for Operation

This chapter discusses software applications for tracking batches and other quantities, monitoring station performance, optimizing energy and facility usage, detecting abnormal operating conditions, and training the pipeline operators with a computer-based training system. These tools can help the pipeline operations staff operate the pipeline system safely and efficiently. It describes these in terms of operating concept, system architecture, and applications.

These applications are classed as "real-time modeling (RTM) system", because they are directly or indirectly interfaced with the host SCADA system for their operations. Depending on the application, some may be used off-line while others in real-time. Even though each of these applications can be used as a stand-alone application separately, this chapter describes them as if they are integrated.

Most operation tools are based on a transient hydraulic model, but a few applications such as pipeline system design on steady-state model due to performance constraints and/or technical limitations. Refer to Appendix 2 for the comparison of a steady-state model with a transient model. Leak detection is one of the real-time applications, but it is discussed in Chapter 7 due to its unique but essential nature of the application.

6.1 Introduction

Pipeline operators are responsible for balancing supply and demand volumes while maintaining a safe and efficient operation. Together with operation engineers, they have to manage receipts and deliveries to achieve the nominated volumes. At the same time they need to minimize equipment changes, minimize energy costs, detect operational problems such as efficiency deterioration and even leaks, and plan for emergencies and contingencies. In order to carry out these responsibilities effectively, the operators need proper support tools. A real-time modeling system can be an effective tool to help the operators to meet these objectives.

Transient simulation models have been widely used for pipeline system design and operation planning. Normally, steady state simulations are initially performed to design a pipeline system with fixed flow profiles, determining an optimum pipe size, station spacing, etc. However, they are not adequate to analyze pipeline system operations under varying operating scenarios, transient simulations are required to analyze pressure surge problems, large changes in load factors, facility commissioning and sudden loss problems, etc. For liquid pipeline system design, transient analysis is needed for pump station control system design, surge control, and pipeline construction economics analysis. For gas pipeline system design, transient analysis is used mainly for capability study, compressor station location optimization, and pipeline availability analysis (1, 2). For various pipeline system design examples, refer to Chapter 6 of "Pipeline Design and Construction: A Practical Approach (3)."

A real-time model may use the same pipeline simulation model as an off-line mode. However, it uses real-time data to drive the model while an off-line transient model uses boundary conditions entered by the user. As a result, the real-time data quality and availability are very critical for the accuracy and applicability of a real-time transient model (RTTM) and subsequently of an RTM system.

A properly designed RTM system can provide the following functionality:

- Operate the pipeline system in a more efficient and profitable manner within its operating constraints.
- Provide the pipeline operators with operational information such as batch or DRA tracking.
- Detect operational limit violations such as over-pressure or under-pressure.
- Advise operators in advance that a transient condition is present in the pipeline system which could lead to a pipeline system upset.
- Provide the operation engineers with a tool to formulate corrective operational strategies for avoiding a pipeline system upset due to a transient.
- Provide leak detection capability in case there is a leak in the pipeline system.

6.2 Fundamentals of a Real-Time Modeling System

A real-time modeling (RTM) system is based on the host SCADA and/or real-time transient model (RTTM). Here, the RTM system is defined as an integrated application system, while the RTTM simulates the hydraulics in the pipeline to represent the current pipeline state. The RTTM takes into account the normal pipeline operations including packing and unpacking. The model has to model the pipeline hydraulics with sufficient accuracy to be able to apply the modeling results to actual pipeline operations.

To simulate a pipeline system for all operating conditions, the model must incorporate three basic conservation principles and an appropriate equation of state for the fluid; momentum, mass and energy conservation laws are used in the model, and also it incorporates composition or batch data. Some models perform extensive thermodynamic property calculations including flash calculations to determine multi-phase flow behaviors. Refer to Appendix 1 for the discussion of the pipeline flow equations.

The RTM system requires flexibility in the selection of modeling time step, so an implicit solution method is preferred (Refer to Appendix 1 for the discussion of solution methods). It may require heavy computational and disk storage requirements. Therefore, a RTM is normally implemented on a separate modeling computer with large disk storage. This arrangement also ensures that RTM applications do not compromise the reliability of the SCADA system.

The RTM system includes not only the real-time systems such as the host SCADA system and real-time transient model (RTTM) but also several on-line and off-line applications. The on-line applications include the leak detection model, line pack management and tracking applications, and off-line applications include the Automatic Look-ahead Model (ALAM), Predictive Model (PM), energy optimization, and training simulator. As part of an off-line application, a training simulator can be integrated into the system, and is used to train the operators in operating the pipeline system. The same transient model is normally used for all applications, each of which can be executed independently. Figure 1 shows a typical RTM system components and their relationship with the integrated database.

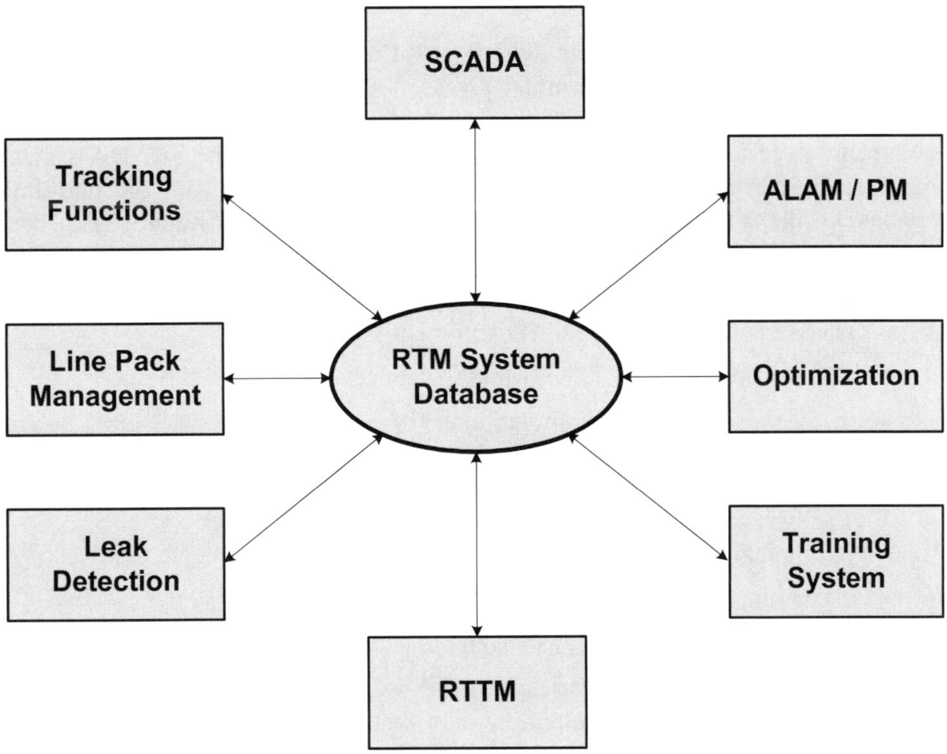

Figure 1 RTM System Database and Components

6.2.1 RTM System Architecture

The RTM system is normally located at the primary control center where the SCADA system that controls the pipeline system is located. The SCADA and RTM systems reside on two separate computer hardware platforms. They are in an open architecture client/server connected on a LAN (Local Area Network). Most SCADA systems are provided with a redundant configuration, while the RTM can be in either redundant or single configuration depending on the criticality of the system. If the RTM is in a redundant configuration, it may have dual redundant RTM servers with hot standby and automatic failover.

Occasionally, there is a backup control center, which is in a different location from the primary control center. The backup SCADA system, which resides in the backup control center, is connected via a WAN (Wide Area Network) and its database is synchronized to the primary system. If the RTM system is also installed in the backup control center, it has to be interfaced with the backup SCADA and the RTM database has to be refreshed with the latest data received from the primary RTM system at the same time as the SCADA database is updated. Refer to Chapter 1 on SCADA system architecture.

6.2.2 Real-Time Data Transfer

The host SCADA system collects field data and refreshes its real-time database at regular (polling) intervals, determined by the "scan rate" of the host. Since the RTTM and its direct applications run in sync with the host, the RTTM time has to be synchronized to the SCADA polling interval. Each scan, the SCADA system transfers the current polled data with time tag information, from the real-time database to the RTM database. The polling cycle dictates the frequency of data transfer between the SCADA and RTM databases. In general, the following tasks have to be performed after the end of one poll and before the next poll begins:

- Data transfer from the SCADA to the RTM database
- Data access by the RTTM from the database
- Completion of RTTM simulation
- Data transfer from the RTTM to the database
- Data access by the direct applications from the database
- Completion of the execution of these applications
- Data transfer from the applications to the database
- Data transfer from the RTM database to the SCADA.

RTTM simulations always start after the SCADA data transfer has been completed. As a result, the simulation start time is one scan behind. The non-direct applications such as ALAM, optimization and training system need not complete their execution.

6.2.3 Real-Time Data Validation

The high quality of real-time data is crucial to the proper operation of a real-time model and is central to the success of the RTM system as a whole. SCADA data received by an RTM system is neither always available nor reliable and accurate. Measurement problems include measurement accuracy, instrument locking, instrument malfunction, and instrument drift. SCADA problems include data out of scan, communications outage, and others. Section 7.5.6 describes the detailed analysis of the data related problems and other factors affecting the RTM performance.

One of the ways of ensuring consistent accuracy of real-time data and of minimizing modeling error is to use a data validation processor. The RTM system requires data validation beyond the level of validation performed by the SCADA system. A data validation processor analyzes field measurement data and status in order to detect measurement problems. Normally, it takes into account the status and data quality flags in the modeling process, in order to improve the quality of model results. The validation processor detects errors in flow rates, pressures, temperature, and compositions as well as checks real-time data against their measurement ranges. Data that are out of scan or failed are not used in the modeling process.

More advanced validation schemes take into account relationships with other variables at the same data point. It checks each measurement data for consistency with redundant data, data at prior times, and other related variables and status. Also, a known interrelationship between variables at several data points can be used to check the validity of the data. For example, the RTTM can cross-check measurements (e.g. a pressure reading with a flow rate) by means of hydraulic calculations.

6.2.4 RTM Database

At the core of a RTM is the database. The RTM system database contains not only the common pipeline system configuration data for the models and the individual applications, but also real-time data for the RTTM model and the outputs of the applications. Modeling applications such as a PM can generate huge amounts of data depending on the simulation period and number of operating scenarios. If a training system is also integrated into the RTM system, the database may contain extra data such as the computer-based text material and student test records.

An RTM system generates large quantities of data and has to process real-time data, requiring its own database to be suitable for the real-time modeling and applications. Therefore, the RTM database is designed to have fast processing data capability as well as the capacity to store large amounts of data. As a result, the RTM system will have both real-time and historical databases, storing its own dedicated historical data independent of the SCADA historical database. This is consistent with keeping the operating SCADA system separate in all regards from the RTM system. Generally, archiving of the RTM data is not done, but this will be determined by each user's IT procedures.

The RTTM data will contain the following:

- Calculated hydraulic profile for last scan
- Calculated hydraulic profile for current scan
- Various alarm and event messages directly coming from the RTTM applications
- Batch or composition tracking data

The database structure for an ALAM is the same as the PM database structure. However, the database size of the latter should be much larger because it uses various initial states and generates data for many different operating scenarios. The ALAM and PM data will contain the following:

- Future operation data such as set points and intermediate and final states
- Initial states from various sources
- Scenario output data (hydraulic profiles, alarms, set points, etc.)

In addition, the RTM database contains optimization data such as unit selection and station set points as well as the data related to leak detection.

6.2.5 Data Interfaces

As shown in Figure 1, data transfer within the RTM system takes place through the RTM database that is interfaced with the host SCADA and other RTM applications. The data transfer is bi-directional among the RTM models and applications, including the host SCADA. The RTM system requires the following interfaces with the database:

- SCADA Interface – real-time data and data quality indications are received from, and key results from various applications are sent to, the host. Normally, alarm and event messages, line pack management, and tracking information are sent to the SCADA automatically, while other information such as optimum set points is sent on demand.
- RTTM Interface – real-time SCADA data is processed and modeled to generate the current pipeline state that is stored in a real-time portion of the RTM database. Pipeline state alarms such as pressure violations are sent to the SCADA via the database.
- Interface with Leak Detection – the current pipeline state generated by the RTTM or other state estimation module (if no RTTM is available) is sent to the leak detection module to detect abnormal conditions such as leaks, together with historical states and the abnormal condition results are sent to the RTM database.
- Interface with Line Pack Management – current and historical RTTM

data or data generated by other line pack calculation method (if no RTTM is available) are used to analyze line pack changes and change rates.

- Interface with Tracking Functions – current RTTM data or data generated by other tracking method are used to track batches, compositions, and/or product anomalies, etc.
- Interface with Optimization – current and historical SCADA or RTTM data are used to determine optimum unit operating points and display operating history.
- Interface with ALAM – current pipeline state is sent to the Look-ahead Model at regular intervals. The data transfer is under the control of the ALAM, and the data is transferred after the RTTM finishes its modeling cycle and refreshes its database.
- Interface with PM – current pipeline state is sent to PM on a demand basis. The data transfer mechanism of the PM is similar to that of the ALAM.
- Interface with training system – simulated field data generated by the training simulator are sent to the RTM database, from which the operator sends control commands to the training simulator. Also, operation instructions are sent to the training simulator and the instructor can view the trainee responses.

6.3 Real-Time Transient Model (RTTM)

An RTTM (4) is the core model to achieve the objectives of a real-time modeling system. It continuously synchronizes to the actual pipeline state through real-time measurement data received from the host SCADA, to determine current pipeline state in the entire pipeline system on a real-time basis. It provides the operators with the information for the analysis of the pipeline system performance and other RTM applications with a starting pipeline state.

The RTTM runs automatically each scan, calculating the current pipeline states in real-time over the entire pipeline system. It takes the previous pipeline state and current measured data received from the host SCADA system to simulate forward in time to the current time. The measured data includes flow rates or volumes, pressures, temperatures, densities, batch or DRA launch information, and measurements and valve status. For a gas pipeline, it receives gas compositions from the SCADA, which are either entered manually or fed automatically by gas chromatographs. The current pipeline state is expressed in terms of flow, pressure, temperature, and density or gas composition profiles along the pipeline. The model provides batch tracking and other tracking information such as DRA movement. The state includes information about line pack and packing or drafting rate. In addition, the state may include operating points of all operating pump or compressor stations.

The model should be able to track batches or gas compositions unless a single product is transported. When a batch is launched, a batch launch signal with its batch identifier is passed to the model, and for gas pipeline gas compositions should be automatically measured by a chromatograph at receipt points or manually entered and passed to the real-time model. If product blending takes place, a proper mixing algorithm is required to calculate the mixed properties accurately.

The RTTM can determine the current operating points of all the operating pumps or compressors. The operating point includes the unit flow rate, speed and efficiency as well as head, horsepower and fuel consumption. These points can be plotted on the operating pump performance curve or compressor wheel map. Refer to Section 6.4.2 for further discussion.

For gas pipelines, the model can detect not only condensation and dew point conditions but also hydrate formation if gas compositions and vapor data are available. For liquid pipelines, it should be able to identify the segments where slack flow or two-phase conditions would occur. Multi-phase flow models perform thermodynamic and physical property calculations including phase equilibrium to predict mass transfer between vapor and liquid phases. Since multi-phase behaviors are more complex and the models are not accurate for real-time applications, the multi-phase real-time model implementation is limited (5).

6.3.1 RTTM Requirements

The key requirements of an RTTM are to calculate the pipeline hydraulics accurately and to run in a robust manner. Without accuracy and robustness, the model has limited applications, especially for leak detection, which requires the highest accuracy. Improving the quality of real-time data from the host SCADA system through a validation process will enhance robustness. The accuracy of the RTTM system is improved by the above validation process as well as short simulation time and distance steps. Refer to Appendix 1 for a discussion of solution techniques. To maximize the accuracy of the RTM hydraulic calculations it should include the ability to:

- Simulate the pipeline hydraulics either on an individual leg basis or on an entire network basis, including the partial differential equations of continuity, momentum and energy as well as accurate equations of states appropriate to the fluid in the pipeline. Refer to Appendix 1 for a detailed discussion. These equations include: the pipe diameters with wall thickness and pipe roughness and the elevation profiles along the pipeline.
- Simulate the effects of pipe wall expansion on pipeline transients due to changes in pressure and temperature.

- Simulate the energy transport of the fluid in the pipeline and heat transfer from the fluid to the pipeline surroundings. The energy transport should include the Joule-Thomson effects for light hydrocarbon fluids and friction terms for heavy crude.
- Simulate the movement of batches or fluids with different compositions along the pipeline system by taking into account fluid property changes. The blending of multiple fluids with different properties is included in the model.
- Simulate the movement and effects of drag reducing agent (DRA) for pipeline systems with DRA operation.
- Perform tuning and state estimation to estimate pipeline states more accurately.

In addition, the RTTM has to function during actual pipeline operating conditions which can include:

- All types of hydraulic behaviors (change in flow, pressure, temperature, etc.)
- Pump or compressor station start-up and shut-down
- Pipeline system start-up and shut-down
- Shut-in conditions
- Valve operations: open, close and in-transit
- Batch operations: full and side stream injection and delivery
- Gas streams with multiple compositions: gas property mixing
- Pigging operations
- DRA injection at multiple locations
- Reverse flow operations
- Slack flow conditions: it is not easy to simulate the hydraulics of slack conditions accurately.

Since real-time data is not always available for various reasons, the RTTM must run in a robust manner under adverse conditions such as communication outages and measurement problems. Even if communication outages occur or the real-time database is corrupted, the model should be able to run with limited available data in a degraded mode. It should generate a message of model degradation to advise the operator of limited reliability during such adverse conditions. When such a condition persists for a long period, a warm start is invoked to allow the model to adjust to the new normal operating conditions.

The model should produce accurate results able to be used for actual operations. In order to improve simulation accuracy and robustness, the model normally pre-processes real-time data before it is used for modeling and corrects for calculation errors through a tuning process after the modeling cycle is completed. Some models use state estimation techniques such as a Kalman filter to improve the model accuracy and robustness. Such techniques try to find a state that satisfies the hydraulics, while keeping the differences between the measurements and the modeled values within measurement accuracy. In other words, a state estimation technique imbedded in an RTTM works as an automatic tuning filter.

The tuning process should be smooth to avoid any sudden change in line pack calculations. Tuning parameters include pipe roughness and measured temperature (6). If calculated pressures differ significantly from the measured pressures or temperatures, the model can identify measurement problems such as bias or failures. (Refer to Section 7.5.8.5 for a discussion of the tuning tasks).

6.3.2 RTTM Outputs

In general, the RTTM does not generate a large amount of data, because it is concerned mostly with the current state. An RTTM may be able to generate the following data:

- Hydraulic profiles of flow, pressure, temperature, and density or gas composition
- Violations of operating constraints
- Line pack and its packing or drafting rate for gas pipelines
- Compressor or pump performance including efficiency and horsepower
- Anomaly detection such as slack flow for liquid pipeline, hydrate formation for gas pipeline, etc.

Normally, the complete RTTM data is displayed on a separate console except for a limited amount of key information such as leak alarms. The main reasons are as follows:

- The RTTM outputs can be too large to display on SCADA consoles.
- The RTTM system is important for efficient operations but not considered as critical as SCADA.

6.3.3 RTTM Degradation

The model is placed in a degraded mode of operation when it is not in perfect working condition due to measurement or modeling problems. The problems that cause model degradation include such things as: temporary measurement unavailability, communication outages, and SCADA problems. The type of measurement and how it is used influences the degree of degradation. In other words, the variables used as boundary conditions have severe impact on the model accuracy and robustness, and as a result the model is degraded more severely than non-boundary variables. Also, lack of pressure data will degrade a model more than lack of temperature data will, because missing or unreliable pressures affect the model accuracy and reliability more than missing temperatures.

If the degraded condition is severe and persists for a long time, the model may enter into a disabled mode of operation. The RTTM should take the operating modes (enabled, degraded and disabled) into its calculations of operating parameters such as line pack and leak detection.

6.3.4 RTTM Operation

The RTTM needs cold, warm and hot start functions. The hot start is invoked with each scan as long as the required data is available. The modeling status is displayed to advise the operator of limited capability during normal operating conditions.

The cold start function is intended to initialize all parameters required by the model. When the cold start is invoked, the model checks to see if enough information is available to estimate the current pipeline state. The cold start invoking decision is based on the following criteria:

- The model is being brought on-line without prior pipeline state.
- The pipeline configuration has been modified.

The decision to invoke a warm start is made if initial pipeline states may not be accurate or reliable due to extensive communication outages or other abnormal situations.

6.4 Applications

An RTM system has three types of applications:

1. The first type is a direct application of the RTTM results.
2. The second type performs facility performance monitoring based on the measured and modeled values of facilities such as compressors and pumps.
3. The third type uses the pipeline states calculated by the RTTM to generate future pipeline states for operational analysis. An ALAM or PM, linked with the RTTM, is used to estimate the future pipeline states.

6.4.1 RTTM Applications

The RTTM can provide the operator with alarms for abnormal operations other than leaks. It can detect pressure limit violations by comparing the calculated pressures against the maximum and minimum allowable operating pressures. It can identify line blockage problems as well as violation of line pack limit and its change rate. An RTTM is applied to the following operations:

- Batch tracking for liquid pipelines

- Composition tracking for mostly gas pipelines
- Line pack management for gas pipelines
- Content tracking for both gas and liquid pipelines
- DRA tracking for liquid pipelines
- Slack line flow detection for liquid pipelines
- Instrument analysis for gas and liquid pipelines

It was reported (7) that some pipeline companies use a compensated mass balance leak detection system for some of these operations. Each of these operations is discussed briefly below.

6.4.1.1 Batch Tracking

Batch tracking monitors each batch for it's volume, origin, current location, destination, and estimated time of arrival to designated locations. A batch is defined as a contiguous entity of uniform fluid properties which moves through the pipeline system as a single entity. For example, a batch is assumed to have constant density, compressibility and viscosity.

Real-time batch tracking information helps the operators to reduce unnecessary downgrading of product or contamination of product in tanks. In addition, up-to-date batch tracking information is useful in improving the accuracy of short-term batch schedules. (Refer to Chapter 5 for practical applications of batch tracking function to actual pipeline system operations).

In an RTM sense, the batch tracking function is batch modeling in the RTTM. Batch movements along the pipeline are modeled, assuming that each batch has uniform fluid properties. An RTTM may include product mixing at the interfaces between two batches and/or at each lifting point. The batch modeling alone is useful for model integrity but not sufficient for actual batch tracking operation. This is because discrepancies in batch positions between actual and modeled batch tracking do occur in actual practice. This difference cannot be simply adjusted in the RTTM model because the volume adjustment can violate mass conservation. This problem can be resolved by maintaining two sets of batch data: one set for modeling batch movements with mass being conserved and the other for accounting batches with allowance for changes to batch volumes. The display in the next page shows a typical batch tracking information.

The batch tracking must be able to perform the following main functions:

- Determine and update the positions of the batch interfaces with each scan.
- Maintain batch volumes in the pipeline.
- Calculate estimated time of arrival (ETA) of batch interfaces at designated locations.

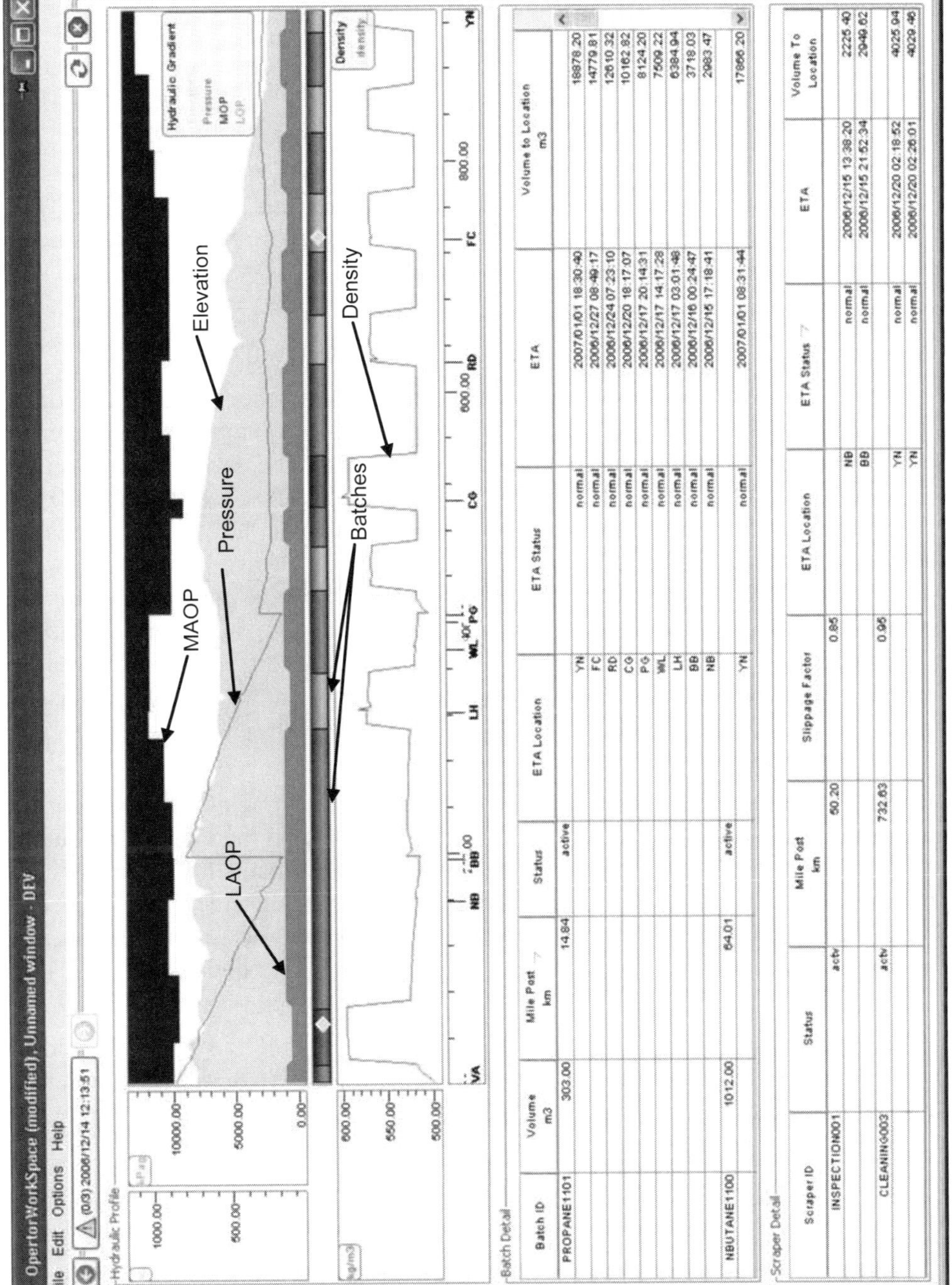

Figure 2 Batch Tracking Display (Courtesy of CriticalControl)

- Calculate batch overages and shortages in the pipeline.
- Estimate interface mixing lengths and volumes.
- Detect an actual interface arrival automatically at a batch interface detector such as densitometer.
- Adjust batch volumes and interfaces automatically according to the specified rules in the event that a batch interface is detected, and provide the operator with the capability to modify batch volumes, batch positions, or batch ID manually.
- Alert the operator of batch arrivals.

Batch launches can be triggered by an indication from SCADA, a change in density, a change in valve status, or can be based on a schedule. Batch volumes are updated based on injection and delivery volumes obtained from metering locations along the pipeline. The interface positions can be determined, given the order and volume of each batch in the pipeline and the line pack in each pipe leg computed by the RTM. Given pipe leg flow rates and interface positions, estimated times of arrival (ETAs) to the designated downstream locations can be determined. Upon completion of delivery and removal of the batch from the pipeline, an over/short volume is calculated and stored. The over/short reflects the difference between metered injections and deliveries along the pipeline as well as any manual adjustment that may have been made along the way. Caution must be exercised when a manual adjustment is made to a RTTM model because changing line pack violates mass balance.

If a side stream injection takes place, two different batch tracking problems arise: the injected product is the same as the flowing product and the injected product is different from the flowing product. The former case maintains the same batch ID but the size is different and the flow rate downstream of the side stream injection point increases by the same amount as the injection rate. However, if a different product is injected into the flowing product, then the following changes take place:

- Two products are blended and the properties of the blended product should be determined for modeling,
- The batch size on the upstream side of the injection point reduces and eventually the batch disappears, and
- The blended product becomes a new batch downstream of the injection point, and its size grows.

Batch tracking may be integrated with a Batch Scheduling system, to determine an up-to-date batch schedule; this is accomplished by comparing actual batch tracking data with scheduled injection and delivery volumes and times. Current batch volumes and positions can also be used to update short-term batch schedules.

6.4.1.2 Composition Tracking

Natural gas can have varying compositions, depending on the source of the gas and the gas processing. An important application of an RTTM to gas pipeline operation is composition tracking. A composition tracking function tracks composition data from the measuring location to a designated downstream delivery point or an in-line measurement location where a new measurement would take place. The composition of the blended product is calculated and tracked from the locations where more than one product stream is mixed.

Composition tracking is required to correct flow rates accurately at the meter station, calculate pipeline state and line pack, and track gas quality. If a gas processing upset takes place, the injected gas stream is either contaminated with prohibited components or enriched with heavier components. In such upset conditions, the composition tracking information is required to handle the contaminated or enriched components at certain points in the pipeline network. Composition tracking capability allows the RTTM to determine condensation and hydrate formation if heavier hydrocarbons or free water vapor are present in the gas stream.

The gas composition is normally determined at injection and delivery locations, where flows or volumes are measured. The composition is determined automatically by on-line gas chromatographs or manually from lab measurements of gas samples. If gas composition data is not available at a delivery location, the data tracked by the RTTM can be used for heating value calculation, assuming that composition tracking is accurate (8). Tracked gas compositions may be used to correct delivery flow if the chromatograph is temporarily unavailable or may be used for measurement at small volume locations to avoid the installation and operating cost of a chromatograph. The composition tracking accuracy is dependent on the accuracy of both the composition measurement and model. Such an application to gas volume transactions requires regulatory approval if it is to be used for billing purposes.

Similar to batch tracking, the RTTM calculates the movement of composition interface points by utilizing local gas velocity, which varies with location and time. Another method of composition tracking is to solve the transport equation. If gases from different sources are blended at a junction, the outlet composition is averaged with the inlet compositions and weighted by the standard flow rate or volume fractions of each composition.

The RTTM can track other physical qualities such as heating value and sour components as well as non-physical qualities such as ownership and source of the gas. The heating values can be calculated using the tracked gas compositions, and the tracked quantities and components can be easily displayed to gain operational benefits.

6.4.1.3 Line Pack Management

The line pack management calculates the amount of gas volumes resident in the pipeline system and makes the line pack information available to the pipeline operators. Since the pipeline is a pressure vessel and used as a conduit, operation efficiency and safety can be achieved by operating between the maximum and minimum line pack limits. By proper use of the line pack information, the operator can respond to demands with minimum changes in pipeline operations. Therefore, the main objective of the line pack application is to increase operational efficiency and safety of a gas pipeline system. The management of line pack is not important and practical to use for liquid pipelines, because the compressibility of liquids is so small compared to the compressibility of gas that the change in line pack is much smaller and the time of line pack change short.

Since the RTTM calculates the pressure and temperature profiles and tracks gas compositions along the pipeline system, it can calculate the gas volumes or line pack and their changes over time accurately, together with pipeline physical configurations, calculated pressure and temperature profiles, and an appropriate equation of state. Line pack is normally expressed in standard volume units.

The line pack information is used to check receipts and deliveries and often included in operating and billing reports. The gas pipeline operator compares the calculated line pack with the predefined target value and adjusts receipts and deliveries to achieve the desired level of gas inventory. The desired level has to be maintained to use compression facility efficiently, and to avoid contract violations and delivery shortfalls. For example, if the line pack is drawn down, then the delivery is greater than the supply; the difference should be made up before the line pack reaches the minimum allowable level. If the line pack is packed, then the supply is greater than the delivery and thus the line pack needs to be controlled before it reaches the maximum allowable level. The line pack data is required at the end of the gas day to balance daily billing. The information can also be used in volume balance leak detection.

Even though the RTTM can calculate the line pack and its change accurately, it is not absolutely required to estimate these values. If an RTTM is not available, the line pack and its changes over time can be approximated using steady state average pressures and temperatures. This approach can result in large line pack calculation errors during transient operations. An appropriate filtering technique can be applied to reduce the errors by compensating for these conditions. This simpler approach is acceptable in most normal operating conditions. Some problems may occur when the line pack starts to drop significantly due to an extended period of reduced supply or excessive delivery and as a result line pressure may drop below the limit.

The current line pack and its change over time require real-time receipts and delivery flow data as well as pressure and temperature measurements. The future line pack and its changes can be estimated using the gas supply and demand data.

A supply forecast model may be used to estimate the gas supply, while a load forecast model is used for the gas demand in the near future. However, gas nomination data can provide the gas supply and demand data with sufficient accuracy.

Line pack values are generally grouped to provide several levels of detail, dividing the entire pipeline system into several sub-systems and if necessary each sub-system into smaller pipeline segments. A sub-system may be bounded by two compressor stations or individual laterals. Typical outputs of this application include system line pack and its rate of change, sub-system line pack and its rate of change, and/or segment line pack and its rate of change, together with their associated time stamps and target values. Typical displays of total and segment line packs over time are shown below.

This application should include both graphic and tabular displays. The line pack and its change rate can be displayed in color on the pipeline configuration graphically. The tabular displays list all pipeline sub-systems sorted by pipeline and sub-system, and include for each pipeline sub-system data such as service status, current line pack, change in line pack since the last hour, and change in line pack since the beginning of the gas day. Subtotals and totals of these quantities need to be displayed for each defined sub-system as well as the entire system.

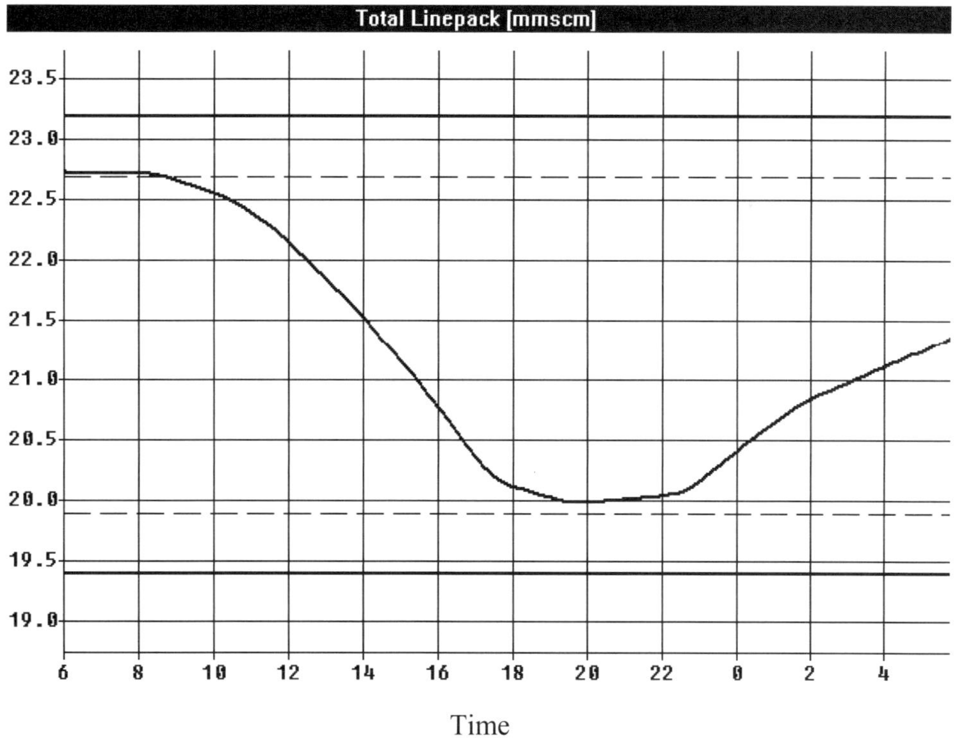

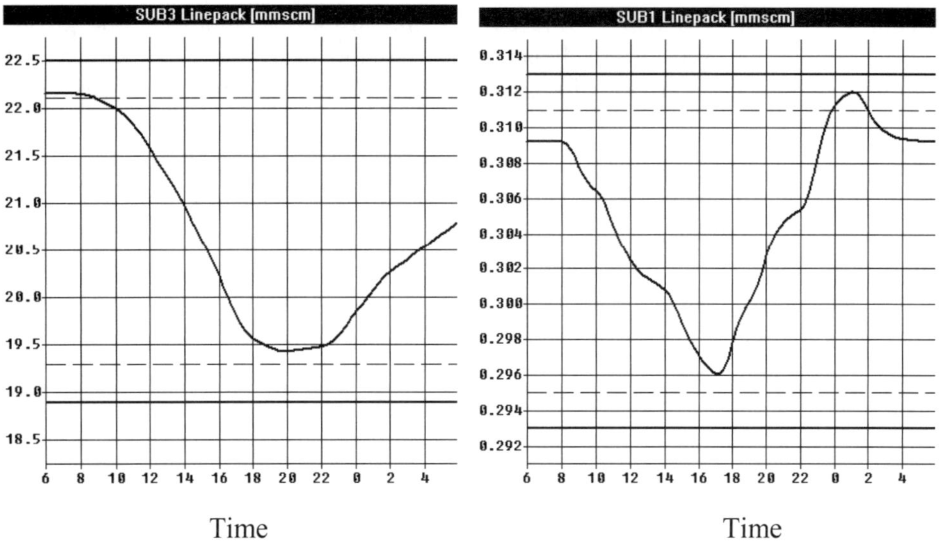

Figure 3 Line Pack Changes over Time (Courtesy of Liwacom)

6.4.1.4 Content Tracking

The RTTM can track any product characteristic in real-time. These values are tracked in a fashion similar to the way batches are tracked in the model, but the tracking functions listed below have their unique tracking requirements. The common tracking functions include:

- Scraper tracking – The RTTM can track the movement of scrapers or pigs in the pipeline from their introduction to their removal at the receiving trap. Pigs enter the pipeline at an upstream pig launcher and are removed at a downstream pig trap. Pig launch and receipt for tracking can be triggered by telemetered or manually activated signals. Pigs usually do not travel at the same velocity as the product in the pipeline, slipping as product passes by the pig. The pig tracking function calculates the location and estimated time of arrival and combines it with the calculated local fluid velocity and manually entered pig slippage factor. The pig tracking function helps the operators to schedule their pig recovery operations.
- DRA concentration tracking – The RTTM can track both sheared and non-sheared DRA concentrations. DRA is injected into the pipeline at the DRA skid located downstream of the pumps. A DRA injection rate is used with measured or calculated product flow rate to calculate the DRA concentration. When a DRA passes through a pump, it is sheared and no longer active. The DRA tracking function tracks the sheared and active

DRA concentrations and checks the concentration against the maximum DRA concentration allowable in the product. For example, DRA is not allowed in jet fuel and thus its concentration should be checked against zero concentration level. A graphic view of the DRA contents within a pipeline can show active, sheared and total concentration of DRA in the product as well as the positions relative to DRA injectors or pump stations.

- Anomaly tracking: - the RTTM tracks the anomaly, once initiated, with flow of product until fully delivered or removed from the pipeline. Anomalies include excessive amount of BS&W in a liquid pipeline, or H_2S and other sour gas content in a gas pipeline. If an anomaly is entered into the pipeline, the pipeline operator would enter the anomaly attributes including anomaly concentration, ID and description, location entered, and specified downstream location to be delivered. The tracking function determines the current location and estimated time of arrival at the specified location.

Other product characteristics such as color, haze and flash can also be tracked. The values can be displayed in a profile graph or table in combination with other hydraulic profiles. This function provides the user with a much clearer view of product ownership and quality.

6.4.1.5 Slack Line Flow Detection

The phase of a fluid turns from liquid to vapor in a liquid pipeline whenever the pressure at a given temperature drops below the vaporization point of the fluid. A slack line is the condition wherein a pipeline segment is not completely filled with liquid or is partly void, and often occurs near high elevation drop points when the pipeline back pressure is low. Since a real-time flow model calculates the pressure and temperature profiles along the pipeline, slack flow can be detected by monitoring pressure and temperatures for slack conditions (9).

As pointed out in reference 9, the slack line condition is frequently encountered in liquid pipelines as one or more of the following cases:

- Slack line flow occurs always in certain segments of the pipeline.
- Slack line flow occurs occasionally during normal pipeline operations in certain segments of the pipeline.
- Slack line flow occurs only during abnormal pipeline operations.

Sometimes it is unavoidable to operate the pipeline in a slack mode, particularly in the case where the back pressure cannot be increased beyond the designed pressure level. However, slack operations need to be avoided for efficiency and safety reasons. The problems caused by slack line conditions include:

- Very large pressure drop due to constriction in slack regions
- Increase in batch interface mixing length, resulting in unnecessary extra

cost

- Pressure surges caused by the collapse of slack line regions
- Increase in metal fatigue rate

The modeling of slack line flow behaviors is difficult, because it requires both accuracy in the pressure and temperature calculations and well-defined phase behavior of the fluid. Therefore, caution should be exercised in interpreting the modeling results.

6.4.1.6 Instrument Analysis

As part of the RTTM and leak detection process, potential instrumentation problems are detected by comparing the actual measurement against the value expected by the RTTM or analyzing the measurement trend. The following problems can be detected and reported to the pipeline operators:

- Abnormal imbalance produced by failed measurement, if not detected by the host SCADA
- Flow meter bias
- Erratic measurements

6.4.2 Optimization of Facility Performance

This section discusses how to make the best use of pipeline facilities by optimizing the pipeline system and/or monitoring compressor/pump performance. The compressor/pump performance is monitored by displaying the current and past operating points on the compressor/pump performance curves along with operating efficiency. System-wide optimization requires more sophisticated mathematical approaches, such as pipeline network simulation and optimization algorithms.

6.4.2.1 Compressor Station Monitoring

A compressor station consists of compressors and drivers, coolers or chillers (arctic application), measurement and control systems, and various other ancillary facilities including station yard piping, valves and auxiliary units. (Refer to Chapter 3 for further discussion of station components and automation.)

The compressor station monitoring functions may include compressor performance monitoring and unit statistics. Operation efficiency of a compressor station can be improved by monitoring unit performance and taking corrective action if required. A compressor performance monitoring function calculates compressor station performance, monitors the trends of each unit's performance, and displays the performance of compressor units including alarms for deviation in performance.

The station's overall operating efficiency is determined by a compressor station model, which uses measured or modeled pressures, flows and temperatures in accordance with the equipment's characteristics, unit alignment and the control system. A compressor station model can consist of four main parts: station model, compressor model, driver model and cooler model (chiller model is not discussed here because it is limited to arctic applications). Normally, a station model treats yard piping as a lumped parameter and valves as flow direction control devices only.

The station model uses the station control system and unit alignments to determine the unit load or flow rate for a parallel alignment and pressure and temperature for series alignment. The compressor model locates the unit operating point with the compressor wheel map using the following values:

- Calculated flow and measured pressure and temperature for a parallel alignment
- Measured flow and calculated pressure and temperature for a series alignment

The operating point on the wheel map shows the head and flow rate, rotation speed, efficiency and power. The driver model in turn uses the calculated compression requirements such as shaft power to calculate the driver efficiency based on driver performance curves and ambient temperature. Finally, the cooler model calculates the cooling efficiency from the cooler temperatures and pressures at the cooler inlet and outlet points.

A compressor wheel map is a relationship drawn in the X and Y coordinates, represented in terms of flow and head and wheel rotation speed or RPM. The map also shows the compression efficiency in terms of flow and speed. The performance curves are bounded by the surge line for low flow rate and stonewall for high flow rate as well as by the lowest and highest speed lines. The wheel map is provided by the compressor manufacturer. If the entire wheel map is not available, it can be generated from the rated performance curve by applying the affinity law and a curve-fitting algorithm.

Each SCADA scan, the station model and compressor unit model calculates the operating points. The operating points with their calculated and measured values are plotted on each compressor wheel map over a specified time period. The plot may provide the following additional information:

- Data quality of the operating point
- Flow conditions for a plotted operating point, showing the state of a recycling or full flow through the unit
- Operating point history, showing the history of the operating points

The compressor unit efficiency and operating points are trended. Efficiency alarms can be generated if the trended efficiencies violate the specified limit repeatedly over a certain period. Surge control alarms can also be generated if the operating point approaches closer to the surge control line. Figure 4 shows a typical plot of operating points on the wheel map of a centrifugal compressor.

The benefits of using the compressor performance plot are significant. It helps the operator to run compressor units more efficiently and to control surge more efficiently.

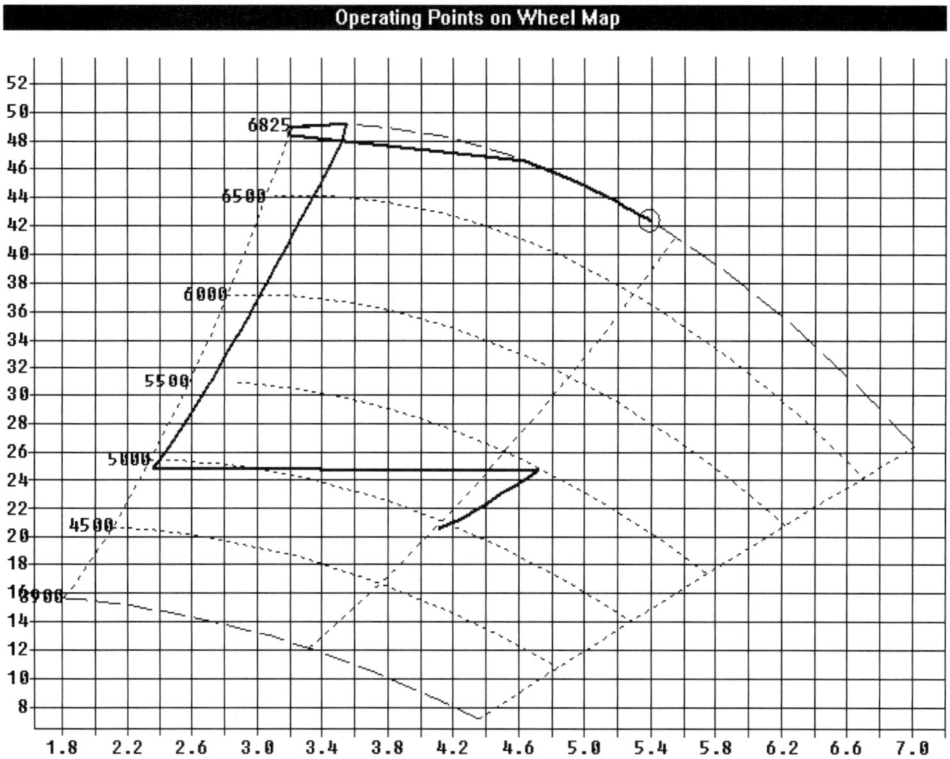

Head vs. Flow Rate

Figure 4 Operation Trajectory on Compressor Wheel Map (Courtesy of Liwacom)

The driver performance is calculated by the driver model using the shaft power, measured fuel, gear box ratio if any, ambient temperature, and other driver data such as RPM. The driver efficiencies are trended to detect efficiency changes. The cooler performance is monitored to detect loss of cooling efficiency due to deposits, corrosion or damage. By maintaining high cooling efficiency, the heat can be removed quickly and the pipeline's designed efficiency can be maintained.

The compressor performance monitor collects not only operating data but also compressor unit statistics. The unit statistics function can be provided by the station PLC or host SCADA system without a real-time model. The compressor unit statistics are useful to run compressors efficiently and safely and to determine the compressor and driver maintenance schedule. The unit statistics may include the following data:

- A count of limit violations such as surge control/surge line, power, etc.
- A count of compressor unit starts and unit total operating hours to check against the allocated number of annual starts for a compressor unit. The count of unit starts is segregated into the number of attempted and successful starts.
- Measured input power, calculated output power and station efficiency
- Accumulated driver operating hours for maintenance purposes

6.4.2.2 Pump Station Monitoring

The pump station monitoring functions are similar to the compressor station monitoring functions. The functions monitor and display the pump unit and driver performances. The pump drivers are mostly electrically driven motors and the electric power is measured.

Most pump stations do not have either coolers or recycling valves for surge control, and their thermodynamic effects on the fluid and pump are insignificant. Therefore, station and pump unit modeling approaches are similar to but simpler than those of compressor stations. Also, pumps driven by fixed speed motors, require a control valve for controlling the discharge pressure.

The pump station monitoring function is concerned mostly with the pump unit operating efficiency. If the driver is of variable speed, then the pump performance curves are bounded by the minimum and maximum speeds, with the efficiency related to the flow, head and speed. On the other hand, the fixed speed pump has a single performance curve with the efficiency related to the flow and the head controlled by a control valve. The station and pump models use the pump characteristics, control logic, fixed and variable speed motor characteristics, and different combinations of pump units to determine the operating point of each operating unit.

Plots of the current and historical operating points are superimposed onto static performance curves which show the minimum and maximum operating ranges. The operators use these plots to operate pumps efficiently. In addition, the efficiency can be trended to identify improper throttling operations or degradation of pump unit efficiency. Such information can be used to determine the operator training and equipment maintenance requirements and to re-rate the pump curves.

The pump unit statistics are similar to the compressor unit statistics and equally valuable for efficient operation and maintenance. The function determines and displays at regular intervals all of the unit statistics and efficiency of the operating stations. The following data may be required for the statistics:

- On-peak/off-peak run time and volume moved
- Number of on-peak starts and total number of starts
- Date and time the unit was last running and started
- Limit violations and their counts
- Measured input power, calculated output power and station efficiency

6.4.2.3 Optimization Model

The purpose of an optimization model is to minimize the pipeline system operating costs. Pipeline system optimization systems can be divided into three main categories: schedule optimization, throughput maximization, and energy minimization. An optimization technique can be applied to minimization of pollutant emissions such as NOx and CO_2 (10).

Schedule optimization refers to optimization of a batch schedule over a long time period, say a few days, one week or even longer. The parameters that are optimized can include energy, interface mixing, throughput, and batch lifting and delivery schedule. The schedule optimization is technically challenging and more suitable for batched liquid pipelines. The objective of throughput maximization is to maximize flow through a set of receipts and deliveries by adjusting the set points for pump/compressor stations and regulators.

This section discusses the energy optimization only. Energy optimization refers to short-term or real-time energy minimization for current pipeline operations and off-line optimization for future operation planning. The results of a real-time energy optimization are typically treated as recommendations and are not generally used for a closed-loop control.

Figure 5 is an example of real-time optimization displays of a gas pipeline system. The first display shows both the pipeline system configuration and compressor stations with set point values selected for energy optimization, while the second display shows only the optimum compressor station selection. The former display allows the operator to visually relate the optimum compressor stations and their set points with the operating pipeline system.

The energy optimization model deals with fuel consumption for gas pipelines and power consumption and DRA usage for liquid pipelines. It determines an optimum compressor/pump station selection and unit line-up as well as pressure set points at the on-line stations so as to minimize fuel and/or power/DRA cost. The model may adjust flow rates to take advantage of lower energy costs during off-peak hours.

An optimization model can provide the following:

- Compressor/pump stations and units to be brought on-line
- Optimum compressor/pump station suction or discharge pressure set points, compressor/pump unit on/off switching schedules, and minimum

fuel and/or power cost for a specified time period.

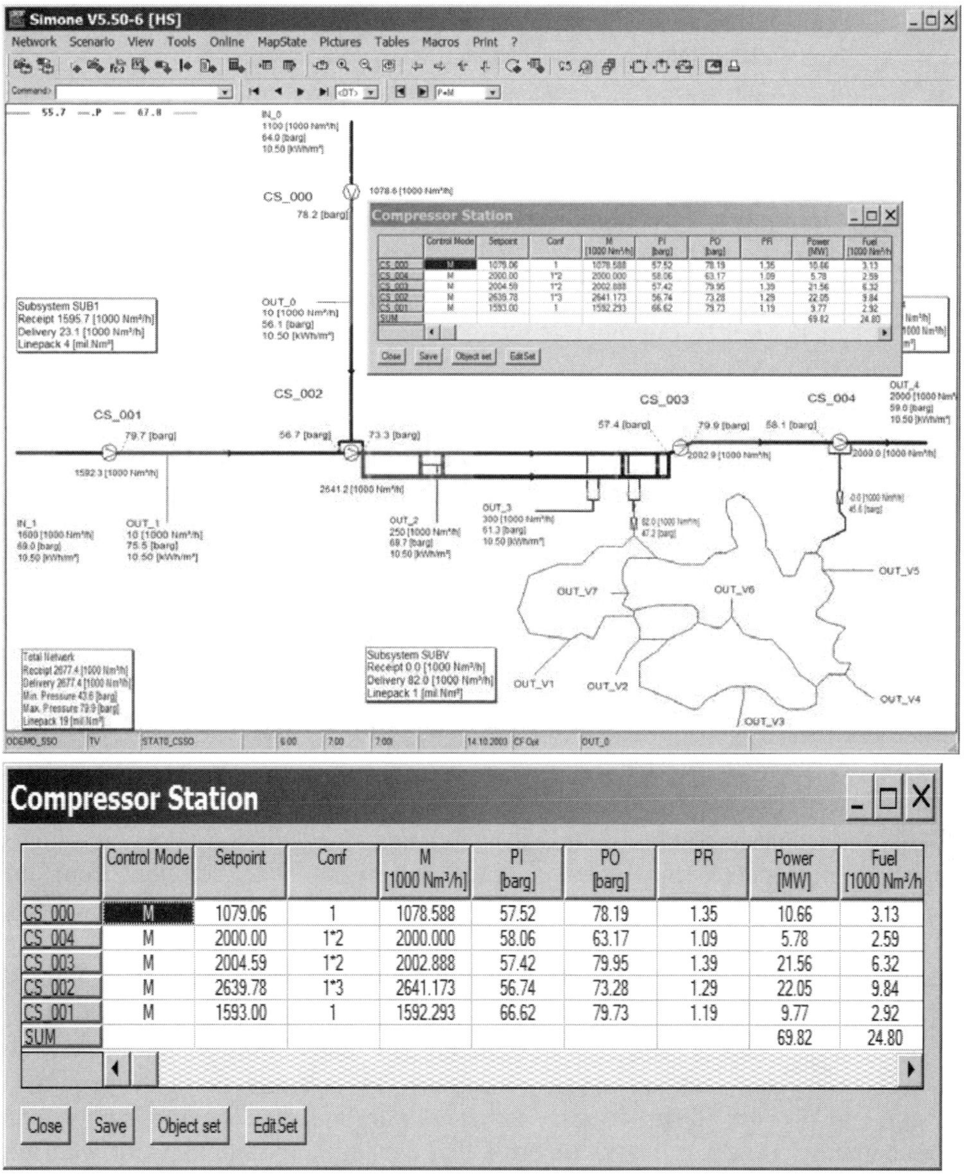

	Control Mode	Setpoint	Conf	M [1000 Nm³/h]	PI [barg]	PO [barg]	PR	Power [MW]	Fuel [1000 Nm³/h
CS_000	M	1079.06	1	1078.588	57.52	78.19	1.35	10.66	3.13
CS_004	M	2000.00	1*2	2000.000	58.06	63.17	1.09	5.78	2.59
CS_003	M	2004.59	1*2	2002.888	57.42	79.95	1.39	21.56	6.32
CS_002	M	2639.78	1*3	2641.173	56.74	73.28	1.29	22.05	9.84
CS_001	M	1593.00	1	1592.293	66.62	79.73	1.19	9.77	2.92
SUM								69.82	24.80

Figure 5 Real-Time Optimization Display (Courtesy of Liwacom)

- Compressor/pump unit line-up and operating point, considering that a station may consist of different compressor/pump units and that the units can be combined in various modes. The operating points, overlaid on the pump performance curve, can be displayed on the host SCADA screen.

- Calculation of the overall compression or pumping costs. When drag reducing additive (DRA) is injected for a liquid pipeline operation, the cost without DRA is compared against the cost with DRA.

In addition, some optimization systems provide the following information to analyze and improve pipeline operation efficiency performed by operation staff:

- Key optimization results and historical records
- Flow rate vs. suction/discharge pressure trends with set point change records
- Flow rate vs. number of compressor/pump units brought on-line and fuel/power consumption
- Cumulative compressor/pump operating records
- Compressor/pump efficiency trends

The model employs the following data in addition to the pipeline configuration and facility data:

- Pipeline hydraulics and equipment such as compressors or pumps
- Pipeline and facility availability data
- Fuel and/or power contract data
- DRA cost for liquid pipeline only
- Line fill and batch schedule data and injection and delivery flow rates for batched liquid pipelines

The primary criterion for an optimization model is to minimize fuel and/or power costs. A secondary criterion is to balance compressor/pump unit operating hours and avoid frequent unit start-ups and shut-downs. The solution from the optimization model should not violate any pipeline and facility constraints. These constraints may include maximum and minimum pressures and flows at certain points in the pipeline network such as minimum delivery flow, maximum and minimum compressor/pump flows and compression ratio, maximum power, etc.

Optimization models can be challenging to apply on complex network configuration and pump/compressor stations. Optimization problems based on these models are difficult due to their non-linearity, non-convexity and discontinuity. However, it was reported that dynamic programming and gradient optimization techniques were successfully implemented for gas (10, 11) and liquid pipeline energy optimization (12, 13).

Dynamic programming is an enumeration technique that starts tabulation at the lifting point and ends at the delivery point and applies the following feasibility and optimality conditions on each stage:

- Feasibility condition: The pressure should be between the maximum and minimum operating pressures. At each station, the maximum discharge and minimum suction pressure should not be violated, the minimum head has to be maintained, the maximum available power should not be exceeded, and the minimum flow should be maintained.
- Optimality condition: A set of discharge pressures in the pipeline system should not consume more fuel than the already established minimum fuel up to that stage.

A stage is defined as a pipe segment between two stations, between the lifting point to the next station, or between a station and a delivery point. When the calculation is finished in the last stage of the pipeline system, optimum suction and discharge pressure combinations are determined at each station, and the minimum fuel cost is determined at the delivery point. By backtracking from the delivery point to the first lifting point, the optimum suction and discharge pressure combination is selected at each stage.

The dynamic programming approach is an optimization technique for certain types of optimization problems, which is relatively simple to implement. The technique is easily applicable to straight pipeline systems which do not have pumps or compressors on laterals. It may be possible to apply the technique to more complex pipeline networks, but other solution techniques are more suitable to solve their problems. Another weakness of this technique is that it provides only one global optimum solution and thus it is not easy to find the next optimum solution if the global optimum solution is difficult to implement on the pipeline.

An energy optimization system can be implemented as a part of the RTM or the SCADA, and requires an interface with the RTM or SCADA system. Through the interface, the SCADA or RTM system sends the current pipeline states required for an optimization run, controls its execution with the data, and receives the optimization results along with alarm and event messages such as new batch lifting and station startup or shutdown. The current states may include the following data:

- Receipt and delivery flow rates
- Compressor/pump stations and units which are on-line and off-line
- Batch and DRA tracking data for liquid pipelines or composition tracking data for gas pipelines
- Batch and DRA injection schedules
- Pipe roughness or efficiency to improve hydraulic calculation accuracy
- Unit utilization data and maintenance schedule

It is advantageous to implement an optimization system as a part of the RTM, because the RTTM constantly checks and improves the accuracy of the hydraulic profiles and compressor/pump characteristics so that the effect of errors are minimized. If it is implemented on the SCADA system, the accuracy of the batch/DRA or gas composition tracking data and friction factor needs to be improved in order to calculate the hydraulic profiles accurately. In order to calculate the pipeline hydraulics accurately, accurate pipe roughness or pipe efficiency along the pipeline may be required. A real-time batch tracking capability can provide more accurate hydraulic calculation. Some optimization models can re-rate pump performance curves by analyzing recent data automatically to determine actual pump performance and efficiency.

An energy optimization system is typically configured to run at regular intervals as well as on demand by the operator. Running the system at regular intervals ensures that the system will notify the operator of any system changes required due to changes in the pipeline line fill (e.g. batched operation, composition change, etc). When there is a need for flow rate change, the operator will enter the new parameters and obtain new system changes.

6.4.3 Future Pipeline States

The SCADA system or RTTM provides the pipeline operator with the current and historical pipeline states. To operate the pipeline system efficiently and safely, the operator needs to understand future pipeline states. Both ALAM and PM (1, 4) provide the capability to predict future pipeline states, which allow the operator to take corrective actions or to develop better operating strategies. The models are off-line simulators interfaced to an RTTM, which provides them with current pipeline states. In general, an ALAM is used for on-line operation, while a PM is used for off-line operation planning.

6.4.3.1 Automatic Look-ahead Model (ALAM)

The primary purpose of an ALAM is to assist the operator in operating the pipeline system safely, by monitoring future pipeline states continuously and informing the operator of potential near-term operations. The operator may examine the problems that the ALAM identifies in a simulation run, and then respond to the problems by altering the current operating scenario or initiating other remedial actions.

An ALAM calculates future pipeline states using the current pipeline states received from the RTTM and an operating scenario entered by the operator. The operating scenario defines the control set points including injection or delivery flows and pump or compressor units and stations as well as the status of valves and equipment such as pump or compressor. An ALAM assumes the operating scenario remains constant or varies according to a pre-defined schedule. An ALAM may detect that a major system constraint is violated and generate an alert to the operator.

An ALAM can provide the following information:

- Pipeline states for the look-ahead simulation period. The pipeline states include pressure and temperature profiles, flow rates, densities, batch or gas composition tracking, line pack data in the network, power or fuel consumption, and pump or compressor operating information at every station.

- Look-ahead alarms for violation of maximum or minimum pressure or temperature and line pack in certain pipeline segments. In addition, it provides alarms for violation of supply or delivery constraints and pump or compressor variables such as surge, horsepower and speed.
- Survival time determination. This function is needed if the line pack can drop below its specified limit. This problem arises when the line pack is exhausted due to reduced or lost supply and/or excessive delivery.
- Sudden deterioration of pipeline efficiency or abnormal operating conditions such as condensation and dew point problems for gas pipelines and slack flow and their positions for liquid pipelines. If such conditions arise, it produces alarms for analysis.
- Operating events such as arrival of batches or anomalies to designated locations.

The ALAM function depends on reliable and accurate gas or liquid product supply and demand data as well as availability and schedules of operating facilities and other pipeline system parameters. The model needs to be able to execute quickly and produce accurate results.

Normally, an ALAM runs automatically at regular intervals as defined in the system, looking ahead for several hours into the future. In general, an ALAM is more useful for gas pipeline operations, because transient behaviors are slow enough for the operator to respond to and a limited inaccuracy of the results can be tolerated.

Figure 6 shows an example of ALAM displays of flow and pressure trends over time. The flow and pressure trends show both the RTTM history and the ALAM predictions beyond the actual time (13th hour) based on the current pipeline state and control schedule, thus providing a synoptical view from the short-term past into the short-term future.

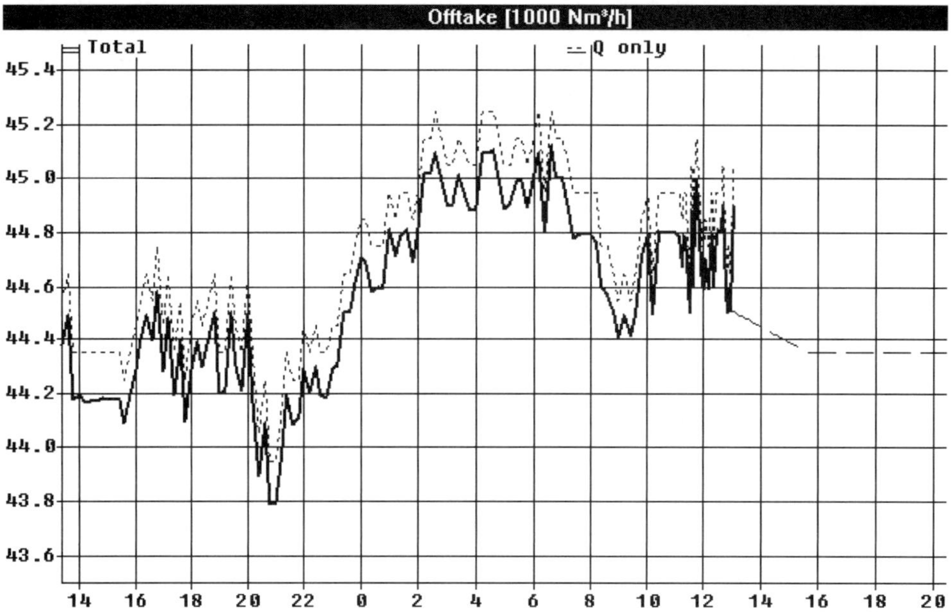

Flow vs. Time

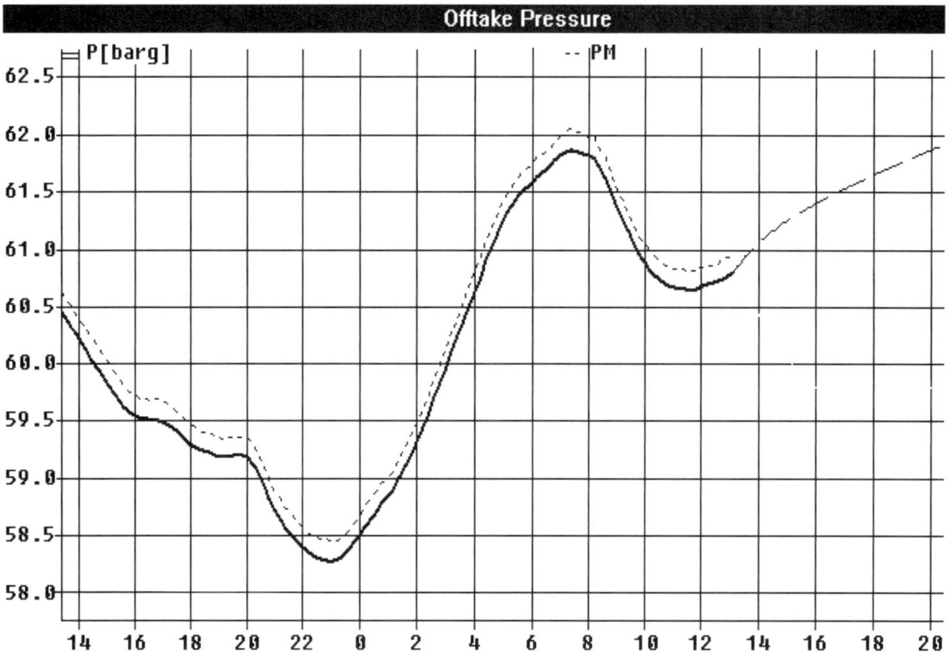

Pressure vs. Time

Figure 6 Look-ahead Flow and Pressure Behaviors (Courtesy of Liwacom)

6.4.3.2 Predictive Model (PM)

The main purpose of a PM is to assist the operator in operating the pipeline system efficiently by providing the information about various operating scenarios. The PM is mainly concerned with efficient operations over a long-term horizon (e.g. one or more days), while the ALAM is concerned with safe operations for a short-term operating period (e.g. several hours). The model enables the operator to:

- choose an efficient operating plan to minimize operating costs,
- develop a new strategy to deal with any unscheduled events, or
- decide on corrective action if a different optimum operating scenario is found or if an upset condition is detected in the pipeline network.

The PM permits the operator to analyze "***what if***" operating scenarios. While the ALAM is constrained to run in a fixed operating condition, the PM is allowed to operate in response to many operating scenarios with different control strategy and commands. Also, the number of predictive simulation runs is large to provide the information on many possible operating scenarios.

The PM calculates the future pipeline states in a similar way to the ALAM. Input data for predictive simulation runs can likely be the current pipeline state generated by the real-time transient model. However, initial state can possibly be a pipeline state under a steady state condition or one from previous predictive runs.

The PM can determine an operating scenario for minimum fuel or power consumption even under transient conditions, by reviewing the results of multiple operating scenarios. Energy optimization discussed in Section 6.4.2.3 is an automated process, while optimization using the PM is a manual process. Specifically, it provides the following information:

- Hydraulic profiles of the pipeline network over the period of the predictive run for all operating scenarios. It provides the information on batch or composition tracking or line pack and packing/drafting rates.
- Pump or compressor operating data such as power or fuel consumption and pump or compressor performance such as operating point and efficiency.
- Maximum throughput or capacity in the pipeline network and equipment required for capacity runs.
- Violations of pre-set operating constraints for certain operating scenarios. Operating constraints are similar to those produced by a look-ahead model.
- Survival time when supplies are limited and/or delivery requirements are excessive. Various scenarios can be evaluated to avoid line pack exhaustion.

- Abnormal operating conditions including condensation and hydrate formation for gas pipelines and slack flow and their locations for liquid pipelines.

Since the predictive model run-times tend to be long and the number of runs is large, the model requires fast execution time and a flexible time step. An implicit solution technique is preferred to satisfy these requirements and at the same time to maintain mathematical solution stability for adjustable time steps. Even though model accuracy is important it is not as critical as the accuracy required by an ALAM, because the PM is primarily used for off-line operation planning.

The PM can generate a large amount of data, which includes hydraulics and throughput data, batch or composition tracking data, pump or compressor operating points and fuel usage, and limit violations. Graphic and tabular tools along with summary and comparative analysis reports facilitate easy analysis of the simulation results. Another important output is trend displays, which plot data such as pressures and flows with respect to time at a given location. The results of the predictive simulation should be saved for use as initial condition for future predictive runs.

In general, a PM should allow a user to:

- Edit the operating scenarios, change the receipt or delivery flows and pressures, control set points, have pump or compressor statuses, etc.
- Set model control variables such as time steps, simulation duration, etc.
- Select variables to be trended or reported.
- Modify pipeline and operating constraints.
- Execute the model in an interactive mode. This allows a user to stop and review the simulation run and user input, modify the operating scenario, or abort the run if errors are discovered.

Running a PM on a complex pipeline system can be difficult for model users because input data preparations are labor intensive and require expert knowledge of the system and model capability. It is suggested (14) that the predictive model runs be automated by using heuristic rules in order to fully utilize the model's capability of predicting future pipeline states.

6.5 Training System

The U.S. Department of Transportation (DOT) requires that pipelines must be operated by qualified operators. The objective of the rules is to reduce the risk of accidents on pipeline facilities attributable to human error. To satisfy the rules, the pipeline companies must describe all covered tasks, list the knowledge and skills required of operators to complete those tasks, identify abnormal operating conditions, and develop a student evaluation method and qualification record keeping plan. API 1161 addresses the operator qualification issues in detail (15) and ASME B31Q (16) lists detailed covered tasks. These standards cover all aspects of pipeline system operation and maintenance. A training program normally describes the standard of training required for pipeline operators, including the knowledge, skills, and abilities required, training program metrics for success and the keeping of training performance records.

A computer-based training system can be very effective in satisfying the training requirements. (Refer to Reference (17) for the experiences in using a pipeline simulator for operator training.) This section is limited to a discussion of computerized pipeline operator training from a viewpoint of pipeline system hydraulics and SCADA, its centralized control system.

Traditionally, pipeline operators have been trained using a combination of on-the-job training and classroom sessions. Such an approach is effective for teaching routine operations but ineffective for dealing with upset and emergency situations because they seldom occur in actual operations. It is also inadequate if the pipeline system changes rapidly such as may occur during pipeline expansion or when new businesses are involved. A computer based training system makes it easier for the trainee to gain experience in upsets and emergencies without putting the pipeline at risk. By responding to rapid changes efficiently, it helps overcome the limitations of more traditional instruction methods. As well, the training system can be used for initial training and qualification of new operators and cross training and re-certification of experienced operators.

6.5.1 Functional Requirements

A computerized training system should provide trainees with a realistic training environment, which includes realistic pipeline hydraulics and equipment reactions to an operator's control actions. This objective can be achieved through a pipeline model, which can simulate pipe, valves and pumps/compressors and their control capability.

A training system is an off-line transient model, but can be used as a standalone system or a part of an RTM system. If the training system is integrated with the RTM system, it can capture the current pipeline state, load an initial pipeline state and begin a simulator session.

There are two types of computerized training systems distinguished by the trainee interface: an integrated training system that is interfaced with an offline copy of the SCADA system used for trainee interaction and a hydraulics training system that has its own trainee interface. This arrangement is analogous to an aircraft simulator wherein the trainee uses the same control system for training that he will use when he is operating the pipeline. A stand-alone hydraulics training system has a trainee interface with SCADA-like screens and is used mainly for hydraulic training. However, an advantage of a stand-alone system is that it can be used anywhere since it is not linked to a SCADA system.

A training system can be run in unassisted self-instruction or instructor-assisted mode. The instructor and trainee can have the ability to select from a list of all possible operating scenarios and run a simulation using an interactive user interface. The training system provides tools to easily build and maintain scenarios. This includes initializing the scenario start-up conditions such as SCADA point values, measurements and line fills.

The playback function is used to playback data previously saved by the host SCADA or to playback data generated by the training system. The playback includes both trainee and instructor initiated events from the recorded session. This function allows the instructor and trainee to review the training session, discuss the responses, and go over errors. It should be possible to increment playback time faster and slower than real-time as well as to rewind to the start of the playback period and to fast forward/backwards to specific playback times.

As a minimum, the training system should allow a trainee to:

- Perform normal operations.
- Respond to abnormal operations including upsets and emergencies.
- Predict the consequences of facility failures.
- Recognize monitored operating conditions that are likely to cause emergencies and respond to the emergency conditions.
- Understand the proper action to be taken.

6.5.2 System Structure and Components

A training system consists of training simulator, trainee interface, and instructor interface. An integrated training system uses the host SCADA as a trainee interface, while a hydraulic training system uses a generic trainee interface. The hardware architecture of an integrated training system is shown in Figure 7.

In general, a complete training system requires four main components and databases to support the functions. The components include:

- Training simulator
- Trainee interface
- Instructor interface
- Record keeping

It also requires the following databases, some of which are unique to the system:

- Hydraulics database, which is the same as the databases of the other RTM modules
- Computer-Based Training database, containing all the required training material
- Record keeping database, containing the training records
- Data playback database, if the playback training function is required

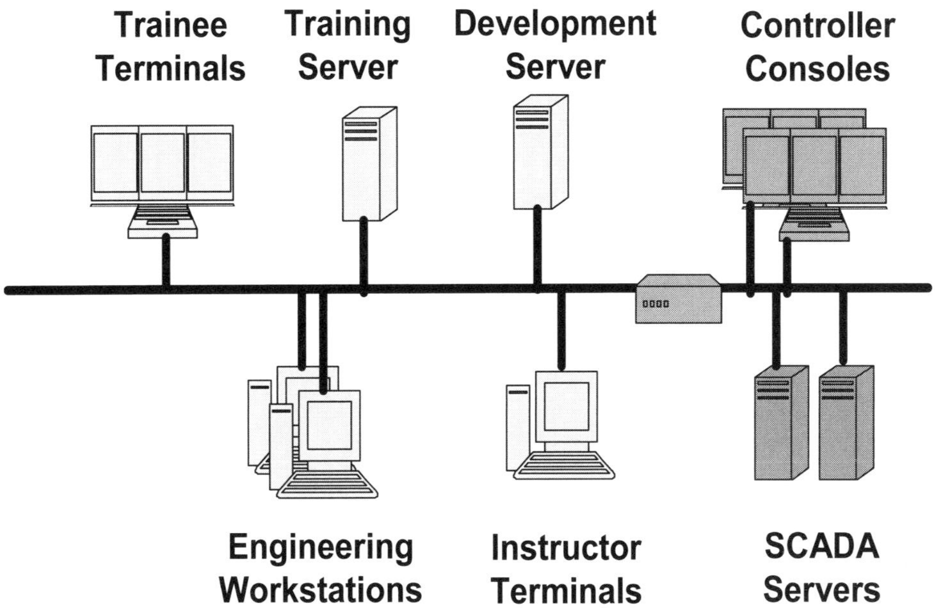

Figure 7 Hardware Architecture of Training System (Courtesy of CriticalControl)

6.5.2.1 Training Simulator

A training simulator requires the capability to simulate pipeline data. The pipeline data simulation uses a transient model, which supplies the SCADA system with data that would normally come from field equipment and remote terminal units (RTU). It mimics the real responses of the pipeline system in terms of hydraulics but also control actions and their responses. In order to emulate the reaction of local stations based on their control logic, it will model all major pipeline system components such as pump or compressor stations, valves and junctions.

For a truly realistic training, the training system should be able to emulate PLC control logic for all field equipment, so that the training simulator replicates the real-time control as closely as possible. The PLC logic for a compressor or pump station includes not only the single and multiple unit control but also station valve control.

In summary, to provide a realistic training environment a training simulator should be able to:

- Generate realistic hydraulic profiles, control responses, alarms, and simulated field data.
- Display hydraulic profiles and other equipment data.

- Simulate RTUs and field equipment failures.
- Simulate a leak, blowdown or a pipeline blockage condition.
- Simulate field equipment failures such as pump or compressor failure.
- Simulate transducer failures.
- Accept changing variables during the simulation from the trainee and instructor consoles.

6.5.2.2 SCADA Trainee Interface

In an integrated training system, the trainee interface is the same as the SCADA's man-machine interface. The integrated training system provides realistic training in an environment where the operators control their pipelines using the same control interface as the real pipeline. It is analogous to an aircraft simulator. This allows the trainee to use the same SCADA HMI displays and commands to execute the scenario as the operating SCADA system.

In addition to hydraulic training, the integrated training simulator helps operators to learn the operation of the SCADA system without interfering with actual pipeline operations. In general, the training efficiency is higher with the integrated training system than with a simpler hydraulic-based training system.

6.5.2.3 Instructor Interface

The instructor interface is used by the instructor to control and monitor training. The instructor, through the instructor interface, can set up new training scenarios including initial states, change controls and measurements, introduce abnormal operations, and monitor the trainee's responses to the changed conditions.

The instructor interface provides the features required to control the training system execution such as start/stop, change simulation speed, rewind, etc. The instructor can run the training simulator in three different modes: slower than real-time, the same speed as real-time and faster than real-time. The instructor can select an initial pipeline state from which a training session starts. The initial state of a training scenario may come from a pipeline state generated previously by the training simulator. This state is loaded on the training system upon the request of the instructor in an interactive mode or by the trainee on a batch mode of operation.

As shown in Figure 8, there is an analogy between the actual operations and training system, in that the training simulator emulates the pipeline devices and the trainee interfaces are the same as the SCADA operator interfaces but the instructor interface doesn't have a directly equivalent component in actual operations.

The instructor interface should be able to change control points and measurements. The following is a partial list of possible changes:

- Station set point change such as suction or discharge pressure set points
- Flow rate reduction below the surge line
- Pump/compressor control such as speed or horsepower change
- Pump/compressor status change such as trip
- Valve status change such as valve closure from open state
- Pressure change to test pressure violation such as MAOP
- Flow measurement change to test flow rate violation such as maximum flow
- Measurement or transmitter failure, communication outage, etc.

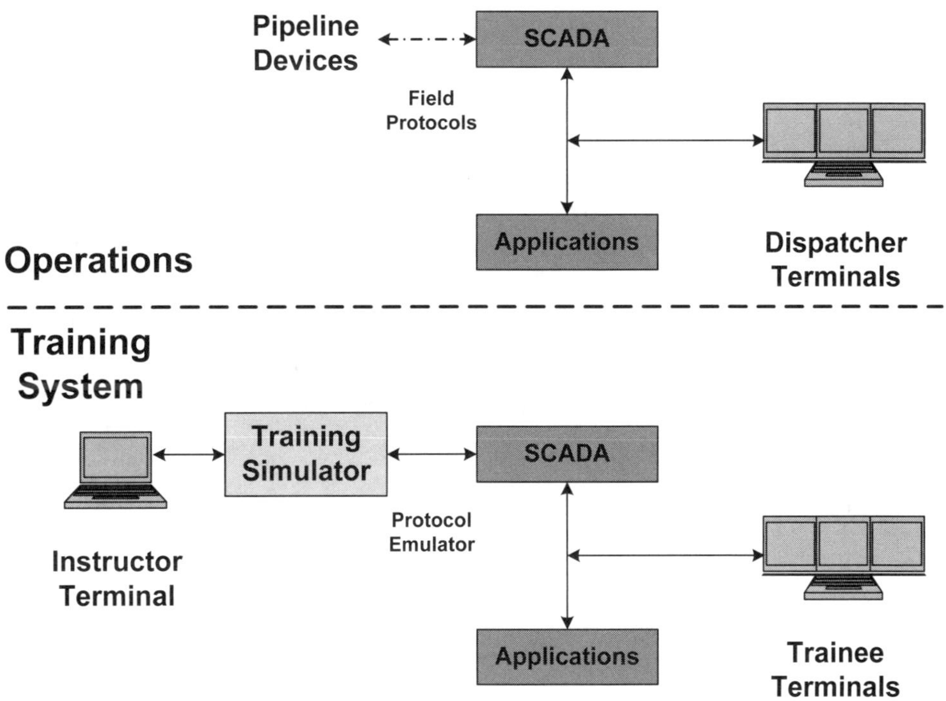

Figure 8 Analogy between Actual Operation and Training System (Courtesy of CriticalControl)

Also, the instructor interface should be able to introduce abnormal operations:

- Leaks
- Line blockage
- Relief valve open

- Blow-down valve open

The instructor interface has the capability to monitor the trainee's responses. A training system can be executed interactively or in a batch mode. The instructor interface allows the instructor to select either a batch mode or interactive mode of training system operation.

The instructor interface looks similar to the trainee interface but it has several additional functions. The scenarios can run without interruption or also allow real-time instructor intervention through a real-time interactive user interface. All instructor interventions must be recorded in the session event log. Figure 9 shows an example of an instructor interface.

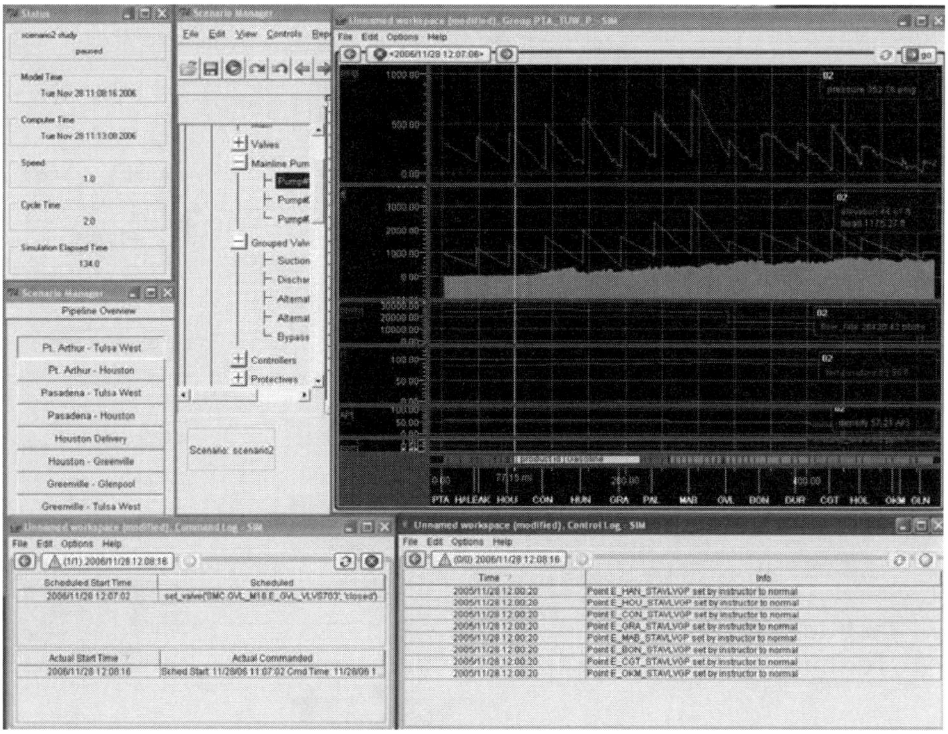

Figure 9 Example of Instructor Interface (Courtesy of CriticalControl)

6.5.2.4 Record Keeping Requirements

The pipeline companies in the U.S. are required to adopt an operator qualification standard based on ASME B31Q. This standard requires that each trainee should be tested and scored and that all the training records be kept to track the training progress of each trainee.

Each pipeline company may have a different set of scoring requirements, and thus the scoring system and training metrics should be established to measure training performance. Metrics may include a record of course modules completed, test scores, history of performance in training sessions, such as responses to emergencies and abnormal operations. Each lesson begins with a base score, and this score can be adjusted as the lesson proceeds. Each time a score is changed, a comment has to be entered into the trainee log and can be displayed on the trainee console. At the end of the lesson, a final score and the log should be available to the trainee.

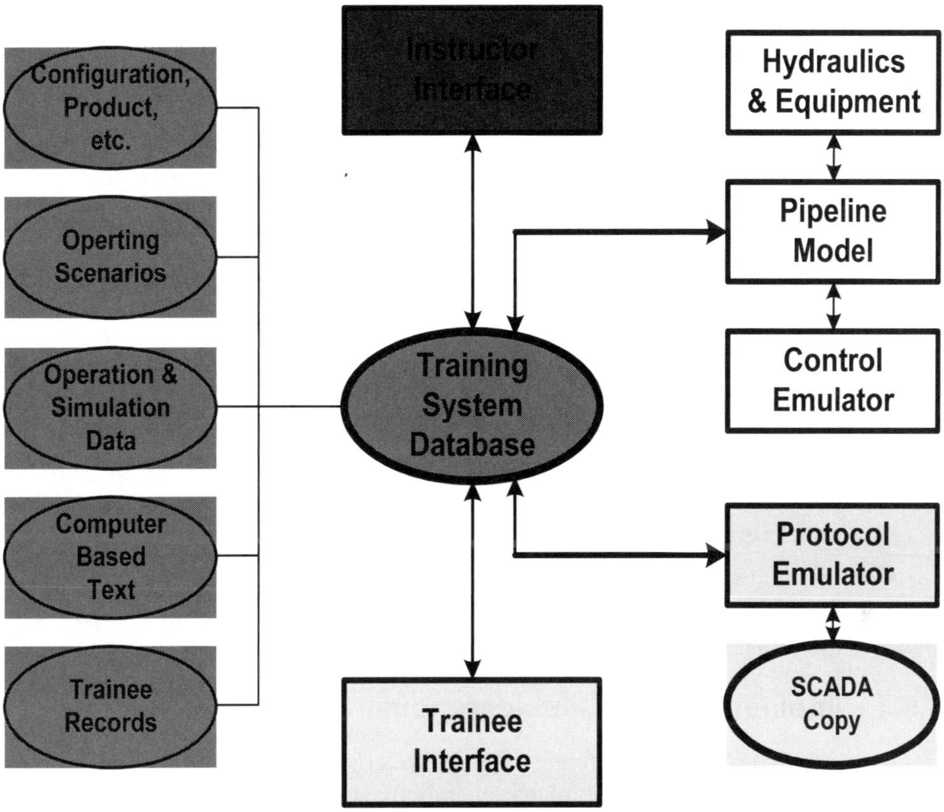

Figure 10 Training System Components

A well designed training system will incorporate the ability to store and track training records and logs when each training module is completed. It logs all student initiated actions, instructor initiated actions, SCADA events and scenario results such as alarms received, SCADA responses, etc., so that the instructor can enter and save comments into the log. The training record lists every training session conducted, training module completed, training session results, etc. This tracks the status of operator training and documents the training for internal and external personnel including the regulatory agency. The record keeping function has the capability to report the training activities and progress for each trainee, which includes training session results and training modules completed. Figure 10 shows the software components and structure of an integrated training system.

In addition to the above components, a computer-based text would be beneficial. A computer-based text is the presentation of textbook material and the performance of tests on the training computers. It does not have any pipeline model but rather is a self-taught classroom environment. A computer-based text can provide trainees with various operating scenarios, even covering abnormal operations. It includes training material on pipeline operations, hydraulics, equipment and facility operations, SCADA, and other relevant topics. Each training lesson will include a test on the subject material. The training material will typically include at least the following:

- Hydraulic knowledge
- Pipeline control for system balance
- Normal operations
- Emergency and upset handling

6.5.3 Modes of Operation

A computerized training system can be operated in two different modes:

- Batch mode: a pre-programmed operating scenario is loaded and no commands are changed during the training period. The instructor is not required in this mode of operation. If the training session is a long one, or if an instructor is not available, a batch training run may be more effective.
- Interactive mode: the instructor interactively changes commands, introduces upset conditions, or overrides measurements in order to help the trainee better understand pipeline operations.

6.5.4 Implementation Considerations

Generally the computer used for a training system is separate from the SCADA computer in order to avoid any interference between the SCADA and the training system. The transient model for the training simulator and the RTM system are usually the same model; this ensures uniformity and simplifies maintenance.

6.5.5 Benefits

A training system can provide the following benefits for pipeline companies:

- Accelerated learning – A training simulator can provide key operations that would not be easily available on-the-job. Many pipeline operations such as valve closure or events such as leaks rarely occur. With a training simulator, every operator is given an opportunity to observe such operations and events.

- Reduced live system risks – Inappropriate operator responses to pipeline operations can cause system disruptions, loss of operating efficiency, and even missed market opportunities. The risk is caused partly because the operator does not recognize a problem and partly because the operator doesn't know how to undertake appropriate actions to meet the operating objectives.
- Accommodation of system changes – As a pipeline system or operation is changed even experienced operators have difficulty adjusting to a new environment. A training simulator allows operators to understand system behaviors before the changed system or operation is actually in use.
- Quantitative evaluation – The training simulator with text and recording modules provides the basis for quantitative evaluation of an operator's performance in a give scenario. The opportunity for the trainee to demonstrate that s/he has mastered the required responses to a scenario allows competence to be assessed.
- Operating efficiency – Pipeline operators tend to make conservative control responses that can lead to less efficient system operation; this in turn translates into higher fuel and operating costs.

6.6 General Requirements

6.6.1 Configuration

Ideally a RTM system will have a common configuration data for all models. By creating and maintaining a single configuration data set, the configuration effort and data error can be minimized without the necessity of entering and modifying the same variable in several models and displays. The configuration files should reside on and be managed through the RTM computer.

Normally, three types of RTM configuration data are required and shared by all the models and applications:

- Pipeline network data defines elevation profile, physical characteristics of the pipeline system including pipes, valves, pump or compressor stations, and measurement related data such as types, locations and limits.
- Product data defines product properties including appropriate equations of state and parameters.
- Operating parameters define the parameters for RTM operations such as model enabling and disabling, alarm thresholds, time steps, and initial and boundary conditions.

6.6.2 Displays

A well designed human-machine interface (HMI) for the RTM system allows the operator to use the system easily. Assuming that the operators are familiar with the host SCADA system, the displays need to have the same "look and feel" and operational characteristics as the SCADA displays. Such features will reduce training time for RTM users and increase the acceptance of the system.

The RTM applications generate extensive data and a graphical presentation of the information is more effective for user interpretation. Graphic displays include hydraulic profiles, batch or composition tracking profile, DRA tracking, etc. Tabular presentation is also useful for detailed analysis.

6.6.3 Alarm Processing

The RTM system provides the capability to process and generate alarms and events. There are three types of alarms:

- Model results – These include not only the violations of limits and constraints but also operation related problems such as slack flow and leak detection. They also include event messages such as anomaly delivery and DRA injection.
- Model data – These include static data such as pipeline configuration and real-time data such as measured data and status.
- Model execution – This includes abnormal or aborted execution alarms.

Since each RTM application has different alarm criteria, each model or application requires its own set of alarm points. These alarms should be uniquely identified with the model or application name such as RTTM alarm or leak alarm. It is recommended that only critical alarms be sent to the SCADA screen.

6.7 Summary

As pointed out in the objectives of an RTM system, it provides the timely and accurate information necessary to help the pipeline companies operate their pipeline systems safely and efficiently. More benefits can be gained if all the information generated by the RTM system is integrated and made available throughout the company. The benefits derived from an integrated RTM system include:

- Increased efficiency
- Greater insight into pipeline operation
- Ability to foresee upcoming changes and have plans in place
- Analyze future events or change of operating conditions

The implementation of an entire RTM system can be time consuming due to its complexity. Sometimes, the results can be unreliable, because they depend heavily on real-time data quality and availability. Therefore, careful consideration of the costs and potential benefits as well as the impact on pipeline operations should be made before a decision is made to install an RTM system.

References

(1) Janzen, T., "Real Time Model as a Business and Operations Mission Critical System", Pipeline Simulation Interest Group (PSIG), 2002

(2) Stoner, M. A., Richwine, T. E. and Hunt, F. J., "Analysis of unsteady flow in gas pipe lines" Pipe Line Industry, 1988

(3) Mohitpour, M., Golshan, H., Murray, A., "Pipeline Design and Construction: A Practical Approach", ASME Press, New York, N.Y., 2004

(4) Klemp, S., et al, "The Application of Dynamic Simulation for Troll Phase I", Pipeline Simulation Interest Group (PSIG), 1993

(5) Griffiths, G.W., Willis, D. J. and Meiring, W. J., "The Woodside Trunkline Management System " Proceedings of OMAE Conference, Vol. 5, ASME, 1993

(6) Price, R.G., et al, "Evaluating The Effective Friction Factor and Overall Heat Transfer Coefficient During Unsteady Pipeline Operation", Proceedings of International Pipeline Conference, ASME, 1996

(7) Hagar, K., Young, B., Mactaggart, R., "Integrity Monitoring – Not Just Leak Detection", Proceedings of International Pipeline Conference, 2000, ASME, New York, N.Y.

(8) Seeliger, J. and Wagner, G., "Thermal Billing Using Calorific Values Provided by Pipeline Simulation", Pipeline Simulation Interest Group (PSIG), 2001

(9) Nicholas, R. E., "Simulation of Slack Line Flow: A Tutorial", Pipeline Simulation Interest Group (PSIG), 1995

(10) Jenicek, T., "SIMONE Steady-State Optimization – Money and Pollution Savins", SIMONE User Group meeting, 2000

(11) Grelli, G. J. and Gilmour, J., "Western U.S. gas pipeline optimization program reduces fuel consumption, trims operating costs", OGJ, 1986

(12) Short, M. and Meller, S., "Elements of Comprehensive Pipeline Optimization", Proceedings of International Pipeline Conference, 1996, ASME, New York, N.Y.

(13) Jefferson, J. T., "Procedure allows calculation of ideal DRA levels in

products line", OGJ, 1998

(14) Wheeler, M. L. and Whaley, R. S., "Automating Predictive Model Runs for Gas Control", PSIG, 2001

(15) "Guidance Document for the Qualification of Liquid Pipeline Personnel", API 1161, 2000

(16) "Pipeline Personnel Qualification Standard" a draft version of ASME B31Q, 2005

(17) Wike, A., et al, "The Use of Simulators to Comply with Legislated Pipeline Controller Proficiency Testing", Proceedings of International Pipeline Conference, 2002, ASME, New York, N.Y.

7 Pipeline Leak Detection System

7.1 Introduction

This chapter discusses various aspects of pipeline leak detection. Emphasis has been put on the widely accepted computational pipeline monitoring (CPM) techniques and their implementation considerations. The techniques and implementation of leak detection are presented objectively to enable engineers to make informed decisions. Specifically, it is intended to provide the following information on the:

- most widely used leak detection techniques and their working principles
- evaluation and selection method of a leak detection system, particularly the computational pipeline monitoring (CPM) methodologies described in API Publication 1130.
- various aspects of the implementation of CPM
- brief discussion of emerging leak detection technologies

Pipeline leak detection is only one aspect of a pipeline leak management program; it encompasses leak prevention, detection and mitigation procedures. In order to minimize the consequences of a leak, pipeline companies require a comprehensive leak management program. A leak detection system by itself does not improve on a pipeline's integrity nor reduce potential failures of a pipeline system. However, such a program will not only help prevent and monitor the degradation of a pipeline that may eventually lead to failure, but will also minimize the consequences of pipeline leaks if they occur.

Pipeline companies minimize leaks through a leak prevention program. The main causes of leaks are: third party damages such as excavation equipment hitting the pipeline, geophysical forces such as floods and landslides, improper control of the pipeline system, and pipe corrosion. Proper control of third-party damage is achieved through: marking of the right of way; education of employees, contractors, and the public; and effective use of systems such as "One-Call." Geophysical forces cannot be controlled but can be monitored and their effects can be mitigated. Corrosion control and defect assessment are significant subjects and are discussed in separate volumes in this monograph series.

Leak mitigation is the attempt to reduce the consequences of a leak when it occurs and there are many ways to do this. If a leak can be detected quickly and isolated quickly, the spillage can be minimized. This requires that the leak alarm and its associated information are reliable and accurate. Having effective procedures in place and the proper resources and tools to enact them are critical in addressing the leak mitigation issues efficiently. This chapter discusses leak confirmation and isolation issues as part of leak detection. It does not deal with spillage management issues such as cleanup procedures and manpower mobilization, as each legal jurisdiction has its own regulations and each company its own requirements.

Historical data indicates that leaks were predominantly detected by local operation staff and third parties. Successful detection by means of a single leak detection system was random. This was because no single leak detection system could detect leaks quickly and accurately or provide reliable leak detection continuously and cost-effectively. Therefore, more systematic approaches to leak detection were required, such as a combination of line patrol, sensing devices and SCADA-based systems with automated leak detection capability.

Since SCADA systems have become an integral part of pipeline operations, particular emphasis has been placed on leak detection methods that are easily implemented on the SCADA system. API Publication 1130 addresses various Computational Pipeline Monitoring (CPM) methodologies, integrated with a host SCADA system (Note that throughout this chapter, SCADA can also include DCS systems).

The general principles and evaluation criteria for each leak detection technique are discussed here. And while reasonable efforts have been made to present all the relevant features of each technique, more detailed information can be obtained from published articles or from equipment or system vendors. API Publication 1149 and API Publication 1155 are briefly discussed with respect to how they are used for specifying and evaluating leak detection performance.

It is assumed that the reader has a clear knowledge of his pipeline system, its operations and pipeline hydraulics. This chapter therefore does not include topics such as pipeline system configurations, fluid properties, hydraulic behaviors and system operations, as important as they are to the functioning and requirements of a leak detection system. For the same reason, subjects such as batch operations and DRA tracking are not elaborated.

The information presented here has been gathered from the published articles listed at the end of this chapter and from the author's operation and implementation experiences. General references on the subject are "A Study of the Pipeline Leak Detection Technology" (1) and API Publication 1130 "Computational Pipeline Monitoring" (2).

7.2 Pipeline Leaks

7.2.1 Definition of a Leak

This chapter uses the definition of leaks as defined in "Petroleum Pipeline Leak Detection Study (3)." There are two types of leaks: an incipient leak and an actual leak. "Incipient leaks" are defined as those that are just about to occur. Certain incipient leaks can be discovered by inspecting the pipeline and dealt with before they become actual leaks.

If fluid is leaking out of a pipeline system, it is called an "actual leak." Leak detection methods can be used to determine and isolate either incipient or actual leaks. Some inspection methods such as hydrostatic pressure testing and visual inspection can be applied as a leak detection method if a leak happens to appear during a test.

7.2.2 Leak Phenomena

All pipeline leaks are associated with certain external and internal phenomena. External phenomena include the following:

- Spilled product around the pipeline
- Noise generated from leakage at the hole in the pipe
- Temperature changes around the hole

Internal phenomena include:

- Pressure drops and flow changes
- Noise around the hole
- Temperature drop for gas pipeline

All leak detection systems take advantage of the presence of one or more leak phenomena.

7.3 Leak Detection System Overview

In North America, a leak detection system is normally required on new liquid pipelines, but not on existing pipelines unless mandated otherwise by the appropriate regulatory agency. In general, there is no leak detection requirement on gas pipelines other than a few new gas pipelines. The same is true of multi-phase gathering pipelines.

Pipeline companies are using various leak detection methods with varying degrees of success. Pipeline companies, the pipeline service industry and academe have put significant efforts into the development of an ideal leak detection system. So far, no single method truly stands out as an ideal system able to detect the wide ranges of leaks with absolute accuracy and reliability, and having low installation and operating cost. Some are accurate and reliable but too expensive, and some are economical but unreliable.

This section will discuss the following aspects of a leak detection system:

- Objectives of a leak detection system
- Leak detection system selection criteria to help engineers in evaluating competing techniques
- Classification of leak detection methods

- Available standards and references

7.3.1 Standards and References

As illustrated in the previous section, different countries have developed different leak detection regulations and practices. A few references and standards are introduced below. However, in general, the codes and standards on pipeline leak detection are not well defined.

7.3.1.1 API Standard References

American Petroleum Institute (API) has published several reference books on pipeline leak detection. They are listed below:

- API 1130 "Computational Pipeline Monitoring" addresses the design, implementation, testing and operation of Computational Pipeline Monitoring (CPM) systems. It is intended as a reference for pipeline operating companies and other service companies. The publication is used as a standard by regulatory agencies in many parts of the world.
- API 1149 "Pipeline Variable Uncertainties and Their Effects on Leak Detectability" discusses the effects of variable uncertainties and leak detectability evaluation procedures for a computational pipeline monitoring methodology. This publication describes a method of analyzing detectable leak sizes theoretically using physical parameters of the target pipeline. It can be used for assessing leak detectability for new and existing pipelines.
- API 1155 "Evaluation Methodology for Software Based Leak Detection Systems" describes the procedures for determining CPM's leak detection performance. Unlike API 1149, this publication addresses the performance evaluation procedures based on physical pipeline characteristics and actual operating data collected from pipeline operations.

7.3.1.2 Canadian Standards Association (CSA) Z662

Canadian Standards Association (CSA) is responsible for developing Canadian Codes and Standards. The Canadian standards applicable to oil and gas pipelines are specified in Z662, "Oil and Gas Pipeline Systems." Section 10.2.6 of Z662 specifies leak detection for liquid hydrocarbon pipeline systems, and Section 10.2.7 for gas pipeline systems.

The specifications in Section 10.2.7 for gas pipeline systems states:

> "Operating companies shall perform regular surveys or analyses for evidence of leaks. It shall be permissible for such leak detection surveys or analyses to consist of gas detector surveys, aerial surveys, vegetation surveys, gas volume monitoring analyses, bar-hole surveys, surface

> detection surveys, mathematical modeling analyses, or any other method that the operating company has determined to be effective. Operating companies shall periodically review their leak detection programs to confirm their adequacy and effectiveness."

This specification describes several techniques for leak monitoring and detection, and operating companies have several options as long as the leak monitoring program works adequately.

The specifications in Section 10.2.6 for liquid pipeline systems states:

> "Operating companies shall make periodic line balance measurements for system integrity. Operating companies shall periodically review their leak detection methods to confirm their adequacy and effectiveness. Installed devices or operating practices, or both, shall be capable of early detection of leaks. Measuring equipment shall be calibrated regularly to facilitate proper measurement."

The specification states that using this technology for multiphase pipelines may be limited or impractical and thus other techniques shall be used to confirm system integrity.

The title of Annex E is "Recommended Practice for Liquid Hydrocarbon Pipeline System Leak Detection." The annex describes a practice for leak detection based on computational methods, particularly material balance techniques. It does not exclude other leak detection methods that are equally effective. The annex emphasizes that operating companies shall comply as thoroughly as practicable with the record retention, maintenance, auditing, testing, and training requirements, regardless of the method of leak detection used.

The annex describes measurement requirements and operational considerations to perform material balance calculations. "All pipeline segment receipts and deliveries should be measured. Under normal operating conditions, the uncertainty in the receipt and delivery values used in the material balance calculation, including uncertainties attributable to processing, transmission, and operational practices, shall not exceed 5% per five minutes, 2% per week, or 1% per month of the sum of the actual receipts or deliveries." It also specifies that pipeline equipment shall be installed to ensure that only liquid is present in the pipeline unless the material balance procedure compensates for slack-line flow.

The pipeline service fluids are divided into two types of liquids: high vapor pressure (HVP) liquids and low vapor pressure (LVP) liquids. Also, applicable pipelines are both transmission and gathering lines. Depending on the class location, normal flow rate, and the types of liquids and pipelines, the intervals for data retrieval, maximum calculation intervals and recommended calculation windows are different.

7.3.2 Objectives

Implementation of leak detection systems can range from simple visual inspection of the pipeline to sophisticated monitoring of the pipeline by means of hardware and software. Although no single leak detection system is applicable to all types of pipelines, pipeline companies can select a suitable system for use in a wide range of applications.

It is essential that the objectives and requirements for employing the leak detection system are defined and that the system can fulfill them. The objectives of the leak detection system are to assist the pipeline operators with (4):

- Reducing spillage of product and thus reducing the consequences of leaks,
- Reducing operator's burden by detecting leaks quickly and consistently without relying heavily on operator experience,
- Providing operational advantages such as additional useful information for responding to emergencies and other operational situations reliably,
- Satisfying regulatory requirements.

Spillage can be reduced in several ways:

- Detecting and locating the leak quickly,
- Confirming and isolating the leak rapidly,
- Reducing block valve spacing with remote control capability.

A leak would be initially detected and located by the leak detection system and then confirmed by some means such as visual inspection. After, or even before the leak is confirmed (depending on the company's leak response procedures), the leak must be isolated by closing block valves adjacent to the leak. After the leak is isolated, a significant volume of product can be lost depending on the leak location and terrain of the leaked pipeline section. The spillage during the detection phase is often relatively small compared to potential total spillage. Therefore, the importance of rapid detection time as a valuable feature of a detection system should not be over-emphasized.

7.3.3 Leak Detection System Selection Criteria

It is important to define a set of selection criteria for use in assessing the performance and selection of various leak detection systems. Typical performance criteria are listed below (5, 6):

Criteria	Description
Detectability	Detectability of leaks is measured in terms of leak detection time and range of leak size. Some leak detection methods depends on leak size, others not. Some systems can detect a

	wide range of leak sizes.
Sensitivity	Sensitivity is defined as the minimum leak size that the leak detection system can detect.
Reliability	Reliability of a leak detection system is defined in terms of false alarm rate. If the frequency of false alarms is high, the operators may not trust the leak detection system. This can increase the confirmation time and thus increase spillage volume.
Robustness	Robustness is defined as a measure of the leak detection system's ability to continue to operate and provide useful information in all pipeline operating conditions, including less than ideal situations such as instrumentation or communication failures.
Applicability	Some leak detection systems can be applied to both single and multiple phase pipelines, but others cannot. Many flow lines and off-shore pipelines operate in multi-phase.
Operability	The leak detection system needs to operate not only continuously but also in all operating conditions (shut-in, steady state and transient state). In addition, the system should not interfere with normal operations.
Accuracy	Accuracy is defined as a measure of the leak detection system's ability to estimate how close the estimated leak location and size is to the actual leak location and size.
Cost	The cost includes the installation and operating costs of a leak detection system, including instrumentation or sensing devices.
Other Applications	Some leak detection methods can perform operations other than detection; many model based systems for example, can track batch movements

The purpose of any leak detection system is to detect leaks, not to prevent them. An effective system helps pipeline operators mitigate the risks and consequences of any leak. It can shorten leak detection time, increase reliability (not miss actual leaks and at the same time not produce false alarms), and reduce leak confirmation and isolation time with accurate leak location estimates. Simply, overall cost can be reduced using an effective leak detection system. However, there are costs to implement and operate a leak detection system.

Therefore, the decision making process of implementing and operating a leak detection system can be looked at from a cost-benefit point of view by assessing potential risks. In other words, the decision is made by balancing the risk and consequences of possible leaks against the cost of a leak detection system and mitigation program. The following process may help in analyzing potential risks of leaks in terms of cost and the cost savings resulting from the implementation of a leak detection system:

- Estimate likely probabilities of various leaks and thus the potential number of leaks.
- Estimate the direct and indirect costs of leaks over a period of time without a leak detection system by using historical data for the consequences of the leaks.
- Assess attainable leak detection performances of various leak detection systems by applying all the above criteria.
- Determine the costs of implementing and operating these leak detection systems over the same period of time used in the cost calculation above.
- Estimate potential cost savings from the use of a leak detection system.

7.3.4 Classification

Broadly, there are three different types of leak detection methods: Inspection Methods; Sensing Devices; and Computational Pipeline Monitoring Methods. This chapter focuses on the discussion of the Computational Pipeline Monitoring methods. The other two methods are described in Appendix 3.

- Intermittent Inspection methods are very accurate, sensitive and generally reliable. Particularly, ultrasonic and magnetic inspection techniques can detect both actual and incipient leaks by determining the pipe wall thickness. However, internal inspection methods are very expensive requiring specialized tools and expertise, and a pipeline cannot be inspected continuously. Due to the nature of intermittent operation, only leaks that occurred prior to the inspection will be detected and any occurring after will remain undetected until the next survey.
- Sensing Devices continuously sense particular characteristics of leaks such as sudden pressure drop, noise, electrical impedance or other signals caused by a leak or interference around a pipe. Some sensing devices can detect not only leaks but also third party interference around the pipeline system. Traditionally, these techniques have been relatively unreliable and impractical. There are several emerging technologies in sensing devices such as fiber optics that are showing increasing promise. Certain techniques such as specialized fiber optic cables can be expensive for existing pipelines, as the pipeline has to be retrofitted with the cable or sensing devices.
- Computational Pipeline Monitoring (CPM) methods are based on mathematical or statistical computations of certain quantities using commonly available measured values such as flows and pressures obtained through the host SCADA system. In general, the cost is

relatively reasonable but the sensitivity is lower than other methods.

7.4 Computational Pipeline Monitoring Methods

The American Petroleum Institute (API) has introduced the term, "Computational Pipeline Monitoring" or CPM in its publication, API 1130. Many liquid pipeline companies utilize one or more of these CPM methods which are discussed in this section.

Any pipeline monitoring system that continuously checks for leaks can be considered a real-time leak detection system. All CPM methodologies are classified as real-time leak monitoring systems. Real-time leak detection as discussed in this section includes only the methods based on leak detection software operating in conjunction with a host SCADA system. SCADA systems are discussed in Chapter 1.

Any CPM system consists of the following components:

- Field instrumentation and RTU which sends the field data to the host SCADA
- SCADA system, which collects the field data, sends them to the real-time leak detection system, and annunciates event and alarm messages. The SCADA system requirements for leak detection are discussed in Section 7.7.2.
- Hardware and software interfaces which integrate the functions of the host SCADA and real-time leak detection system
- Real-time leak detection computer and software

The key advantage of the CPM methods is that they seldom need additional instruments and equipment to those that already exist for normal pipeline operations. As a result, the implementation and operating costs are typically lower than the costs for inspection and sensor methods.

API Publication 1130 defines the following eight CPM methodologies:

- Line balance technique
- Volume balance technique
- Modified volume balance technique
- Compensated mass balance technique
- Real-Time transient model (RTTM) method
- Flow/pressure monitoring method
- Acoustic/Negative pressure wave method
- Statistical technique

Each of these eight methodologies is discussed in terms of fundamental principles and equations, required data and instrumentation, implementation approaches, and its strengths and weaknesses with respect to the leak detection system selection criteria (Refer to Section 7.3.3 for the discussion of the selection criteria). The first five methodologies are based on mass balance principle and will be discussed in that context. Since the RTTM methodology requires other principles and solution techniques, a separate section is devoted to the discussion of this methodology. The last three methods are discussed separately.

7.4.1 Mass Balance Leak Detection Methodologies

The mass balance principle applied to a pipeline means that the difference between the amount of fluid that enters and leaves the pipe over a given time must be the same as the change in fluid inside the pipe over the same period of time. This principle is expressed mathematically as follows:

$$V_{in} - V_{out} = \Delta LP$$

Or

$$\mathrm{Im}\, b = V_{in} - V_{out} - \Delta LP$$

where

V_{in} = mass or corrected volume entering the pipeline over a fixed time interval

V_{out} = mass or corrected volume leaving the pipeline over the same time interval

ΔLP = change in line pack over the same time interval

Imb = imbalance

In theory, the imbalance must always be zero by mass balance principle, unless there is a leak or unaccounted flow in the pipeline section. In practice, however, the imbalance is not precisely zero. The non-zero imbalance can be attributed to a number of factors including measurement errors and line pack calculation errors. It is interpreted as a leak if the imbalance is positive beyond a predefined limit, and as an unaccounted flow if it is negative, assuming that the measured flows are accurate.

The flows that go into and out of the pipe are measured quantities and line pack changes are calculated quantities. Depending on how line pack changes are treated, the mass balance method has several forms. API Publication 1130 includes four different variations of the mass balance techniques: line balance, volume balance modified volume balance, and compensated mass balance.

7.4.1.1 Line Balance (LB) Method

API Publication 1130 defines the line balance methodology as follows:

> "This meter-based method determines the measurement imbalance between the incoming and outgoing volumes. The imbalance is compared against a predefined alarm threshold for a selected time interval. There is no compensation for the change in pipeline inventory due to pressure, temperature or composition. Imbalance calculations are typically performed from the receipt and delivery meters, but less timely and less accurate volumes can be determined from tank gauging."

Line balancing can be accomplished manually because of its simplicity. This methodology uses only the measured flows into and out of the pipeline system, ignoring line pack changes. It assumes that a leak may have occurred if more fluid enters the pipeline than leaves the pipeline over a certain time period. It generates a leak alarm when the measured flow differences accumulated over a certain period exceed a defined limit. This methodology cannot provide leak locations because it doesn't calculate the pressure profile.

At a minimum, this method requires flow sensors at all fluid injection and delivery points. The higher the accuracy of the flow meters, the better the long term leak sensitivity. Pressure and temperature sensors are needed only if the measured flows are to be corrected to a standard condition. Any other measurements along the pipeline are not used in this method for line pack calculation. The accuracy of the pressure and temperature is neither critical for leak sensitivity nor does it improve overall leak detection performance. Since this methodology ignores the line pack change term in the mass balance equation, it does not require pipeline configuration, product data, or a method to calculate line pack change.

Long-term sensitivity of the LB method depends only on the flow measurement accuracy and short-term sensitivity depends on the amount of line pack changes. In general, leak detection sensitivity increases as the time interval increases, because the line pack changes are reduced to near zero over longer intervals because any pipeline transient will have died out. In addition, sensitivity can be increased if the measured flows are corrected to a standard condition with pressure, temperature and a proper equation of state.

Tuning requirements for this method are simple, because only flow measurements and if required, flow corrections are involved. The flow or volume difference between injection and delivery meters is analyzed in terms of long-term bias and short-term random errors. If they are clearly evident, these errors are corrected during a tuning period. If flow or volume correction to a base condition is required, the measured pressures and temperatures at all injection and delivery points have to be checked for their availability and accuracy. As well, the selected equation of state has to be checked for accuracy.

Figure 1 shows a flow difference trend that uses this method. It was measured during a pipeline operation, which included pump start-up and shut-down. As shown in the trend graph, the flow differences are large and thus line balance may not be feasible during transient operations. This problem during transient operations will last longer if the pipeline size and length is larger than in the example, and/or if the product is lighter (i.e. more compressible).

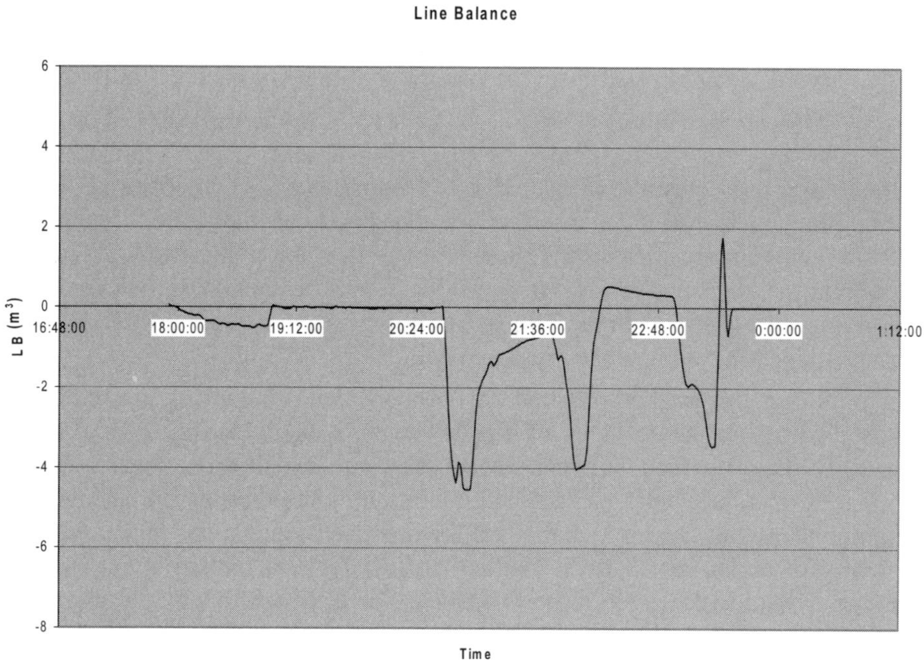

Figure 1 Line Balance Trend (Courtesy of CriticalControl)

Line balance was widely used because it is simple and the computational requirement is not extensive. It is a suitable method for pipeline systems with small pipe size and short pipe length and if reliable flow measurements are available at both injection and delivery ends. This technique is less popular now because:

- It is applicable to few operating situations and small pipeline systems.
- It depends entirely on flow measurements and their accuracy.
- Detection time is long.
- The low computing requirements no longer offer a significant advantage.

In order to reduce the dependency of flow measurement accuracy, a few companies have developed a statistical technique. This technique is discussed in Section 7.4.5.2.

7.4.1.2 Volume Balance (VB) Method

API Publication 1130 defines the VB methodology as follows:

> "This method is an enhanced line balance technique with limited compensation for changes in pipeline inventory due to temperature and/or pressure. Pipeline inventory correction is accomplished by taking into account the volume increase or decrease in the pipeline inventory due to changes in the system's average pressure and/or temperature. It is difficult to manually compensate for changes in pipeline inventory because of the complexity of the imbalance computation. There is usually no correction for the varying inventory density. A representative bulk modulus is used for line pack calculation."

The VB technique (7) uses both the flow difference and line pack change terms in the mass balance equation, adding the line pack change to the Line Balance discussed in the previous section. In other words, this method compensates for the difference between the volumes into and out of the pipeline with line pack changes over a certain time period. Imbalance beyond a set limit is interpreted as a potential leak. Since imbalance includes both flow difference and line pack change, a leak can be detected by increased flow differences, line pack changes, or both. Therefore, leak detection using this methodology is faster than that using the line balance methodology but it does not provide leak location because it doesn't calculate the pressure profile.

Line pack change depends on fluid properties such as compressibility and thermal expansion, pressure, and temperature, and pipe data such as size and length. Since line pack change is a dynamic quantity, the fluid compressibility, pressures and temperatures are the most important parameters in calculating line pack change.

This technique doesn't calculate line pack change rigorously. Normally, the line pack and its changes are calculated by using the measured pressure and/or temperature together with an average fluid density. Assuming that the pipeline is in a steady state condition, average pressure and/or temperature are estimated and the resulting average density is calculated using a representative bulk modulus and thermal expansion coefficient. This calculation is made even when there are multiple products in the pipeline. The average density is then multiplied by the pipeline volume to obtain the line pack. The line pack change is the difference between the line packs at the current time and the previous time. Since this technique assumes a steady state to estimate average pipeline pressure and temperature, the line pack calculation error can be large if a transient condition is severe, pipe size is large, pipe length is long, or the product is more compressible.

In line pack calculation, the product properties are very important. They can either be measured or correlated based on the known parameters such as crude API gravity or specified product name such as gasoline. The properties include the product density, compressibility or bulk modulus, and its' thermal expansion coefficient. They are expressed in an equation of state in terms of density, pressure and temperature.

At a minimum, this method requires flow and pressure sensors at all fluid injection and delivery points. The measured flows or volumes are used directly to calculate the flow difference in the pipe section bounded by flow meters. The higher the accuracy of the flow meters, the better the long term leak sensitivity. Unlike with the line balance leak detection method, the measured pressures and temperatures, particularly at the injection points, are used to calculate line pack as well as to correct measured volumes or flows to a standard condition.

The inclusion of line pack change in the mass balance equation can reduce the imbalance error, particularly during packing and unpacking operations. However, the steady state assumption under all operating conditions and simplified calculations of line pack changes can result in a large line pack calculation error.

Figure 2 shows an imbalance trend using this method, for the same pipeline and operations as those for the line balance example. As shown in the trend graph, the imbalances during the transient operations are reduced compared to the flow differences as shown in Figure 1.

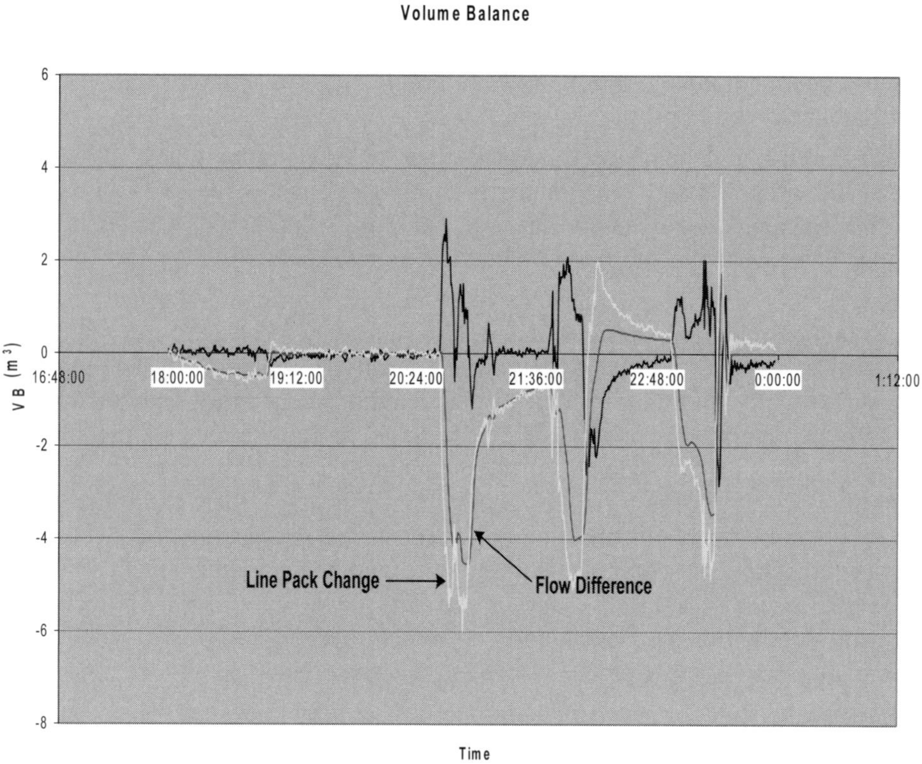

Figure 2 Volume Balance Trend (Courtesy of CriticalControl)

Tuning effort for this method is more involved than for a LB method, because product compressibility is required to estimate line pack changes. During a tuning period, product compressibility may need to be adjusted from a theoretically known value in order to minimize the calculated imbalances in transient operations. In addition, more tuning effort is required for a batch pipeline with high compressibility.

The VB technique has been widely applied to heavier hydrocarbon liquid pipelines with small pipe size and short pipe length, because it:

- can detect leaks relatively quickly for a wide range of leak sizes.
- has a relatively low false alarm rate.
- requires basic instrumentation such as flow, pressure and temperature, and doesn't require instrumentation such as viscometers.
- does not require a high level of expertise to maintain, thus has a lower operating cost.

However, a basic VB method has limited leak detection capability for the following pipelines due to potentially large line pack calculation errors:

- Long pipelines containing large line pack
- Pipelines carrying light hydrocarbon liquids
- Pipelines carrying multiple products in batch
- Pipelines whose temperature profiles change significantly

7.4.1.3 Modified Volume Balance (MVB) Method

API Publication 1130 defines the MVB method as follows:

> "This meter-based method is an enhanced volume balance technique. Line pack correction is accomplished by taking into account the volume change in the pipeline inventory utilizing a dynamic bulk modulus. This modulus is derived from the bulk moduli of the various commodities as a function of their percentage of line fill volume."

A MVB method is a modified version of a VB leak detection method; it differs in that it uses a more accurate accounting of product properties. As discussed in the previous section, the VB method is not accurate when dealing with product movements such as batching and blending. Instead of using one representative bulk modulus or product compressibility for the whole pipeline, as in the VB method, the MVB method tracks batches along the pipeline and calculates the average bulk modulus dynamically in each pipe segment. The segment bulk modulus is applied to calculate the segment line pack and its changes, and the whole line pack and its changes and then all changes are added together.

Most comments made for the VB method are valid for this method, except that it generally calculates the line pack change more accurately than the VB for batch pipelines, as seen in Figure 3. The MVB method requires extra tuning effort in establishing batch tracking and other product movement related data.

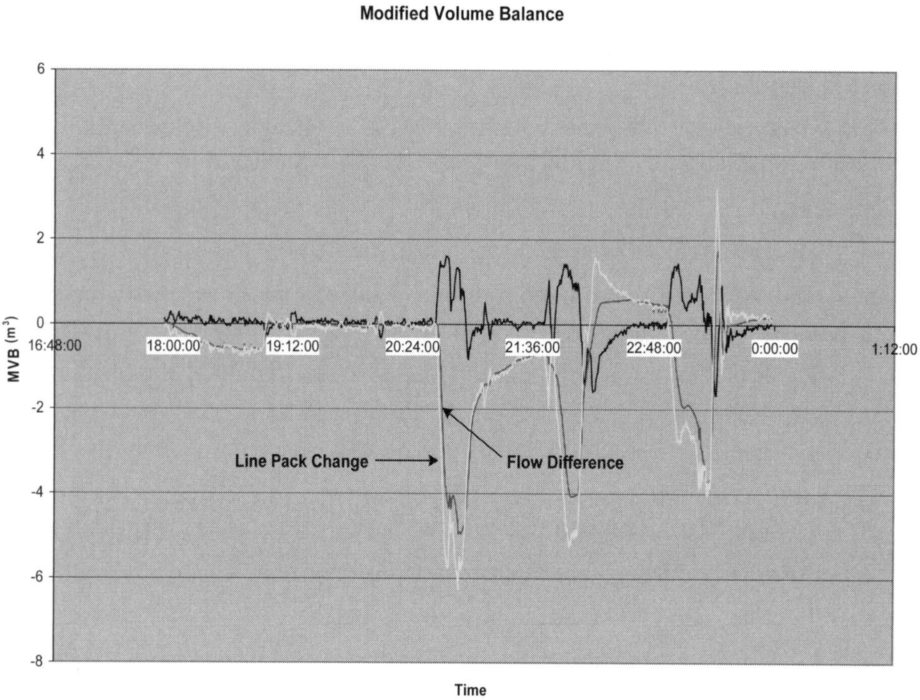

Figure 3 Modified Volume Balance Trend (Courtesy of CriticalControl)

7.4.1.4 Compensated Mass Balance (CMB) Method

API Publication 1130 defines the CMB methodology as follows:

> "As a further enhancement to the MVB method, this volume balance technique models pipeline conditions between measurement points to more accurately determine pressure and temperature profiles as input for the line pack calculation. The pipeline is sub-divided into a predefined number of segments based on available instrumentation, elevation characteristics, and the desired level of sensitivity. In addition, inventory locations are determined through batch/DRA/hydraulic anomaly tracking. Volume imbalance is typically monitored over a number of time periods (e.g., 15 minutes to 24 hours, also weekly and monthly) to detect commodity releases of different sizes."

As defined above, a CMB method is an enhanced version of the MVB (8). The CMB method calculates temperature profiles along the pipeline by solving an energy equation with the temperatures at the injection as a boundary condition.

The temperature model includes heat transfer with the ground, the transportation of energy with the movement of the product, frictional heating, and possibly the Joule-Thomson effect.

The CMB method takes into account the fluids movements including batch, fluid blending, and product characteristics together with anomaly tracking such as basic sediment and water content (BS&W). It monitors the volume and product properties of each batch. Batch volumes are updated based on injection and delivery volumes obtained from metering locations along the pipeline from the host SCADA. Batch launches can be triggered by an indication from SCADA, a change in density, a change in valve status, or can be based on a schedule. At junctions where a side stream injection occurs, blending of the products is modeled with mass and energy conservation in the blending process.

The CMB method uses measured pressures, elevation profiles, product densities and batch positions to calculate pressure profiles along the pipeline. It doesn't solve a momentum equation directly, because solving the momentum equation requires data such as viscosity and other detailed fluid properties, which are sometimes difficult to obtain. The calculated pressure and temperature profiles are used to calculate line pack and its change. Since the pressure and temperature profiles are calculated based on the assumption that the pipeline is in a steady state, the calculated line pack in a near steady state is accurate.

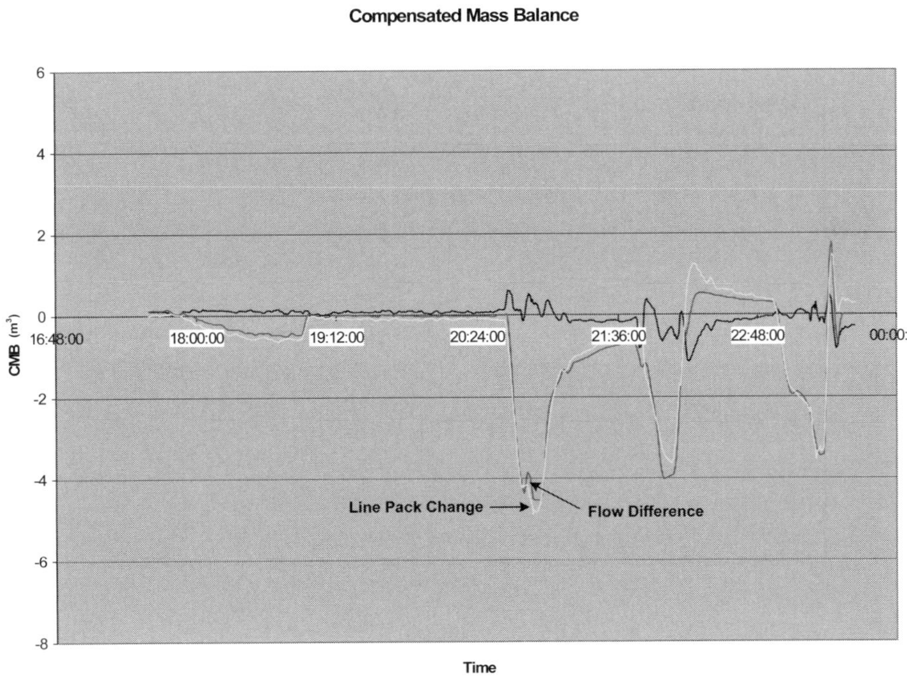

Figure 4 Compensated Mass Balance Trend (Courtesy of CriticalControl)

However, the calculated line packs and its changes under transient operations cannot be very accurate because of the steady state assumption. To reduce line pack calculation error during transient operations, a filtering technique can be applied to line pack changes. A filtering parameter can be determined by the ratio of the distance the pressure wave can propagate over a scan time to the distance between two pressure measurements. This filtering technique can dampen the growth of line pack change errors caused by various factors during transient operations. This technique is particularly useful in minimizing line pack change errors for lighter hydrocarbon product pipelines with a large line pack.

At a minimum, this method requires flow and pressure sensors at all fluid injection and delivery points. If temperature modeling is required, at least the measured injection temperature should be made available to the model. If the pipeline transports multiple products in a batch mode, either a densitometer or batch launch signal with a batch identifier is required.

Figure 4 shows an imbalance trend of this method for the same pipeline and operations as those for the previous two methods. As shown in the trend graph, the imbalances during the transient operations are even more reduced compared to the imbalances shown in Figure 3. Since the imbalances during severe transients are relatively small, there is only small performance degradation under transient conditions.

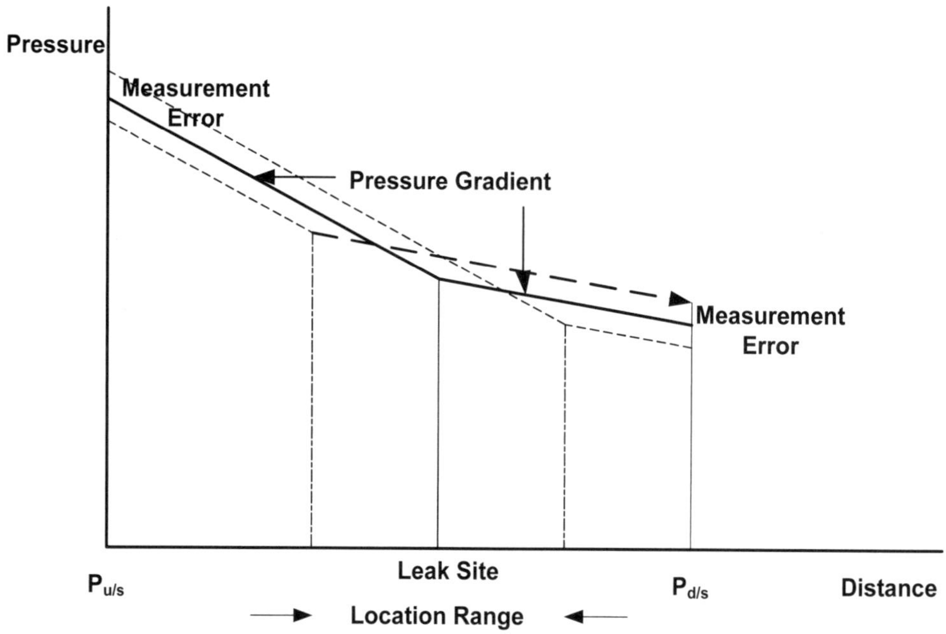

Figure 5 Leak Location and Accuracy

This methodology can determine leak location using pressure gradients, assuming that the pipeline is in a steady state during the leaking period. Pressure gradient can be calculated from a measured flow rate or from two measured pressures with elevation correction. The pressure gradient on the upstream side of the leak is higher than the gradient on the downstream side. It can be safely assumed that pressure gradient of heavy hydrocarbon liquids is linear, if the pipeline is in a steady state and the pressure profile is corrected with fluid temperature and elevation. The leak location is determined at the intersection of the two gradients.

In general, the leak location can be reliable in a steady state condition, but unreliable when the pipeline is in a transient state. The location accuracy depends on operating conditions, leak size and measurement accuracy (9). Figure 5 shows the gradient leak location method and the range of leak location as a factor of measurement accuracy.

This method may not be able to produce a leak location for very small leaks because the difference in the pressure gradients is too small to find an intersection point, nor for very large leaks because the pipeline is likely to be shut down before a steady state is reached.

Tuning effort for this method is more involved than that for a volume balance method, because in addition to the data required for the volume balance method, temperature profile calculations and batch tracking need to be included. A temperature profile calculation requires soil conductivity, product specific heat and other data such as Joule-Thomson coefficient. Batch tracking requires data related to batch operations, including batch launch and delivery, product density and compressibility. Also, a line pack change filtering coefficient needs to be determined. During a tuning period, appropriate adjustments have to be made to minimize imbalance errors.

A MVB technique offers the following advantages over other methods:

- It can detect leaks quickly and reliably for a wide range of leak sizes.
- It has a low false alarm rate.
- It requires basic instrumentation such as flow, pressure and temperature, and doesn't require specialized instrumentation such as viscometers.
- It does not require a high level of expertise to maintain, thus has a lower operating cost.
- It is simple to install and easy for the operators to make quick informed decisions.

One of the key disadvantages of this technique is leak detection sensitivity. As with the Volume Balance technique, it introduces line pack error that may not disappear even in a long-term window. Also, it may not be applied easily to detect small leaks in large gas pipeline systems.

7.4.1.5 Implementation and Data Requirements

1. Time Windows

Even under normal operating conditions, imbalance is not always close to zero mainly because measurement and line pack calculation errors can be large for a short period. To avoid this short-term problem, the calculated imbalances are accumulated over several time intervals, called windows. This accumulation technique is intended to increase the signal to noise ratio in order to reduce the false alarm rate and to detect large leaks in a short time and small leaks in a longer time. In other words, it can reduce false alarm rate and to increase leak detection sensitivity.

Each window has a different time period and will accumulate a series of imbalances calculated for that time period. For example, a 5 second window will have an accumulated total of imbalances calculated for successive 5 second windows.

The multiple samples in each window can be statistically analyzed or the total value of the samples in each window compared against the predefined threshold for the corresponding window.

2. Data Requirements

The data requirements for the three methods are summarized below.

Data	LB	VB & MVB	CMB
Pipeline configuration data			
- Pipe size	Not required	Required	Required
- Pipe length	Not required	Required	Required
- Pipe wall thickness	Not required	Required	Required
- Elevation profile	Not required	Not required	Required
- Pipe roughness	Not required	Not required	Optional (3)
- Depth of burial	Not required	Not required	Optional (4)
- Soil conductivity	Not required	Not required	Optional (4)
Product data			
- Equation of state	Optional (1)	Required (2)	Required
- Viscosity	Not required	Not required	Optional (3)
- Vapor pressure	Not required	Not required	Optional (5)
- BS & W contents	Not required	Not required	Optional (6)
- Other product specifications	Not required	Not required	Optional (6)
Facility data			
- Instrumentation location	Not required	Required	Required
- Injection point	Required	Required	Required
- Delivery point	Required	Required	Required
- Batch launch	Not required	Not required (7)	Optional (6)
- DRA launch	Not required	Not required	Optional (6)

(1) An equation of state is needed if the measured flow or volume is corrected to a base condition.

(2) Product compressibility is required to calculate line pack and its change.

(3) There are two ways of calculating pressure profile; one using measure pressures and elevation and the other using elevation and friction pressure drop calculation equation. These values are required to calculate pressure profile using the latter method.

(4) These values are required to calculate temperature profile.

(5) The product vapor pressure is used to check if the pipeline pressure causes a slack flow condition.

(6) The launch signals are used for batch or DRA tracking.

(7) A batch launch signal is required for the modified volume balance method.

3. Instrumentation Requirements

The instrumentation requirements are not necessarily the same for the three methods.

- **Flow rate or volume:** Both inlet and outlet flow or volume quantities are essential data in all mass balance techniques. Unlike the line pack change term in the equation, the flow difference term is accumulated over time, and is required for both short-term and long-term balance calculations.
- **Pressure:** Pipeline operation changes cause pressure and flow changes, creating fast and large transients. These changes result in rapid line pack changes in the pipeline.
- **Temperature:** Temperature variations in liquid pipelines are so gradual that temperature change effects on the line pack vary slowly. If the inlet temperature changes, the temperature profile along the pipeline gradually moves at the speed of the fluid and so does the line pack change. For heavier hydrocarbon liquids, temperature effects take place over a longer period compared to a leak detection time.
- **Fluid properties:** Compressibility and thermal properties may change if different grades of crude or refined products are transported in batches in the same pipeline. When a new stream of fluid enters the pipeline, a new fluid profile is established. In the batch or blending process, the pressure and fluid properties will vary along the pipeline. The speed of line pack change is directly proportional to pressure and fluid compressibility. Since a leak causes fast transients, both pressure and compressibility are important parameters for leak detection.

Instrumentation	LB	VB & MVB	CMB
Flow or Volume	Required	Required	Required
Pressure	Optional (1)	Required	Required
Temperature	Optional (1)	Optional (3)	Optional (4)

Density	Optional (2)	Optional (2)	Optional (2)
Measurement status	Required	Required	Required
Valve status	Required	Required	Required

(1) If pressure and/or temperature measurements are available at the flow measurement points, they can be used for volume correction to a base condition.

(2) If a densitometer is available at injection points for a batch pipeline, it can be used for batch identification and tracking as well as volume correction with its appropriate equation of state.

(3) If a temperature measurement is available at injection and delivery points, it can be used for correcting volume and also for estimating temperature in the pipeline.

(4) If a temperature measurement is available at injection and delivery points, it can be used for correcting volume and also for calculating temperature in the pipeline with an energy equation.

In addition to the above analog values, the status of each of the values is also required. Possible statuses include good, failed, old, stale, etc.

7.4.2 Real-Time Transient Model (RTTM)

API Publication 1130 defines the Real-Time Transient Model based leak detection methodology as follows:

> "The fundamental difference that a RTTM provides over the CMB method is that it compares the model directly against measured data i.e., primarily pressure and flow) rather than use the calculated values as inputs to volume balance. Extensive configuration of physical pipeline parameters (length, diameter, thickness, pipe composition, route topology, internal roughness, pumps, valves, equipment location, etc.), commodity characteristics (accurate bulk modulus value, viscosity, etc.), and local station logic (e.g., pressure/flow controllers) are required to design a pipeline specific RTTM. The application software generates a real time transient hydraulic model by this configuration with field inputs from meters, pressures, temperatures, densities at strategic receipt and delivery locations, referred to as software boundary conditions. Fluid dynamic characteristic values will be modeled throughout the pipeline, even during system transients. The RTTM software compares the measured data for a segment of pipeline with its corresponding modeled conditions."

Each scan, an RTTM receives an updated set of SCADA data and sends a set of the modelled results back to SCADA through the SCADA interface software. Normally, an RTTM model processes the real-time data received from the host SCADA checking the data quality including availability and validity. Some models even filter the received data.

The RTTM performs the following functions after the real-time data is processed each scan:

- Modelling of the pipeline hydraulics, determining the pipeline state in terms of flow, pressure, temperature and density profiles along the pipeline. The pipeline state includes batch movement and thus batch tracking information.
- Automatic tuning to reduce the difference between the measured and modelled values
- Leak detection

Theoretically, the RTTM approach of real-time modelling and leak detection can provide the most accurate modelling and leak detection sensitivity results. However, many companies have attempted to make this methodology work in actual operations for more than 20 years (10, 11, 12) with limited success.

This section focuses on leak detection and location methods using an RTTM, and their limitations in actual implementation and operation. Appendix 1 describes the governing principles, solution methods, and applications and operational benefits of a real-time modelling system.

7.4.2.1 Leak Detection with Pressure-Flow Boundary

Two sets of boundary conditions are the most popular for real-time applications including leak detection: boundary values using measured upstream pressure and downstream flow, and boundary values using measured upstream pressure and downstream pressure. The selection of the boundary condition determines leak detection and location methods. For example, the upstream flow deviation and downstream pressure deviation are used for leak detection if the upstream pressure and downstream flow are used as boundary conditions. If the upstream pressure and downstream pressure are used as boundary conditions, the flow deviations and/or line pack changes are used for leak detection.

This modeling approach uses the upstream pressure and downstream flow as a set of boundary conditions (13). Firstly, the last segment is modeled using the upstream station discharge pressure and the flow at the delivery point. The flow at the upstream station and the pressure at the delivery point are calculated. Since the measured flow is not usually available at every pump station, the next upstream segment is modeled using the calculated flow rate at its' closest downstream station, with the discharge pressure at the upstream station used as boundary values. This process continues on up to the first pipe segment. The first segment is modeled using the pressure at the injection point and the calculated flow at the next downstream station. Both measured and calculated flows are available at the injection point, while both measured and calculated pressures are available at the delivery flow location. Figure 6 illustrates the modeling approach using pressure-flow boundary.

Theoretically, in normal operating conditions, the calculated flow is the same as the measured flow at the injection points, and the calculated pressure is the same as the measured pressure at the delivery point. In practice, the calculated and measured flows and pressures do not exactly match. The differences between the calculated and measured values are a result of inaccuracies in product properties, pipe configuration, and instrumentation as well as of calculation errors. These errors can be reduced by a tuning process. The purpose of tuning is to produce a more realistic modeled approximation of a true pipeline operation. Normally, tuning involves the adjustment of pipe roughness to reduce pressure deviation and the adjustment of the injection flow rate to reduce flow deviation. Tuning is performed every scan during normal pipeline simulation, but the adjustments should be very small in order to avoid large changes in the model results and the possibility of tuning out a real leak.

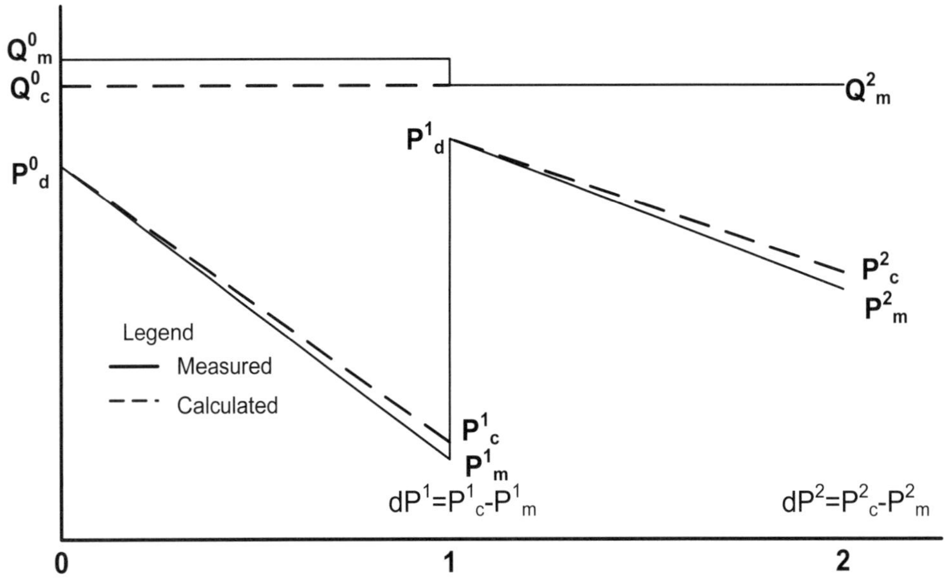

Figure 6 Model with Pressure-Flow Boundary

When a leak develops, the calculated flow will deviate from the measured flow at the injection point, and the calculated pressure from the measured pressure at all the pressure measurement points downstream of the leak. The flow and pressure deviations due to a leak occur because the flows upstream and downstream of the leak are different. A leak detection system based on this set of boundary conditions uses these pressure and flow deviations for leak detection and location. If flow measurement is not available at every pump station the only leak signal is the pressure deviation at every pressure measurement point downstream of the leak and flow deviation at the injection point.

The leak is assumed to occur in the upstream segment of the first largest pressure deviation. The leak location is then determined by using the deviation between the measured and calculated values. The location formula given below can be used if a steady state is reached after the leak.

$$L_x = \frac{(P_u - P_d) + g\bar{\rho}(h_u - h_d) - (2 f_d \rho_d L q_d^2 / D^5)}{2(f_u \rho_u q_u^2 - f_d \rho_d q_d^2)/ D^5}$$

where

L_x = leak location from the upstream flow measurement point
P = measured pressure
q = measured or modeled flow rate
g = gravitational constant
h = elevation
ρ = fluid density available from the RTTM
$\bar{\rho}$ = average fluid density available from the RTTM
f = friction factor available from the RTTM
L = distance between two pressure measurements
D = pipe inside diameter
Subscript u = upstream designation
Subscript d = downstream designation

Another leak location method is to compare the measured pressure behaviors with modeled pressure behaviors. In this method, the estimated leak rate is used to simulate the pressure behaviors at various locations where a leak is assumed along a suspected pipe segment. The assumed location, which results in the best match of the measured with the simulated pressure behaviors, is assumed to be the true leak location.

Made in parallel with the pressure and flow deviation analysis is the volume balance calculation. The model calculates pressure, temperature and density profiles along the pipeline each scan, which determines line pack changes. Leak detection can thus be performed by applying both the mass balance principle and the pressure and flow deviation method. Most users find volume balances easy to understand, but the main advantage of using deviations is that they provide leak detection capability even when a non-boundary flow meter is unavailable.

The following problems can be encountered in this method of leak detection:

- it is heavily dependant on the downstream flow meter. If a meter is unavailable, the model's ability to detect leaks is degraded or disabled for the entire pipeline system.
- if the calculated flow used as a boundary is unavailable at a given station, the calculated pressures and flow in the rest of the segments will deviate

from the measured pressures and flow even though there is no leak. Once the above deviations are created in the absence of a leak, it takes a long time for the model to settle in a lock-step with the running pipeline system because the deviations will have occurred in a large portion of the pipeline system.

- it is time consuming to tune the model, because discrepancies are difficult to identify if modeled pressures are different from their corresponding measured pressures.
- if any boundary pressure is unavailable, the model and leak detection is degraded or disabled for the entire pipeline system.
- the magnitude of pressure deviation depends on the leak location. The closer a leak is to the upstream boundary pressure, the smaller the pressure deviation and the more difficult it is to detect the leak.
- the concepts of flow and pressure deviations are not easily understood by most operators.

Due to the foregoing this approach generates frequent false alarms and is not likely to function properly under transient operating conditions. The estimated leak location can be inaccurate because pressure and flow deviations are used to estimate the leak location. In short, this method of leak detection is not suitable for pipeline systems with intermediate pump stations.

7.4.2.2 Leak Detection with Pressure-Pressure Boundary

Flow meters are expensive and not required for pipeline control except for injection and delivery locations, where custody transfers are used. Pressure measurement devices are not expensive and are used for control at pipeline facilities and at critical points along a pipeline system. Therefore, there are more pressure transducers available on a pipeline system than any other measurement devices. A real-time model with pressure-pressure boundary takes advantage of this reality of pipeline instrumentation (10, 14).

The entire pipeline network is first divided into linear pipeline segments bounded by two pressure measurements. The linear segments are then simulated to determine the pipeline state. This modeling approach requires a pressure measurement from each end of each linear segment and at every side-stream injection or delivery point. A flow measurement is required at every injection and delivery point – that is the flow is calculated at every pressure measurement point in each linear segment. When two linear segments meet at a common pressure measurement point, two flows are determined; one flow on the upstream side and the other flow on the downstream side of the pressure measurement point. The measured flow is on the upstream side and the calculated flow on the downstream side at an injection point, while the measured flow is on the downstream side and the calculated flow on the upstream side at a delivery point.

The imbalance at each point is used to detect a leak and estimate the leak location. If there is a leak in a section, the point imbalances at upstream and downstream pressure points will increase according to the size and location of the leak. Figure 7 illustrates the modeling approach with pressure-pressure boundary. Note that the point imbalance may not be zero at each pressure measurement point all the time.

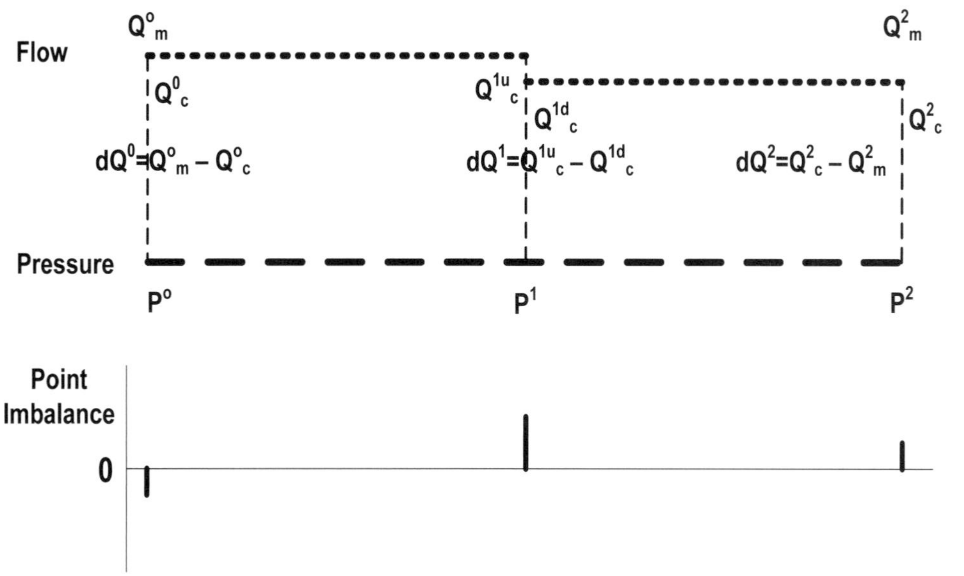

Figure 7 Model with Pressure-Pressure Boundary

Therefore, flow deviations can be determined at every pressure measurement point: the deviation between the measured and calculated flows at the injection and delivery points and the deviation between the calculated flows of the upstream and downstream segments. Theoretically, the flow deviation at each pressure point should be zero in the absence of a leak in the segment. In practice, the flow deviations are not zero, because of errors in product properties, measurements and computation.

The non-zero flow deviations imply that the calculated flow in the upstream segment is different from the calculated flow in the downstream segment. This difference in segment flow results in different batch flow movement in different segments. In order to determine consistent flows for batch movement, some RTTM models perform state estimation on the assumption that net flow should be balanced in a section bounded by flows.

Like the modelling approach with flow-pressure boundary, measured and simulated values are not always identical. The discrepancies between the two values are reduced by means of a tuning process. Normally, the tuning process adjusts the pipe roughness to reduce differences between the pipeline flows calculated by the model and those measured at the flow measurement points. Tuning is performed every scan during normal pipeline operation, but very slowly in order to avoid tuning out true leaks.

There are two ways of detecting leaks; volume balance within a pipeline section bounded by flow measurements and the other the two largest flow deviations at two consecutive pressure measurement points. They both are founded on the principle of mass balance but the former is applicable to the entire pipe section bounded by all injection and delivery flow measurements and the latter to a local pipe segment bounded by pressure measurements. A volume balance method can be easily applied to leak detection because line pack changes are determined from the pressure, temperature and density profiles and flow differences from the measured flows. If a leak occurs in a pipe segment, sustainable flow deviations are detected at the upstream and downstream pressure measurement points because the flow upstream of the leak is different from that downstream of it. Normally, these two methods are used together to detect leaks.

In practise, the flow deviations and volume balance are not always zero in the absence of a leak. Therefore, several techniques are used to improve leak detection capabilities while minimizing the number of false leak alarms. These techniques may use such variables as:

- multiple alarm accumulation time windows with varying durations,
- dynamic alarm thresholds during transient operations, (a consecutive number of conditions under which leaks could occur)
- the removal of constant bias by using a long-term volume balance, and
- analysis of behaviours in flow differences and line pack changes over time.

When the leak alarm conditions are satisfied, the leak location is estimated by using the two consecutive large flow deviations. The leak location is determined as follows:

$$L_x = \frac{FD_d}{FD_u + FD_d} * L$$

where FD_d = flow deviation at the downstream pressure measurement point
FD_u = flow deviation at the upstream pressure measurement point
L = segment length between two pressure measurement points
L_x = leak location from the upstream point

This approach overcomes many difficulties encountered by the pressure-flow boundary modeling: the model settling time is short due to short boundary pressure spacing, even under transient states; degradation and disabling is localized; no significant tuning effort is required; leak signal to noise ratio can be amplified; and leak signal response time is quick.

However, the calculated flow deviations are sensitive to flow rate because the pressure drop is small for lower flow rates while the pressure measurement errors may not be. Therefore, this leak detection method is more prone to false alarms at low flow rates and location accuracy may not be as accurate as that estimated at high flow rates.

As discussed in the previous sections, a real-time model with pressure-pressure boundary can simulate the flow in the pipeline and provide line pack changes accurately. Combined with the measured flows, it can give fast and reliable volume balance, assuming that the required data is accurate. This leak detection system, combined with flow deviation analysis, can detect leaks faster and function more reliably than other volume balance approaches discussed in the previous sections, assuming that the model is properly implemented and the required instrumentation works well.

This method needs more data than other software based methods, requiring:

- flows, pressures and temperatures at inlet and outlet points
- additional pressure measurements at several points along the pipeline
- fluid composition or properties including viscosity

In addition to the above data, other pipeline data are required to configure the pipeline model: pipe size and length, elevation profile, ground temperature, etc. Discrepancies between the assumed parameters and actual pipeline values can generate inconsistencies which result in potential deterioration of the leak detection performance.

This leak detection system is theoretically promising. Most RTTMs can provide a wealth of information on the pipeline state. In practice, however, real-time data quality and availability are often not sufficient for reliable operation of this leak detection approach, and certain values such as viscosity are not readily measurable on-line. In addition, modeling in transient conditions sometimes increases uncertainty when data quality is questionable.

Disadvantages of the technique include:

- requiring a high level of expertise to implement and maintain resulting in high operating costs,
- taking a long time to install and tune the model on operating pipelines,
- strong dependency on instrumentation for reliability and robustness,
- high false alarm rate if fluid components or operating conditions change,
- amplification of measurement errors and attendant loss of reliability with

the use of a mathematical model.

7.4.2.3 Data and Instrumentation Requirements

Most leak detection system based on an RTTM also uses multiple time windows to improve leak detection capability and reliability in a method as described in 7.4.1.5.

1. Data Requirements

The data requirements for the RTTM are summarized below:

Data	Requirements
Pipeline configuration data	
- Pipe size	Required
- Pipe length	Required
- Pipe wall thickness	Required
- Elevation profile	Required
- Pipe roughness	Required
- Depth of burial	Required
- Soil conductivity	Required
- Insulation thickness and conductivity	Required (1)
Product data	
- Equation of state	Required
- Composition or density	Required
- Viscosity	Required
- Vapor pressure	Optional (2)
- Pour point	Optional (3)
- BS & W contents	Optional (4)
- Other product specifications	Optional (4)
Facility data	
- Injection point	Required
- Delivery point	Required
- Block valve location	Required
- Pump/compressor station location	Required
- Pressure control valve location	Required
- Surge or relief tank location	Required
- Instrumentation location	Required
- Batch launch for batch operation	Required (5)
- DRA launch for DRA injection	Required (6)
- Pig launch and trap	Required (7)

(1) Insulation data is required only for insulated pipelines.

(2) The product vapor pressure is used to check if a slack flow condition exists in the pipeline by comparing the pipeline pressure against the vapor pressure.

(3) The pour point of a product is used to check if the product in the pipeline is colder than the pour

point.

(4) This data is used for tracking purposes.

(5) The launch location and signal data are used to model and track batch movements.

(6) DRA is launched at pump stations and sheared at their next stations. The location and launch signal are used to model and track DRA movement and shearing.

(7) Pig launch location and signal data are used for pig tracking. The pig tracking capability is not directly related to modeling and leak detection.

2. Instrumentation Requirements

Instrumentation requirements are:

- measurements to drive the real-time model
- measurements required for leak detection
- measurements to improve performance

In general, the RTTM leak detection method requires more measurements than other CPM methods (15).

1) Measurements Required to Drive the Real-Time Model

As discussed in the previous section, a real-time transient model requires pressure-flow pair measurements at the ends of each pipeline segment, depending on the selected boundary condition. Temperature measurements are required at the upstream end of each segment. If temperature measurement is not available at an upstream pump station, it needs to be calculated using pressure head, product heat capacity and pump efficiency. The pump station measurements must be on the pipeline side of any pump station equipment such as a controlling valve. If the station is bypassed, at least one pressure measurement must be outside the isolation valves.

If the product property is assumed to be consistent, no density measurement is required. For batch pipelines or pipelines transporting products of variable density, however, a density measurement is required at every location where batches or products are lifted. The measured density is used to track batches and to identify the batch and its appropriate equation of state. If the density measurement is not available, the batch identifier or product name should be manually entered at the time the batch is lifted.

The status of block valves is also required. Valve statuses can be manually entered. However, if the model does not have the correct valve status, the calculated flows will be wrong from the time the valve status changes. The instrumentation requirements to drive the model are summarized below.

Instrumentation	Pressure-Flow Boundary	Pressure-Pressure Boundary
Injection flow or volume	Not required (1)	Not required

Delivery flow or volume	Required	Not required
Discharge or injection pressure	Required	Required
Suction or delivery pressure	Not required	Required
Intermediate pressure	Optional (2)	Required
Temperature	Required	Required
Ground temperature	Optional (3)	Optional (3)
Density	Optional (4)	Optional (4)
Measurement status	Required	Required
Valve status	Required	Required

(1) If the injection point is a delivery location of an upstream section, then it is treated as a delivery flow.

(2) They are required if intermediate pressures are used as boundary.

(3) Ground temperatures are required if measured ground temperatures are used to calculate temperature profile.

(4) The densitometer at every injection point is recommended for batch pipeline systems.

In addition to the analog values noted above, the status of each of the values is also required. Possible statuses can include good, failed, old, stale, etc.

2) Measurements Required for Leak Detection

As discussed in the previous sections, the RTTM uses the following values for leak detection:

- Deviation of the calculated flows from the measured flows and the calculated pressure from the measured pressure at the non-boundary points; this is called the pressure-flow boundary data. All injection flow and non-boundary pressure measurements are required for this method of leak detection.
- Deviation of the calculated flow from the measured flow at every flow measurement point is used for the pressure-pressure boundary method. The flow deviation at every pressure measurement point, including all fluid injection and delivery points, is the only data required for this method of leak detection.

3) Measurements Required for Tuning

The real-time flow model with pressure-flow boundary is tuned by comparing calculated pressures with the measured flows. The same pressures used for leak detection are used for tuning.

Similarly, the real-time flow model with pressure-pressure boundary is tuned by comparing the calculated flows with the measured flows. The flow measurements required for leak detection are used for tuning as well.

The thermal model is tuned by comparison of downstream model temperatures with measured temperatures. Therefore, temperature measurements are required at all compressor station suctions and all delivery points. However, temperature increase at a pump station is normally small and temperature effect on line pack change is accordingly small in the pipe segments downstream of the second pump station. So, temperature measurements downstream of the second pump station are optional. Temperature measurements at delivery points are normally required for volume correction.

In general, modeling accuracy and thus leak detection capability are improved with extra measurements such as in-line flow meters and pressures at block valves. However, extra instrumentation increases installation and maintenance costs, so cost-benefit has to be reviewed carefully when extra instrumentation is considered.

7.4.3 Pressure/Flow Monitoring Technique

API Publication 1130 defines this technique as follows:

> "Pressure/flow values which exceed a predetermined alarm threshold are classified as excursion alarms. Initially, excursion thresholds are set out of range of the system operating fluctuations. After the system has reached a steady-state condition, it may be appropriate to set thresholds close to operating values for early anomaly recognition."
>
> Pressure/flow trending is the representation of current and recent historical pressure and/or flow rate. These trends may be represented in a tabular or graphical format on the Control Center monitor to enable a pipeline controller to be cognizant of their parameter fluctuations. Pressure/flow trending can be used to display operating changes from which a pipeline controller can infer commodity releases.
>
> Rate-of-change (ROC) calculates the variation in a process variable over a defined time interval. The rates at which line pressure and/or flow changes over time are the two most common forms of ROC used in pipeline operations. The intent of this approach is to identify rates of change in pressure and/or flow outside of normal operating conditions, thereby inferring a commodity release if operating hydraulic anomalies cannot be explained.

In general, there are four types of pressure/flow monitoring techniques used on liquid pipelines to indicated unusual conditions and potential leak conditions:

1. Pressure/Flow Limit Monitoring – ensures that measurements stay within predefined operating conditions and emergency limits.
2. Pressure/Flow Deviation Monitoring – ensures that measurements stay within a predefined tolerance of an expected operating value. Often, separate deviation limits are established for active and inactive conditions and for positive and negative deviations.

3. Pressure/Flow ROC Monitoring – ensures that any rapid measurement change, above a predefined value per defined time period, is made known. Often, separate ROC limits are established for the positive and negative directions.
4. Pressure/Flow ROC deviation – modified version of the Pressure/Flow ROC Monitoring, that projects expected ROC values during transient conditions. Often, separate ROC deviation limits are established for positive and negative directions."

This monitoring methodology monitors rapid or unexpected changes in pressure and/or flow rate, depending on their availability. The first and third types of monitoring techniques compare the current measurements and rate of changes against pre-defined operating limits.

The second and fourth monitoring types project the next expected values (pressure and/or flow rate) using a specified number of measurements. Mathematically, a projected value is expressed in terms of the following linear regression to predict the next pressure or flow rate using a specified number of pressures or flow rates collected over a specified period:

$$P = a + b * t$$

where

P = expected pressure or flow rate at time t

$$b = \frac{n\Sigma P_i t_i - (\Sigma P_i)(\Sigma t_i)}{n\Sigma t_i^2 - (\Sigma t_i)^2}$$

$$a = \frac{\Sigma P_i}{n} - b\frac{\Sigma t_i}{n}$$

n = the number of sample points

P_i = measured pressure or flow at time t_i

In principle, if the current measurement drops outside a predefined threshold from the predicted value, an alarm condition is satisfied. Normally, a second violation check is performed with the next value in order to avoid generating frequent alarms. If a second consecutive violation is detected, pressure and/or flow rate violation alarms are generated.

In order to reduce the frequency of alarms caused by rapid changes in pressure and/or flow rate, there are a few variations to the above simple violation check. If both pressure and flow rate are available at a measurement point, then two separate checks are performed: first, the pressure violation is checked and then the flow rate one. An alarm is generated only if the violation conditions are satisfied at both measurements. This check is intended to reduce false alarms caused solely by measurement problems.

Another refinement of this approach is the checking of pressure changes at both upstream and downstream points as well as the difference between the two. When the upstream or downstream pressure changes beyond the threshold, the change in the other measurement is checked. An alarm is withheld until rapid changes on both sides are detected.

This method requires measurement of pressure, flow or both. Measurements of density or temperature are not used, because they are not directly related to sudden pressure or flow changes, unless these sudden changes are related to batch changes.

The correct threshold setting is critical: it may not be able to detect a leak if the threshold is too high, while it may produce frequent false alarms if the threshold is too low. Acceptable thresholds and measurements should be determined by analyzing historical operating data.

The Pressure/Flow Monitoring system is normally disabled when communications are lost or measurements are failed. The system is enabled upon the restoration of communications or measurements from the failed state. When they are restored, old pressure/flow measurement data is cleared and new data sets are accumulated for calculation.

This method is simple and easily implemented on the host SCADA system. The main difficulties with this method are as follows:

- Normal operations can produce rapid changes in pressure and flow rate that do not necessarily indicate a leak.
- Pipeline pressure increases can mask a leak.

This method may be useful for detecting unusual events or ruptures. For leak detection purposes, it is normally used in conjunction with other leak detection methods.

7.4.4 Acoustic/Negative Pressure Wave Method

API Publication 1130 defines this method as follows:

> "The acoustic/negative pressure wave technique takes advantage of the rarefaction waves produced when the commodity breaches the pipe wall. The leak produces a sudden drop in pressure in the pipe at the leak site which generates two negative pressure or rarefaction waves, travelling upstream and downstream. High response rate/moderate accuracy pressure transmitters at select locations on the pipeline continuously

> measure the fluctuation of the line pressure. A rapid pressure drop and recovery will be reported to the central facility. At the central facility, the data from all monitored sites will be used to determine whether to initiate a CPM alarm."

An acoustic leak detection system can listen for the natural sound (16) or detect pressure changes (17) created by fluid escaping through the leaking hole. This section discusses a system based on pressure changes or rarefaction waves, while an acoustic system based on identifying leak sound is discussed in the appendix.

A leak generates a negative pressure or rarefaction wave that propagates at the acoustic velocity of the fluid in the upstream and downstream directions from the leak site. In this leak detection method, pressure sensors detect the negative pressure wave generated at the onset of the leak in the pipeline.

In general there are two sensor types: pressure sensor and sound sensor. Normally, pressure sensors detect the negative pressure wave produced by a sudden "blow out" in the pipe, while sound sensors detect the acoustic noise generated by a leak. Therefore, an acoustic system based on pressure sensors can miss a leak if the sensors are not operational when the negative pressure waves reach them at the onset of a leak.

Pressure wave sensors are installed along the pipeline. Sensor spacing depends on the fluid type and desired response sensitivity and time. The sensors are connected to the signal processing computer via a communication network, which processes the signals collected from the sensors and correlates them in order to distinguish leak signals from operational noises. An effective acoustic system should be able to reduce pumping and fluid noises and identify the direction from which the leak wave originates. The following factors should be taken into account in installing an acoustic leak detection system:

- sensor type and spacing
- leak noise or negative pressure change detection ability
- noise reduction technique to be used
- directional or non-directional capability
- leak signal through fluid or steel pipe

The pressure wave attenuates due to energy dissipation as it propagates along the pipeline. The attenuation increases with pipe bends, constrictions in the pipeline, and two-phase conditions (vapor in liquid when the pressure drops below the vaporization pressure of the fluid). In addition, noise is generated not only by a leak but by normal pipeline operations such as flow rate and equipment changes. Therefore, the maximum span between acoustic sensors is dependant on the fluid in the pipeline, attenuation, background noise, and the minimum leak size required to be detected. Typical sensor spacing in use is in the order of 15 kilometers for gas lines and 30 kilometers for liquid.

A schematic of a typical acoustic monitoring device setup is given in Figure 8. The acoustic sensor and signal processor with communication capability may be installed at the same location.

A leak is located by the relative arrival time at different pressure wave sensors. The acoustic monitoring system calculates the location of the leak by using the known acoustic speed in the pipeline fluid. For example, a pressure wave travels

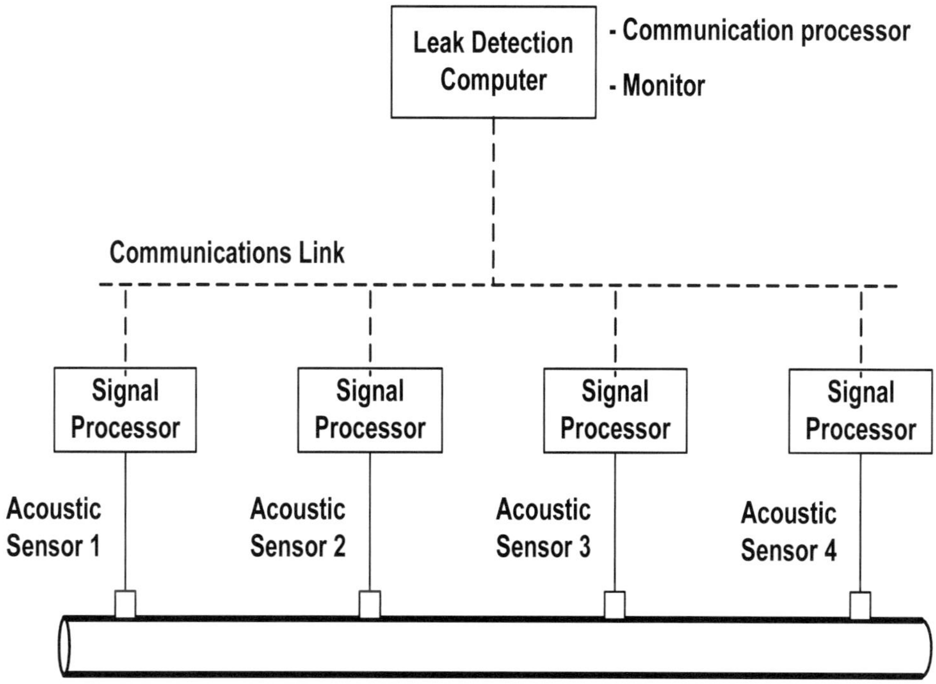

Figure 8 Schematic Diagram of Acoustic Monitoring System

at the speed of about 1 km per second in a crude oil pipeline. An acoustic sensor, which is located 10 km from a leak, can detect a leak signal in 10 seconds, assuming that the pressure wave attenuation is small. The leak location is determined by the following equation:

$$L_x = \frac{d + a * (t_1 - t_2)}{2}$$

where

L_x = distance of the leak from the sensor where the negative pressure wave is first detected

d = spacing between the upstream and downstream sensors

a = acoustic velocity of the fluid between two sensors in the pipeline

t_1 = time of first arrival of the negative pressure wave at a sensor

t_2 = time of arrival of the negative pressure wave at the other sensor

The system provides a fast response following the leak occurrence, which is mainly determined from the time required for the negative pressure wave to reach a sensor and by filtering the noises created by the pressure wave. Leak sensitivity depends on the sensor spacing and pipeline pressure, and is reported to be in the order of a few percentile of the pipeline nominal flow.

Just like any other method, threshold setting is critical for the proper operation of this system. Even though pressure changes can be detected with great sensitivity and more sophisticated data analysis techniques are available, this technique may not be able to detect small leaks due to various operating conditions that create large pressure drops, unless the pipeline operations are steady and pressures seldom change rapidly.

The main advantages of an acoustic monitoring method are:

- rapid detection of relatively large leaks, assuming that the leaks occur rapidly, the sensor spacing is close and operating pressure is high
- accurate leak location

The main disadvantages are:

- Leaks which do not generate rapid pressure drops cannot be detected because this method does not use cumulative effects of pressure drops over time. Therefore, any existing leaks cannot be detected, because pressure waves generated at their onsets already passed away.
- Normal pipeline operations including pumping and even pigging operation can produce pressure wave changes. Therefore, frequent false alarms can be generated.
- Many sensors are required to detect a leak in a long pipeline, making it costly to install and maintain.
- Installation of sensors on a sub-sea pipeline is not possible.
- If existing pressure sensors are not adequate and thus new sensors have to be installed, sensor mounting may have to be intrusive, unless the existing valve assembly allows sensor mounting without intrusive tapping. If a power supply is not available, costs can be high. In addition, sensor spacing needs to be short enough, in the order of 30 km, to detect small size leaks.

There are several techniques that can detect change points: wavelet transform, filtering of leak signal, and statistical analysis of leak signal. One commercially available technique is pressure point analysis (PPA). Since the PPA uses a statistical technique, it is detailed in the next section.

7.4.5 Statistical Analysis Method

API Publication 1130 discusses this technique as follows:

> "The degree of statistical involvement varies widely with the different methods in this classification. In a simple approach, statistical limits may be applied to a single parameter to indicate an operating anomaly. Conversely, a more sophisticated statistical approach may calculate the probability of commodity release against the probability of no-commodity release. Pressure and flow inputs that define the perimeter of the pipeline are statistically evaluated in real time for the presence of patterns associated with a leak. A probability value is assigned to whether the event is a commodity release. The analysis can, with suitable instrumentation, provide intelligent alarm processing which reduces the number of alarms requiring operator analysis. This type of CPM methodology does not require an extensive data base describing the pipeline.
>
> The statistical process control (SPC) approach includes statistical analysis of pressure and/ or flow. SPC techniques can be applied to generate sensitive CPM alarm thresholds from empirical data for a select time window. A particular method of SPC may use line balance data from normal operations to establish historical mean and standard deviation. If the mean value of the volume imbalance for the evaluated time window increases statistically, the CPM system will give a warning. An alarm is generated if the statistical changes persist for a certain time period. Also, it can correlate the changes in one parameter with those in other parameters over short and long time intervals to identify a hydraulic anomaly."

Most CPM methodologies use a particular leak signal of statistical data for various purposes. This section discusses the PPA, sequential probability ratio test, and Bayesian inference method. It should be noted that the statistical analysis technique is not a standalone leak detection approach but is used to augment other leak detection methods discussed in the previous sections.

7.4.5.1 Pressure Point Analysis (PPA)

The PPA system of leak detection (18) is based on pipeline pressure drop as a result of a leak. The PPA technique normally performs a statistical analysis of two data sets - a new pressure data set and an old or previously measured pressure data set. Both average pressures and data variances of the two sets of data are calculated, and the combined data set is put to a Student's *t*-distribution test, assuming that all data is random. The Student's *t*-distribution determines a change point with a small number of leak signal data. A leak is suspected when the statistical tests determine that the mean of the new data set is statistically significantly lower than the mean of the old data set.

Each pressure measurement used in this technique is treated individually. Several partitions may be made using the same data set, and then the partitioned data statistically analyzed in order to maximize both sensitivity and detection speed. The sensitivity can be increased by using a large amount of data for both the old and new data partitions, while a small amount of data in the new data partition can be used for detecting large leaks in a short time. By putting more data in the old data partition and less in the new partition, the affect of the leak on the mean of the new data is felt more quickly and responds to large changes more quickly but sensitivity is reduced. For each partition of data, the PPA determines the probability that the pressure is dropping. When this probability exceeds an established threshold, a change point is confirmed.

Designating the subscripts 1 and 2 for old and new sets of data respectively, the hypothesis is that the average pressures are equal, or $\mu_1 = \mu_2$ against the alternative $\mu_1 > \mu_2$. This means that the new set of pressure data contains pressure drops and is put to the Student's *t* test. To do this, the *t*-distribution of the combined data set needs to be calculated as follows:

$$t = \sqrt{\frac{n_1 n_2 (n_1 + n_2 - 2)}{n_1 + n_2}} \frac{(\mu_1 - \mu_2)}{\sqrt{(n_1 - 1)\sigma_1^2 + (n_2 - 1)\sigma_2^2}}$$

where μ = the sample mean of the pressure data
σ^2 = the sample variance of the pressure data
n = the number of data points in each sample

The *t* value obtained from the above equation is an observed value of a random variable which has a Student's *t*-distribution with $n_1 + n_2 - 2$ degrees of freedom. Using this value and the number of degrees of freedom, the significance level of the test can be found from a table of Student's *t*-distribution.

Because this change can be the result of an operating transient such as a pump stop or valve operation rather than a leak, additional conditions are imposed on the algorithm to avoid false leak alarms. Leak detection is usually disabled during known transients.

The PPA method is simple and requires minimal instrumentation. It works best if the line pack size is small and at the same time is operating mostly in steady flow conditions. This is because the pressure will not change abruptly under a steady state condition unless there is a change in product shipped or loss of product such as a leak. Also, it can be used as an optional leak detection method under limited operating conditions if only one pressure measurement is available.

Since the PPA method uses pressure drop as the only leak signature, it has the following problems:

- Pressure drops can occur in normal operations (for example due to flow rate changes). Therefore, the pressure drop due to a leak can be superposed on pressure changes due to operational changes other than the leak. The pressure changes due to operation changes can mask leaks if they are positive, and generate false alarms if they are negative. In other words, PPA has no way of distinguishing pressure drops caused by a leak from normal operating pressure changes. In pipelines where operating changes are common, the leak alarm thresholds must be loosened to reduce the number of false alarms to an acceptable level, thus reducing the sensitivity of the leak detection system.
- According to published literature, the PPA inhibits leak detection when known transients are present in the pipeline in order to reduce false alarms. If a leak starts while the PPA's statistical test output is nullified or while the leak detection system is not running it may never be detected. This is because the leak signal will show up in both the short and long term data sets that the statistical process compares. This problem becomes worse with passing time.
- A large number of pressure measurements are needed to make the PPA statistically meaningful, and thus a leak has to be sustained for a certain period. A large leak's initial pressure transient will not last long, as it quickly begins to look like a regular delivery point being fed by pipeline flow. Therefore, the PPA may not have enough data to analyze and generate a statistically reliable leak alarm unless fast scan rates are used.
- The PPA alone can neither provide leak location nor leak size or spillage estimate.

7.4.5.2 Sequential Probability Ratio Test (SPRT)

Another statistical leak detection method detailed in this section is a sequential probability ratio test (SPRT) technique to determine an alarm status. It provides a means of making a leak alarm decision by analyzing a time series data statistically. For pipeline leak detection, the SPRT is applied to the time series data of the volume imbalances or flow differences. This technique has been primarily used for equipment fault detection. Shell UK first applied this technique to pipeline leak detection (19).

The primary variable used for leak detection is the volume imbalance or flow difference between the injection and delivery flows. A shift in imbalance or flow difference would signal a leak, given that the imbalance should be theoretically zero. Controlling the percentage of false alarms while having a good probability of detection is a very important part of detection procedures using this method.

This method is based on Neyman-Pearson's probability ratio with sequential testing, or Wald Sequential Probability Ratio Test (SPRT).

The fundamental assumption is that test variables are random or Gaussian. The method involves the testing two hypothesis:

- Null hypothesis: imbalance has a Gaussian distribution with mean "m" and standard deviation "sigma" (no leak) – Alpha (Type 1 error)
- Alternative hypothesis: imbalance has a Gaussian distribution with mean "m + delta_m" and standard deviation "sigma" (leak) – Beta (Type 2 error),

where m is the mean value of the signal (imbalance or flow difference) under normal operations and delta_m a leak size to be detected.

The method uses the natural logarithm of the ratio of the probability of false alarm ("alpha") and the probability of letting a leak go undetected ("beta") to detect a change in mean imbalance from "m" to "m + delta_m". Using a Gaussian distribution, the logarithmic probability ratio using the latest data is determined by the following expressions:

$$PR(t) = \frac{L_a}{\sigma^2}\left[S_0(t) - m(t) - \frac{L_a}{2}\right]$$

where

$PR(T)$ = logarithmic probability ratio

L_a = minimum detectable leak size

σ = current standard deviation

$S_o(t)$ = imbalance or flow difference

$m(t)$ = mean value

The sequential probability ratio can be obtained by adding the current ratio to the previous ratio:

$$\lambda(t) = \lambda(t-1) + PR(t)$$

The log ratio is updated as data is obtained. The Wald test states that if the log ratio exceeds a certain threshold, the alternative hypothesis is accepted and hence a leak alarm is generated. If the log ratio falls below a certain threshold, then the null hypothesis is accepted and everything is assumed to be normal. The two probabilities, alpha and beta, determine the upper and lower thresholds as follows:

$$UpperLimit = \ln \frac{1-\beta}{\alpha}$$

$$LowerLimit = \ln \frac{\beta}{1-\alpha}$$

where *ln* is a natural logarithm function. A leak is detected if the upper limit is violated, while the probability of a leak is extremely small if the log ratio is smaller than the lower limit.

In this technique, three parameters play a distinctive role in leak detection; standard deviation, mean value of imbalance, and minimum detectable leak. The standard deviation affects primarily leak detection time and the mean value the absolute leak detection limit, below which leak detection is statistically impossible. The minimum detectable leak should be the minimum departure from the mean.

The standard deviation of volume imbalances can change as the pipeline operation changes. In general, the standard deviation is very small under a steady state operation and increases under transient conditions due to various uncertainties. In order to avoid a large amount of noise in the imbalance or flow difference during transients, the data is normally smoothed by means of moving averages. Different standard deviations can be used to properly take into account changing pipeline operating conditions.

In trying to catch departures from the mean value, any value that surpasses the mean plus a standard deviation, can be considered a value that is believed to be shifting. This implies that the minimum detectable leak size should be greater than the standard deviation. A minimum value for sigma is necessary because a standard deviation is zero or very small particularly during steady pipeline operations. For example, if the imbalance happened to be constant over the span of a window, then this would result in zero variability and the resulting values of the Wald sequential test would be plus or minus infinity, giving false information. In the actual implementation of this technique, several minimum standard deviations can help to achieve optimum reliability and sensitivity.

Because of errors in the measured and calculated values, the mean values are not always zero, but hover around the zero line. It is known also that in long term operations the instruments might introduce a bias, causing a shift, called instrument drift. The rate at which the bias affects the calculated imbalance or flow difference is much smaller than the rate at which a leak would make a difference in the mean value. Hence, in studying leak detection, statistical techniques involving the probabilistic measure of shifts in the mean value are used. The mean value can be corrected by reducing bias, which can increase leak detection sensitivity.

It is obvious that the instrumentation requirements for each are those implied in the method; flow meters are required if the SPRT technique is applied to a line balance and both flow and pressure measurements are required if it is applied to a volume balance.

The SPRT is a decision making tool using time series data produced by other means. It does not produce the time series data by itself for leak detection. The estimating of leak location also has to be provided by another method like a pressure gradient method.

The SPRT technique can be applied to any time series data. As shown in(b) Probability Ratio Changes Over Time

Figure 9 (a) the pipeline represented was in a transient condition. The probability ratio during the transients shown in (b) doesn't change significantly, but it increases beyond the threshold level when a leak is generated. However, it works best if the time series data is smooth with no anomalies. For example, imbalances are generally smoother than their corresponding flow differences during transient operations, and thus volume balance data is likely to produce better leak detection performance than line balance data.

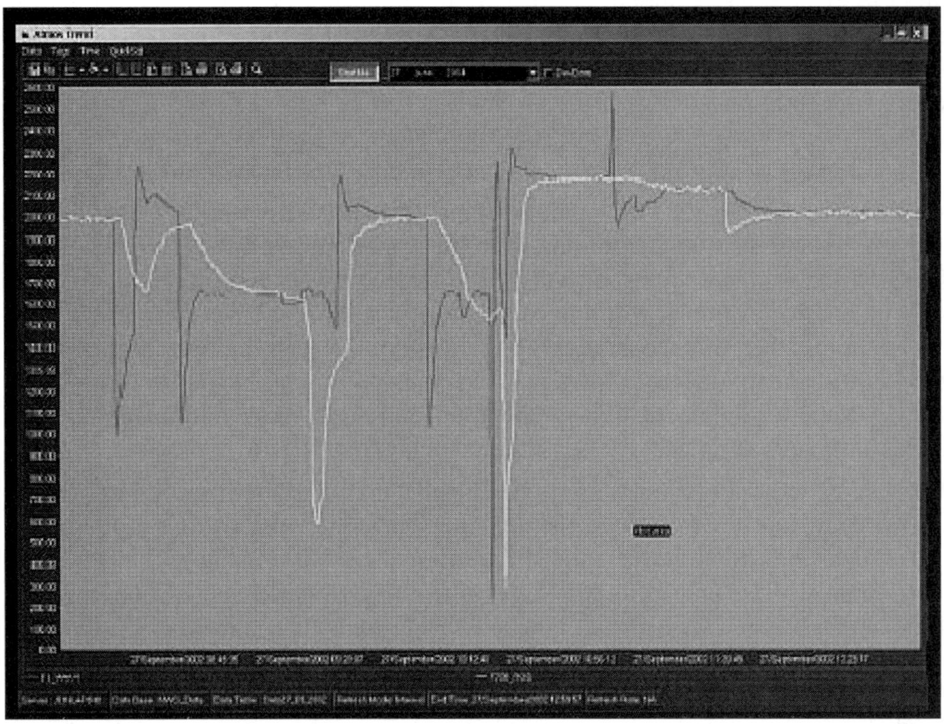

(a) Flow Rate Changes Over Time

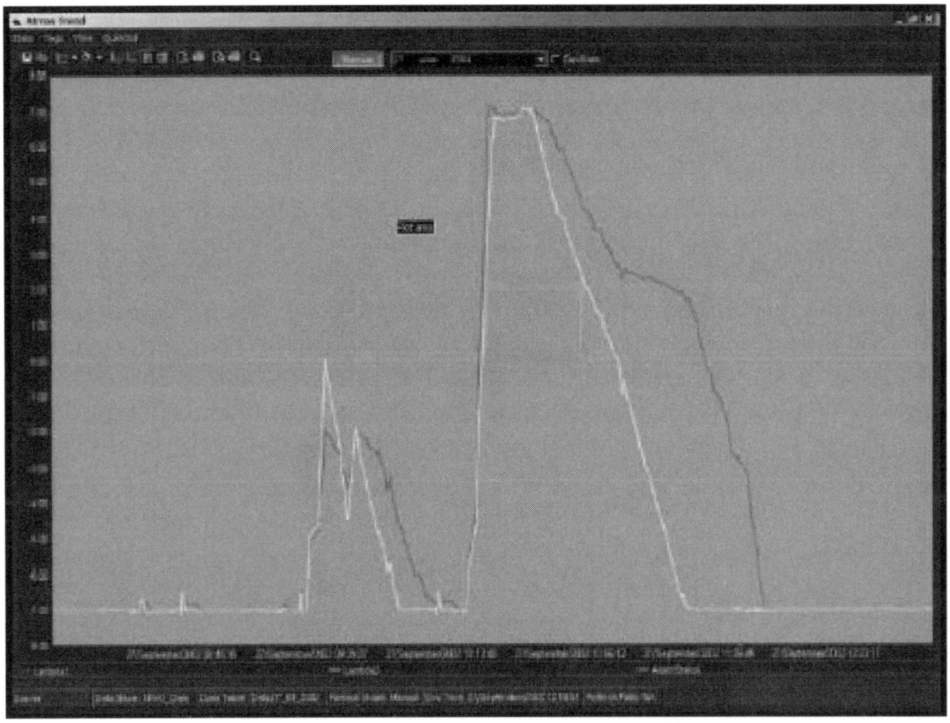

(b) Probability Ratio Changes Over Time

Figure 9 Example of SPRT Display

To achieve reliable and sensitive leak detection performance, test value such as imbalance data, should be reliable and the statistical parameters properly set during a tuning process. A sufficient amount of normal operational data must be analyzed in order to obtain the correct statistical tuning parameters. The tuning parameters include the number of time series data, probabilities that determine the thresholds, leak sizes to be detected with minimum standard deviation, and mean value correction.

The SPRT offers good fault detection capability including pipeline leak detection. The sequential probability ratio test expression includes the standard deviation and mean value terms that indicate variability of the incoming data and inherent measurement bias. Therefore, the equation automatically takes into account the pipeline operations in terms of changes in test values and bias correction. This technique responds to changes quickly, and if properly tuned, it can provide sensitive and reliable leak detection capability.

However, successful operation of the SPRT technique requires that the smooth time series data to be tested be reliable. Since it relies on other calculation methods for its test values such as volume imbalance, the selection of a proper imbalance calculation method is an important factor in achieving good leak detection performance. In general, the SPRT tends to use a lot of test data for proper trending analysis, and thus it may respond too slowly to respond to pipeline ruptures that require immediate leak detection and confirmation.

7.4.5.3 Bayesian Inference Technique

Another option is to use a Bayesian inference technique to make a leak/ no leak alarm decision (20). In other words, assuming known prior probabilities of no leak for a set of no leak patterns, the Bayesian inference technique applies the Bayes' rule to determine the probability of a no leak alarm condition. The same Bayes's rule is applied to a leak condition to determine the probability of a leak occurring. Simply described, this method tries to put certain measured and calculated values into leak and non-leak patterns in a probabilistic sense. If the results fit a leak pattern with a high potential for a leak occurring, a leak is confirmed. This is considered an emerging leak detection technique.

Initially, a large number of operation scenarios including leaks are simulated or alternatively past operational data is used off-line to obtain the leak and non-leak patterns. This data becomes the basis for patterns in the pipeline system. These patterns are then refined with actual operation data obtained while the system is operating in real-time. As a result, reliability and detection sensitivity can be improved as more operating data is accumulated over time and used to refine the pipeline state and probabilities.

A Bayesian inference method has been successfully applied for fault detection, and this pattern recognition technique, at least in theory, can be applied to any pipeline operation. For pipelines with a simple configuration, with no or one intermediate pump station, the probability of a leak and no leak condition may be easily determined. For complex pipeline systems, however, it is time-consuming to determine prior probabilities and establish pipeline operation patterns. Since the Bayesian inference technique needs to build an accurate probability database for almost all possible operations, extensive field and maintenance tuning efforts are required for reliable operation; this may take a long time to acquire for a complex pipeline system.

7.5 Factors Affecting Performance

To successfully implement and operate a real-time leak detection system, leak detection vendors and pipeline operators need to fully understand the factors affecting leak detection performance and their consequences and limitations. As well, leak detection software should be designed to take these contributing factors into account to improve leak detection capability and reduce spurious false alarms. Incorporating field instrumentation and SCADA characteristics correctly in the leak detection system is critical to ensure that a high level of performance is maintained. The issues described in the following can potentially affect leak detection system performance in terms of reliability, sensitivity and robustness (5). These factors are not applicable to pressure/flow monitoring or to acoustic/negative pressure wave techniques.

7.5.1 Pipeline Configuration Data

Pipeline configuration data such as pipe diameters, lengths and elevation profiles are usually well known. However, the following data are often not well defined:

- Soil conductivity
- Ground temperatures
- Pipe roughness

RTTM methodologies are particularly affected by these factors because these parameters can adversely affect the accuracy of the pressure and temperature calculations. Due to these uncertainties, pipe roughness and temperature measurements are often used as tuning parameters.

7.5.2 Measurement Data

The quality of real-time field measurements plays an important role in determining the level of leak detection performance. The measured values derived from SCADA contain any errors originating from both the primary measuring device and associated electronic devices. Most industrial grade instrumentation available today is suitable for use in leak detection applications. However, some instruments do not always behave the way they should due to:

- Bias error in flow measurement - Flow rate (volume) is reported higher/lower than it should be. The flow instrument and its electronic devices have to be re-calibrated along with associated measurements such as pressure, temperature, and composition. Equations of state need to be checked to ensure that they are not introducing any error.
- Random error in flow or pressure measurement - Random error within the instrument specification is acceptable. If the error exceeds the pre-defined limit, the instrument and its associated devices and values need to be checked.
- Flow or pressure measurement locking - The measured flow or pressure remains constant even under a transient condition. A stuck meter turbine or bearing, a power or communication outage, and other conditions can cause this problem.
- Erratic flow or pressure measurement - The measured flow or pressure bounces up and down beyond its repeatability error. This can be caused by vibration or by resonance phenomenon at a pump or pump station.
- Sudden flow or pressure changes - The measured flow rate or pressure

sometimes changes suddenly. If the corresponding pressure or flow does not change appropriately, the measured flow or pressure should be checked.

- Wrong temperature value - If the temperature sensors are improperly mounted, for example, they may be measuring the soil's rather than the fluid's temperature. The temperature sensors can be insulated and mounted in a thermo-well by equipping the pipe with a small flanged riser. This helps ensure accuracy without obstructing pigging operations.

7.5.3 Product Properties

Product density affects volume corrections at flow meters and line pack calculations. The effects of incorrect product compressibility information can be large during transient operations. Errors in product properties constitute the main cause of false alarms for methods using line pack calculations.

The product property problem can be significant in pipelines which transport gas and light hydrocarbon liquids, whose compositions vary significantly. The effect of this problem can be partly reduced by tracking the composition.

7.5.4 SCADA Factors

CPM systems normally receive field measurement data through the host SCADA, which collects and processes the measured data. The particular manner in which SCADA collects and processes the data can impact data resolution and reliability. SCADA related problems are listed below:

- Communication outage – CPM systems can generate a large error if a significant pipeline operation (e.g. transient, fluid delivery) takes place during a SCADA communication outage. This situation is easily identified and taking proper degradation procedures can minimize its effects. Degradation may occur if the required data or its quality is unavailable.
- Long scan time – The reported scan time is longer than expected. If a sudden operation such as a line shutdown takes place between the long scan cycle, the CPM cannot receive proper values and may result in error. This situation is also easily identified and taking proper degradation procedures can minimize effects.
- Wrong status reporting – SCADA can send the wrong status of a measurement to the CPM. The CPM can generate an error if sudden transients are introduced in this situation.
- Measurement override – When a pipeline condition changes, the overridden pressure or flow measurements do not change accordingly. Model results look similar to the locked pressure or flow measurements.

- Batch operation – Sometimes a wrong batch ID is passed by the SCADA to the CPM. Since some CPM methodologies use batch ID to correct injection flow rates and line pack changes, a false alarm condition can be generated with a wrong batch ID and thus wrong product properties.
- Improperly accounted delivery or injection – When a delivery or injection occurs, the volume accounting is delayed or not reported for a certain period. This case is similar to flow bias.

7.5.5 Operation-related Factors

Transient operations can cause more uncertainty in calculating line pack in the pipeline than steady state operations. This uncertainty results from errors in product properties, measurements and data sampling errors. Sometimes, unusual operational features can cause errors:

- Slack flow condition – A slack flow condition occurs whenever the pressure drops below the vaporization point of a fluid, which depends on the fluid's temperature. If the pipeline pressure drops below the vaporization point, vapor is formed around that area of the pipe. It is normally prevented by increasing the back pressure of the affected pipeline segment, unless the upstream pressure is low. Under slack flow conditions, the delivery flow is initially higher than the injection flow and the upstream and downstream pressures do not change significantly. In other words, the line pack change in the pipeline is smaller than the flow difference. When the slack condition starts to collapse, flow and pressure behave in opposite ways. Slack flow conditions are the most difficult situation for any CPM method to deal with. It is difficult to establish that slack flow exists, detect a leak during slack flow and to avoid false alarms under these conditions. The RTTM and modified volume balance methodologies can identify a slack flow condition, and may be able to provide information on its' condition in a display so that the pipeline operators can observe it and avoid unnecessary actions. The performance however cannot be accurate and reliable because the slack volume and its change cannot be accurately calculated.
- Product does not match specification or includes large amounts of BS&W – Just like a wrong batch ID, a product with different than expected properties or excessive amounts of BS&W can potentially result in excessive errors in flow correction and line pack changes.
- Sudden temperature changes – The measured temperature shows an ambient temperature when the temperature transducer is being serviced or when the fluid starts flowing from a shut-in condition, resulting in a temperature error. This will affect the accuracy of line pack calculation.

7.6 Performance Evaluation Methods

It is important to understand the level of expected performance from the leak detection system to be installed and to prepare proper leak detection specifications. API has published two standardized procedures for evaluating leak detection performance and preparing specifications for CPM systems: API 1149 and API 1155. API 1149, "Pipeline Variable Uncertainties and Their Effects on Leak Detectability," provides a theoretical way of estimating detectable leak sizes for mass balance based CPM methodologies, while API 1155, "Software Based Leak Detection Evaluation Methodology," provides a method of analyzing leak detection performance using actual or modeled operating data of the target pipeline.

7.6.1 API 1149

API 1149 is essentially an uncertainty analysis procedure, using physical parameters of the target pipeline and fluids. The API 1149 Procedure provides a theoretical estimate of leak detection performance by estimating the total uncertainty in mass balance with uncertainty analysis of individual parameters such as fluid property and instrumentation. It is applicable to a leak detection system using mass balance principle, and is specific to crude oil and refined products. It is assumed that all receipt and delivery points are metered, pressure and temperature are measured at both ends of the pipeline segment, and the pipeline is operated in a full flow condition.

The API 1149 expresses the maximum likely error in mass conservation in terms of maximum uncertainties in measured flows and line pack changes. Using a root sum square (RSS) process for the independent values, the minimum detectable leak size can be defined as

$$Q_l = Q_{in} - Q_{out} \succ dQ_m + \frac{dV}{dt}$$

where

Q_l = size of the minimum detectable leak

Q_{in} = flow rate into the pipeline

Q_{out} = flow rate into the pipeline

dQ_m = an upper bound of uncertainty in flow measurements

dV/dt = an upper bound of uncertainty in line pack change over a time interval dt.

In other words, a leak can be detected if the size of the leak is greater than the minimum detectable leak size.

Since each flow measurement is independent, a root mean square process can be applied to estimate the total likely uncertainty of the flow measurements, dQ_m. In other words, the total flow measurement uncertainty can be estimated from known individual flow measurement uncertainty as follows:

$$dQ_m = \sqrt{k_{in}^2 + k_{out}^2}$$

where k_{in} = inlet flow meter errors and k_{out} = outlet flow meter errors.

The total uncertainty of line pack change can be estimated from individual factors contributing to line pack change. The API 1149 Procedure takes into account the following factors:

- Pipe volume which depends on pressure and temperature,
- Fluid density in terms of pressure and temperature, using the API equation of state for crude and refined products, and
- Uncertainty in fluid density (which includes those in bulk modulus) and thermal expansion coefficient.

The total uncertainty of line pack change is expressed in two major uncertainties: pipe volume and fluid uncertainties due to pressure and temperature. Using the RSS procedure, line pack change can be expressed as

$$dV = \sqrt{2\sum_{i=1}^{n}(A_0 L_0)^2\left[\left(\frac{\partial I}{\partial P}dP\right)^2 + \left(\frac{\partial I}{\partial T}dT\right)^2\right]}$$

where

A_0 = area of pipe

L_0 = pipe length

I = ratio of the mass contained in the pipe segment to the mass at standard condition

n = number of pipe segment

The API Procedure simplifies these two uncertainties using the API equation of state and relationship between pipe, pressure and temperature. The uncertainty due to pressure is expressed as

$$\frac{\partial I}{\partial P} * 10^6 = a_0 + a_1 P + a_2 T + a_3 PT + a_4 T^2$$

where the coefficients $a_{o,}$ $a_{1,}$ a_2 $a_{3,}$ and a_4 are constants, and the uncertainty due to temperature as

$$\frac{\partial I}{\partial T} * 10^3 = b_0 + b_1 P + b_2 T + b_3 PT + b_4 T^2$$

where the coefficients $b_{o,}$ $b_{1,}$ b_2 $b_{3,}$ and b_4 are constants. These constants are listed in API 1149. They are classified in terms of products, their API gravity, and the ratio of outside pipe diameter to wall thickness.

Combining the uncertainties in measured flows with those in line pack changes, the total uncertainty in mass balance is expressed as:

$$\frac{Q_l}{Q_{ref}} \geq \sqrt{\left(k_{in}^2 + k_{out}^2\right) + \left(\frac{dV}{dtQ_{ref}}\right)^2}$$

where Q_{ref} is a reference or maximum design flow rate. This ratio can be considered as the minimum detectable leak over a time window *dt*.

Note that the above equation is a function of time. This equation shows that the minimum detectable leak size is largely influenced by the uncertainty in line pack change for a short time interval, while the long-term minimum detectable leak size is determined by the uncertainty of the measured flows. It is assumed that the procedure is based on a steady state assumption. The API 1149 does not provide a quantitative estimation procedure for transient operations.

It is clear from the above equation that the calculation procedure takes as input pipe volume with pipe diameter and wall thickness, product group and gravity, flow rate with average operating pressure and temperature, as well as uncertainties of flow, pressure and temperature measurements. Refer to API 1149 for detailed procedures and required coefficients for performance estimation.

7.6.2 API 1155

The API 1149 procedure helps initially in estimating leak detectability if the target pipeline is new or actual pipeline operating data is unavailable. The API 1155 procedure provides a way to estimate the level of performance that can be expected from a CPM system, mostly for existing pipelines. The objectives of performance study using the API 1155 procedure are not only to determine realistic leak detection performance for the pipeline system, but also to improve operator confidence and system reliability. Particularly, it helps operating companies to select a leak detection vendor objectively.

The API 1155 procedure is a way of standardizing data collection and performance metrics. It defines a standard pipeline data file format, provides pipeline configuration file definition, and helps to specify performance criteria and identify extra features and functions. API 1155 recommends the following six step procedure for evaluation of a software based leak detection system:

- Step 1 – Gather information and define the physical pipeline
- Step 2 – Collect data samples and build case files
- Step 3 – Specify performance metrics
- Step 4 – Transmit information to vendors for evaluation
- Step 5 – Have vendors perform data analysis
- Step 6 – Interpret vendor results

Step 1 is time-consuming but intended to provide leak detection vendors with a complete physical description of the target pipeline or system of pipelines in a standard format. The amount of data described in API 1155 for this step is quite extensive. This is because it is meant to cover all possible operating cases. Before collecting all the data outlined in API 1155, it is recommended that the operating company check with the vendors to find out which data they require and which data does not have to be collected.

Step 2 is a key step for this study in order to provide potential leak detection vendors with a realistic snapshot of pipeline operation collected from the host SCADA system. The operating data shall include all major operating scenarios, batch product identification data with batch launch signal, and possibly simulated leak data.

Step 3 is a procedure for specifying leak detection performance. In determining performance metrics, regulatory requirements and operational experience need to be taken into account. The pipeline company may use the sample layout for specification and ranking listed in the document, which defines the performance in terms of sensitivity, reliability, robustness and accuracy. However, the pipeline company should understand that it is difficult to quantify certain values with a limited amount of collected data.

Step 4: The collected pipeline configuration and operating data with performance metrics are sent to all potential vendors that show a willingness to participate in the performance study.

Step 5: Potential vendors analyze the collected pipeline data. During the evaluation process, the pipeline company should expect to clarify incomplete data or operational aspects that could be unclear to the vendors. The vendors prepare a report describing the results of their study. API 1155 describes the vendor report format, including recommendations and performance with respect to the expectations of the pipeline company. The pipeline company needs to discuss the results with participating vendors to ensure they are clearly understood.

Step 6 is the interpretation and comparison of vendor results with the performance criteria specified in Step 3. In addition, other evaluation criteria such as cost, ease of use, and user support need to be considered.

API 1155 provides the pipeline company and leak detection vendors with a framework to evaluate leak detection performance based on operating data of the target pipeline. However, it is costly to the pipeline company, particularly if many vendors participate in the study, and the amount of work required of vendors is significant as well. It should be noted that the document should not be considered as a vendor selection and contracting tool because it doesn't address project requirement issues.

7.7 Implementation Requirements

7.7.1 Instrumentation Specifications

As discussed in previous sections, having adequate instrumentation, particularly for flow and pressure measurements, is critical to the implementation of an effective CPM system. This is because all CPM methods rely on the instrument readings for pipeline monitoring and leak detection; for optimum performance, it is important that the instrumentation be consistent with leak detection requirements.

The instrumentation can be specified in terms of accuracy, repeatability, and precision or resolution. Quite often, instrument manufacturers provide these instrument specifications for the primary devices. Since a CPM system receives the measured data via a SCADA system, a more accurate measurement error should include the extra errors introduced by auxiliary quantities such as equation of state and by secondary devices such as the RTU. The secondary device errors are caused by current-voltage conversion, signal amplification and analog-to-digital conversion. Therefore, the measurement accuracy attainable at the host SCADA should be used to estimate a minimum detectable size using the API 1149 procedures or other estimation method.

7.7.2 SCADA Requirements

A real-time leak detection system is closely integrated with the host SCADA. The final system architecture will depend on the redundancy requirements for the SCADA system and may be one of the following:

- Single SCADA computer and a single leak detection system computer configuration: Both SCADA and leak detection systems can reside in a single computer if the number of data points is small. This configuration is seldom used even for simple, non-critical pipeline systems.

- Dual-redundant SCADA and a single leak detection system computer configuration: The SCADA system is redundant but the leak detection system is not, requiring a separate computer for the leak detection system. This configuration is suitable if the leak detection function is not considered a critical part of pipeline operations.
- Dual-redundant SCADA and dual-redundant leak detection system computer configuration: If the leak detection system is deemed to be a mission critical application, four computers can be arranged in dual SCADA computers and dual leak detection system computers. This configuration ensures high reliability and availability for both the SCADA and leak detection systems. The leak detection system will have dual redundant servers with hot-standby and automatic failover. The leak detection failover arrangement is analogous to the SCADA host failover arrangement; if the active leak detection CPU fails, or is determined to be defective by the standby leak detection CPU, the standby leak detection CPU shall automatically assume the role of the active CPU.
- Dual-redundant SCADA/leak detection computer configuration: The leak detection system can run on each of the SCADA computers, sharing machine resources with the SCADA system. If the leak detection software is simple or SCADA computer capacity is large, this configuration is cost effective. In addition, redundancy of both SCADA and leak detection systems can be maintained with a minimum amount of hardware cost. However, ongoing maintenance of the leak detection system can impact the reliability of the SCADA system.

All SCADA systems have certain properties and uncertainties. The key properties are scan time and time skew; the data uncertainties are due to the dead-band and report-by-exception, which should be minimized. To a certain extent, the shorter the scan time and the smaller the dead band, the better the leak detection performance can be.

7.7.3 SCADA Interface

All CPM systems must interface with a host SCADA system. The interface allows all field data used and data generated by the CPM to be exchanged with the SCADA system. Refer to Chapter 6 for a detailed discussion of the SCADA interface.

The leak detection system interface with the host SCADA system is defined in a Interface Control Document (ICD). The interface requirements should be clearly defined in the ICD at the beginning of the implementation project. A clearly defined ICD is even more critical if the SCADA and leak detection systems are from different vendors. The ICD defines the communication protocol, system control and synchronization, and SCADA data and time stamp. The system control definition includes the startup of functions and data transfer including the connection with the hot SCADA server. The ICD also defines the mechanism of synchronizing the SCADA and leak detection time clocks.

7.7.4 Commissioning and Tuning Tasks

The commissioning and tuning tasks start after the CPM system is installed on site. Upon the successful completion of the FAT, the CPM hardware and software are shipped to the operating site. Integration with the SCADA should be checked, and the field data quality and compatibility with the installed CPM examined. The following tasks are usually performed (the tasks listed below are not necessarily required for all the CPM methodologies):

1. Check the SCADA functions and interface
    - The SCADA functions relevant to the integrated CPM system include scan time, regular and irregular polling, time stamp and skew, and report-by-exception.
    - The SCADA-CPM interface checks include protocol, synchronization, measurement points with their tag names, and point-by-point check of data between the SCADA and CPM databases.
2. Check the following instrumentation for availability, accuracy, repeatability and behaviors:
    - Flow measurement
    - Pressure measurement
    - Temperature measurement
    - Density measurement for batch operation
    - Consequences of missing measurements
3. Check pipeline operations
    - Check the pressure and flow behaviors during both batch launch and movement and lifting and delivery operations in regard to the CPM.
    - Check pressure and flow behaviors during pump or valve operations involving the CPM.
    - Check the pressure and flow behaviors with regard to the CPM when a surge valve is opened.
    - If there are unique operations for the target pipeline, their pressure and flow behaviors should be analyzed with those of the CPM.
4. Check product properties
    - Compare short-term flow differences with their corresponding line pack changes under various transient operational conditions to check the

accuracy of the equations of state, particularly bulk modulus for heavy hydrocarbon fluids.

5. Tune the CPM
    - The tuning process can differ significantly with different CPMs. For example, an RTTM requires significant tuning efforts while those required for a line balance method are minimal.
    - Following the API 1155 procedure, the leak detection thresholds and other parameters are finalized during the tuning period.

7.7.5 Acceptance Tests

Normally, three different acceptance tests are performed:

- Factory Acceptance Test (FAT)
- Site Acceptance Test (SAT)
- Operational Availability Test

All three tests are performed with representatives from the leak detection system vendor and the pipeline's operators in attendance. Each 'test' is a series of structured and unstructured tests. A structured test is conducted in accordance with the pre-approved test procedures, and unstructured tests are performed by the pipeline operator to demonstrate any concerns not addressed or satisfied in the structured testing. Each test is considered successfully completed only when the system has passed all structured and unstructured test items.

1. Factory Acceptance Test (FAT)

The main purpose of the FAT is to verify in a factory condition that all functions have been properly implemented and operate as specified in the statement of work (SOW). The FAT procedures include testing all CPM functions, features and capabilities specified in the SOW. Discrepancies found during the FAT are documented and maintained in the FAT report, and the subsequent corrections are documented and demonstrated by the CPM developer for the operator's approval.

The leak detection system operator should perform the unstructured part of the FAT after the structured test has been successfully completed and all discrepancies corrected. The CPM system software and hardware should be shipped only after both parts of the FAT have been successfully completed and approved of by the leak detection system operator.

2. Site Acceptance Test (SAT)

The main purpose of the SAT is to confirm that all the system's functions and features perform satisfactorily under actual operating conditions. These tests are performed after the CPM system hardware and software have been integrated with the host SCADA and have been commissioned and tuned on site. It is recommended that the SAT be performed after the integrated system has been shown to operate successfully in the field environment for a length of time sufficient to demonstrate its' efficiency.

The SAT should use the real-time data received from the host SCADA and verify that the CPM system functions in all possible operating scenarios. The SAT is conducted using the SAT procedures - similar to those of FAT except that they are also conducted in field conditions. The leak detection operator needs to perform a series of unstructured tests after the structured part of the SAT has been successfully completed and all deficiencies corrected.

3. Operational Availability Test (OAT)

The main purpose of the Operational Availability Test is to confirm that the system performs consistently for a prolonged period while covering all operating conditions in the actual field environment. This test is performed after the successful completion of the SAT and without special test procedures. During the test period, the Owner performs the actual tests including leak tests.

After a specified period of continuous testing, the test records are examined to determine conformity with the performance and availability acceptance criteria. If the test criteria have not been satisfied, the necessary corrections have to be made and the testing continues until the criteria have been met.

7.8 User Interface

The operations staff identifies and analyzes pipeline operation problems via the user interfaces. All CPM systems will generate false alarms with varying frequency. The operations staff has to confirm the leak alarm and make the final decision to isolate and shut down the pipeline in accordance with the pipeline's operating and alarm conditions. A proper response must be made quickly when an emergency such as a leak occurs. Therefore, it is critical to have accurate and timely information in an easy-to-interpret format.

The user interfaces should be consistent for all SCADA and CPM information, implying that the CPM system provides the same "look and feel" across all operator screens. The recommendations API 1130 suggests on building displays are summarized below:

- Displays need to be simple, easy to use and read, and arranged in a systematic way to facilitate easy access of the required information.
- The information presented on the user interfaces should be relevant for diagnosing problems easily.
- The users should be involved in the design of the CPM system so that the pipeline operations staff are satisfied with the layout and design.

The display requirements are different for different CPM methodologies and integration levels between the host SCADA and CPM. If the CPM is tightly integrated, almost all CPM data can be displayed on the SCADA screens and the information from the CPM made an integral part of the pipeline operation. On the other hand, only small sets of key data from the CPM, such as key alarm messages, can be passed to the SCADA system. The following displays are a required minimum:

7.8.1 Alarm Message Display

Alarm messages are critical information that the operations staff must pay attention to. It is strongly recommended to display alarm messages including leak detection alarms on the SCADA alarm display screens. The following features and qualities should be part of the alarm displays:

- Consistent with SCADA system alarms and have an appropriate priority.
- Have different colours for each category of alarm.

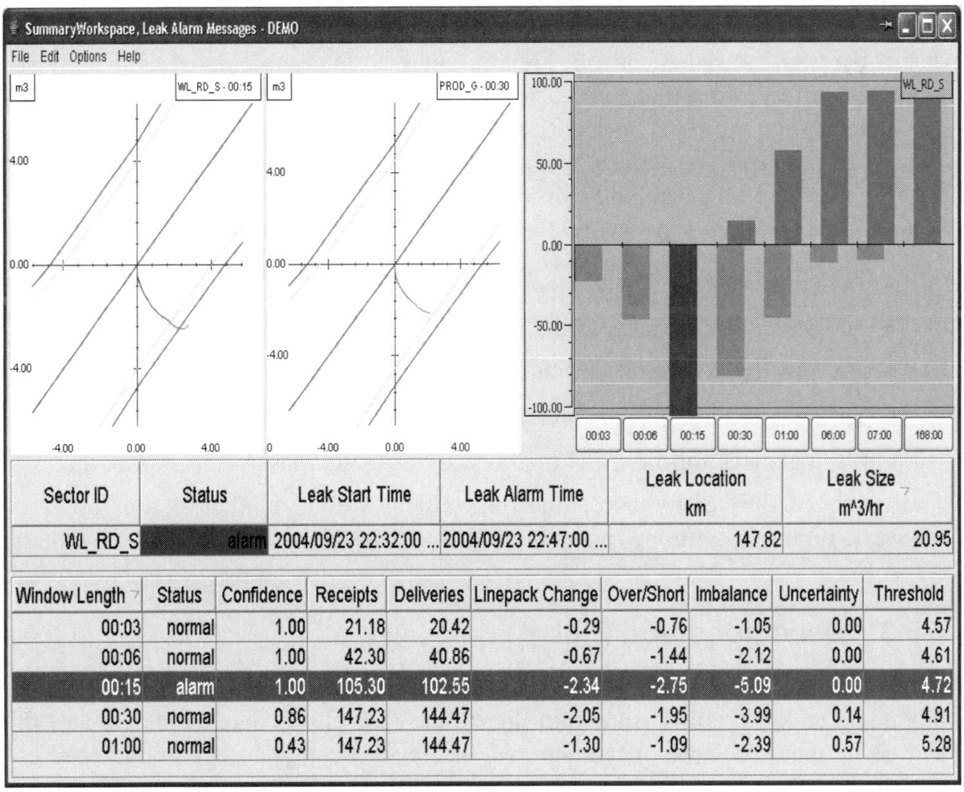

Figure 10 Example Display of Leak alarm Messages (Courtesy of CriticalControl)

- Acknowledged and unacknowledged alarms should be accessible to the pipeline operator in one step. Acknowledged alarms still in the alarm state should remain readily available to the pipeline operator.

- Have a time stamp as part of the displayed alarm.
- Should have both audible and visual cues. Each alarm should have a unique audible tone. Visual cues for any given alarm should persist for a long enough period of time so as not to be overwritten irrevocably by newer alarms.
- Not easily be made defeated, or inhibited without just cause. The use of screen savers or any other screen blanking is strongly discouraged.

An example display of leak alarm messages is shown in Figure 10. It shows the leak alarm status, estimated leak location and size, and other information that helps the operator to quickly identify the potential problem.

7.8.2 Trend Displays

Trending measured and calculated values of the SCADA and CPM system help determine what caused an alarm. Trending may be in graphical and tabular forms: Graphical presentation makes it easier to identify anomalies. The tabular form is useful for analyzing data in detail. API 1130 suggests that a trend cover a long enough duration to see values before a CPM alarm occurred and continue right through to when the alarm ends, or the current time. The following values need to be trended:

- Measured pressures and temperatures
- Measured densities, particularly for batch pipelines
- Measured flow rates and their differences between inlet and outlet flows
- Calculated line pack changes if they are made available
- Imbalances for the CPM methodologies using mass balance principle.

The trends of flow differences, line pack changes and imbalances are shown in Figure 1, Figure 2, Figure 3, and Figure 4 for LB, VB, MVB, and CMB methods, respectively.

7.8.3 Decomposition Plot

Since the imbalance consists of flow difference and line pack change, they can be plotted on an x-y graph as shown in Figure 11. This plot gives an indication of the degree of metered flow difference and line pack change in a pipeline segment bounded by flow meters. It also provides various operating patterns as to the cause of imbalance in the segment over a specific period of time. For example, a flow meter may have a problem if the imbalances were changed by flow differences, while line pack calculations may be wrong if there were no accompanying flow difference. The following figure is an example of a plot, which illustrates the display's features.

The vertical axis represents the change in line pack and the horizontal axis represents the difference in the metered in and metered out volumes. When the pipeline is running in a steady-state condition (i.e., $V_{out} - V_{in} = \Delta LP = 0$), the plotted point should rest at or near the origin. Any operations in the pipeline will appear as a plotted line or locus that starts from the center point of the graph. Normal packing and unpacking in the sector will follow the diagonal line and cross through the point of origin. The locus is a form of trend (a series of connected points plotted over time). Each point represents the accumulation of metered over/short versus accumulation of change in line pack at a particular moment during the window time period.

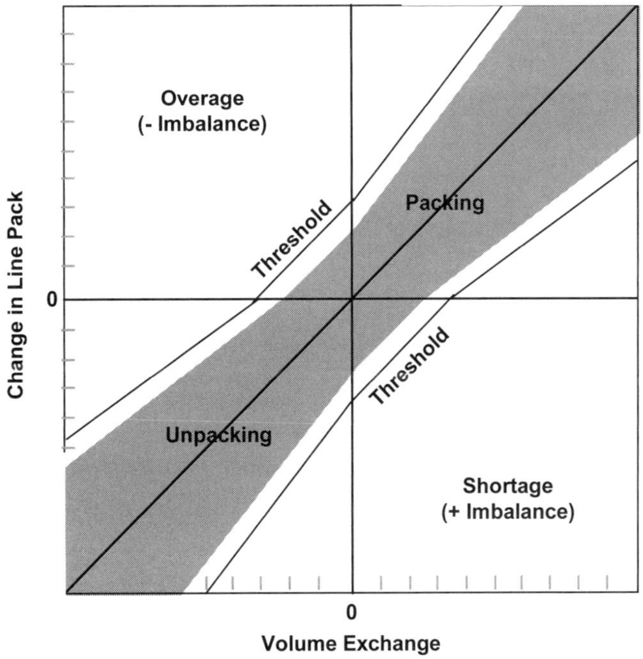

Figure 11 Example of Decomposition Plot (Courtesy of CriticalControl)

The shaded area indicates the allowable bandwidth of residual imbalance (e.g., error due to measurement anomalies, product definition, computational error, etc). While the locus remains in the shaded area, the pipeline is considered to be within normal levels of imbalance. If it crosses from the shaded area into the white, a potential overage or shortage situation exists. On each side of the normal diagonal, a warning threshold bounds the allowable bandwidth. An imbalance exists once the locus crosses the threshold lines.

7.8.4 CPM Operational Status

The CPM operational status should be clearly displayed on the SCADA screen to indicate to the operator the system's reliability and robustness. A CPM system is fully enabled when all the measurements and the integrated SCADA-CPM system are in good working condition. The real-time measurements can sometimes be unreliable or unavailable due to such causes as communication outages or instrument malfunctions. Also, the integrated CPM system may not function to its full capacity when a system hardware or software problem occurs. When some measurements are not available or the system does not function fully, the leak detection system is not fully reliable. Therefore, the CPM system should take into account such problems to minimize the false alarms and warn operators of limitations of the leak detection system.

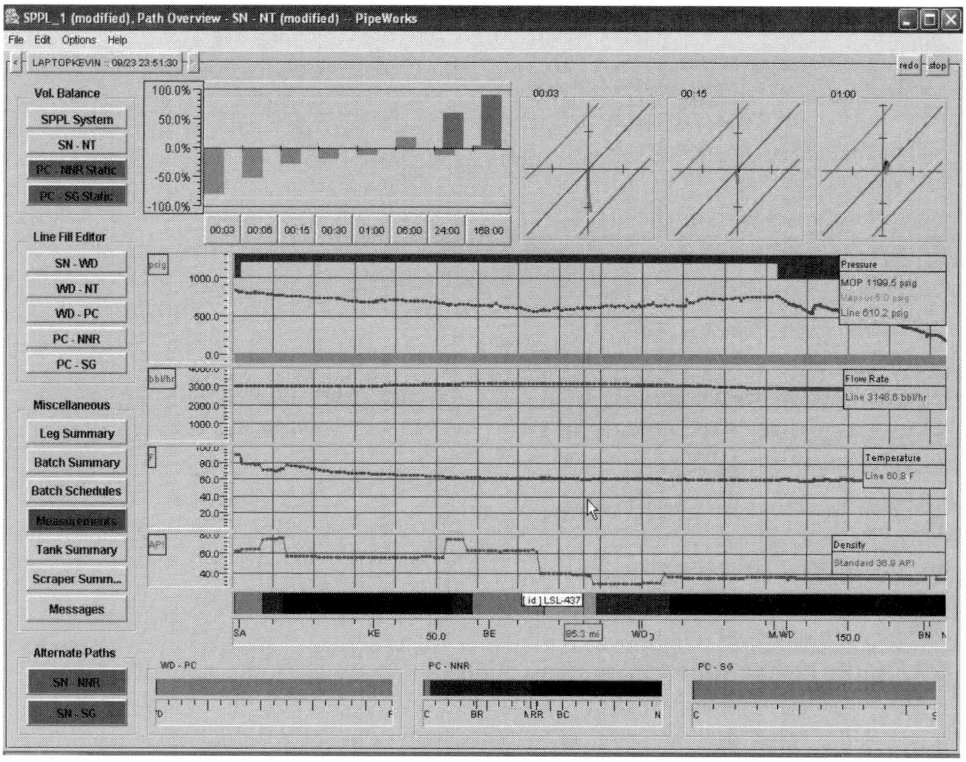

Figure 12 Example of Operation Status (Courtesy of CriticalControl)

In general, pipeline states during transient operations are not as well known as during steady state operations. Although RTTM models can significantly reduce uncertainties in the pipeline states during transient conditions, there are many factors the model has no control over. Because of this, most CPM methodologies incorporate an algorithm to degrade the performance in order to reduce false alarms during transient operations. Some methods apply dynamic thresholds to account for the increased uncertainties that exist during transient conditions caused by operations. Transients created by leaks should not increase thresholds significantly since the leak signals should be distinguishable from operating transients so leaks can be detected during transient operations.

7.8.5 Pipeline Map

A pipeline map is required to efficiently execute emergency response procedures. The map gives detailed information about the names and contact numbers of responsible parties, pipeline route and terrain, population close to the pipeline route, responsible officials including police, critical environmental concerns, etc.

In addition to these two displays, the following information may help the operator to identify and diagnose anomalies.

- Leak detection alarms with their associated diagnostic information such as mass balance
- Leak size and location estimates if they are available
- Data logging displays showing alarm and event messages produced by the SCADA and CPM
- Measured and modelled flow and pressure profiles if available
- Density profile and batch tracking information for batch pipelines

7.9 Operational Considerations and Emergency Responses

7.9.1 Operator Training and Manual

Effective operation of a CPM system requires a thorough understanding of the system installation, operation, capabilities and maintenance. The pipeline operations staff must have extensive training including practical on-line operating experience. Emphasis is placed on how to operate the system effectively and how to analyze the results accurately. The following training courses are a minimum requirement for operations staff:

1. System Operation

This course is for pipeline operators to give them the skills necessary to effectively monitor the CPM system and diagnose anomalies. It covers user interface, display organization, alarms and acknowledgement, data logging, and hydraulic profiling if available. The course must be taken by personnel responsible for operating the host SCADA and managing the CPM system. The pipeline operating staff should also be fully familiar with emergency response procedures required for use of a CPM.

2. System Administration

This course must be taken by pipeline system engineers to give them the skills necessary to effectively manage and maintain the CPM system. It covers system start-up and shut-down, configuration and threshold changes, database configuration and management, hardware set-up, software components, system security, and system performance monitoring.

3. Leak Detection Manual

CSA Z662 Annex E recommends that operating companies have a leak detection manual readily available for reference by those employees responsible for leak detection on the pipeline. It suggests that the manual contains the following information:

- A system map, profile, and detailed physical description of each pipeline segment
- A summary of the characteristics of each product transported
- A tabulation of the measurement devices used in the leak detection procedure for each pipeline segment and a description of how the data is gathered
- A list of special considerations or step-by-step procedures to be used in evaluating leak detection results
- Details of the expected performance of the leak detection system under normal and line upset conditions
- The effects of system degradation on the leak detection results

The RTTM and modified volume balance users need to understand basic concepts of the technology to use the system effectively. In addition, system engineers may need a system maintenance course.

7.9.2 CPM System Testing

The CPM system operating company needs to define leak test policy - including test methods, test frequency and periodic testing. API 1130 recommends the following testing practices:

1. Test methods – Possible methods of testing include:
    - Removal of test quantities of commodity from the line,
    - Editing of CPM configuration parameters to simulate commodity loss,
    - Altering an instrument output to simulate a volume imbalance or a

pressure output to simulate a hydraulic anomaly, or

- Using archived data from a real leak.

CPM tests may be "announced" or "unannounced." An unannounced test is started without the knowledge of the pipeline operator and intended to test the proper functioning of the CPM system as well as the response of the pipeline operator.

2. Test Frequency – Each company is responsible for establishing test frequency.
3. Leak Rates – It is recommended to test using multiple leak rates to assess the system's overall ability to detect leaks.
4. Initial tests – Initial tests a site acceptance test.
5. Retesting – API 1130 recommends the CPM system be tested every 5 years for continued effectiveness, while CSA Z662 Annex E recommends annual leak tests. Some pipeline companies perform unannounced tests regularly not only to check the performance of the CPM system but also to test whether operators follow the company's emergency response procedure.

7.9.3 Emergency Response Procedure

Effective emergency response is one of the key tasks for mitigating leaks when a leak is detected. Emergency response procedures must not only be clearly written but understood and practiced by pipeline operating staff. The procedures may provide for the following:

- Emergency response policy
- Leak confirmation and isolation procedures
- Notification to management, emergency response teams and clean-up/repair crews
- Notification to responsible officials including police
- Management of media and the public
- Local one call system support

7.9.4 Record Keeping and Archiving Data

API 1130 recommends keeping design records, software changes and test records, and specifies the record retention length. Records of tests should include:

- Date, time and duration of the test
- Method, location and description of the commodity withdrawal
- Operating conditions at the time of the test
- Analysis of the performance of the CPM system and, for tests, the

effectiveness of the response by operating personnel

- Documentation of corrective measures taken or mitigated as a result of the test
- SCADA data generated during the test.

It also recommends that records detailing the initial or retest results should be retained until the next test.

CSA Z662 Annex E specifies test record requirements similar to the above. It also recommends that the pipeline companies keep a record of daily, weekly, and monthly material balance results for a minimum period of six months, and that this data be readily available and reviewed for evidence of small shortages below established tolerances.

7.9.5 System Maintenance

To maximize performance of the implemented CPM system, operating companies need to establish a procedure and schedule for maintaining all instruments, communication tools, and hardware and software that affect the leak detection system. API 1130 describes several aspects of a system maintenance and support program. CSA Z662 Annex E recommends internal audits to monitor performance of the leak detection system and if necessary to correct performance degradation.

7.10 Summary

A leak detection system is a tool for mitigating the consequences associated with a leak by fast detection and accurate location. If a leak detection system is effective, it can be good insurance for reducing risks. An appropriate leak detection system can help pipeline companies protect the environment and/or operate their pipeline systems safely.

A SCADA system, integrating computer, instrumentation and communication technologies, is an integral part of daily pipeline operations, particularly real-time operations. The CPM methods of leak detection take advantage of real-time capability and the effectiveness of the SCADA system as a monitoring and controlling tool. As the historical data indicates, the current CPM technologies are far from satisfactory in their performance. They need further improvement in their reliability and leak detection sensitivity. Also, a single CPM system may not satisfy all the criteria of an effective leak detection system. Combining a few CPM methodologies however, can meet not only most regulatory requirements but also effectiveness criteria.

In summary, the current CPM methodologies can satisfy the regulatory requirements but need improvement in performance, particularly leak detection sensitivity. Leak detection performance under transient conditions or when the quality of real-time data used is questionable, remains as a problem area.

Performance can be improved by introducing advanced process control techniques including statistical fault detection methods. At present, a CPM-based leak detection system can be augmented by a sensing technology like fiber optic cable in order to improve reliability and sensitivity. It should be emphasized that CPM systems are not designed to replace pipeline dispatchers but to assist them in detecting pipeline hydraulic anomalies (21).

References

(1) Yoon, M.S., and Yurcevich, J., "A Study of the Pipeline Leak Detection Technology", 1985, Contract No. 05583-00106, Government of Canada

(2) "Computational Pipeline Monitoring", API Publication 1130, 2nd Edition, American Petroleum Institute, 2002

(3) Mastandrea, J. R., "Petroleum Pipeline Leak Detection Study", 1982, EPA-600/2-82-040, PD 83-172478

(4) Yoon, M.S., Mensik, M. and Luk, W.Y., "Spillage Minimization Through Real-Time Leak Detection", Proceedings of OMAE Conference, ASME, 1988

(5) Yoon, M.S., Jacobs, G.B. and Young, B.E., "Leak Detection Performance Specification", Proceedings of ETCE Conference, ASME, 1991

(6) Nagala, D.W. and Vanelli, J.C., "An Evaluation Methodology for Software Based Leak Detection Systems", API Cybernetics Symposium, 1994

(7) Luopa, J.A., "Design and Performance of a Material Balance Leak Detection System With a Lumped Parameter Pipeline Model", Proceedings of OMAE, ASME, 1993

(8) Mactaggart, R. H. and Hagar, K, "Controller 2000", Proceedings of International Pipeline Conference, ASME, 1998

(9) Seiders, E.J., "Hydraulic Gradient Eyed in Leak Location", OGJ, Nov. 19, 1979

(10) Nicholas, R.E., "Leak Detection By Model Compensated Volume Balance", Proceedings of ETCE Conference, ASME, 1987

(11) Blackadar, D.C. and Massinon, R.V.J, "Implementation of a Real time Transient Model for a Batched Pipeline Network", PSIG, 1987

(12) Fukushima, K., et al, "Gas Pipeline Leak Detection System Using the Online Simulation Method", Computers & Chemical Engineering, 2000

(13) Vanelli, J.C. and Lindsey, T.P., "Real-Time Modeling and Applications in Pipeline Measurement and Control", IEEE Petroleum and Chemical Industry

Conf, 1981

(14) Dupont, T.F., et al, "A Transient Remote Integrity Monitor for Pipelines Using Standard SCADA Measurements", INTERPIPE conf., 1980

(15) Wade, W.R. and Rachford, H.H., "Detecting Leaks in Pipelines Using Standard SCADA Information", Pipeline Industry, Dec. 1987 and Jan. 1988

(16) "Leak and Shock Acoustic Detection System", Private communication with 01DB Acoustics and Vibration, France

(17) Baptista, R.M. and Moura, C.H.W, "Leak Detection System for Multi-Phase Flow – Moving Forward", Proceedings of IPC, ASME, 2002

(18) Whaley, R.S., et al, "Tutorial on Software Based Leak Detection Techniques", PSIG, 1992

(19) Beushausen, R., et al, "Transient Leak Detection in Crude Oil Pipelines", Proceedings of International Pipeline Conference, ASME, 2004

(20) Private communication with a vendor

(21) Scott, D.M., "Implementing CPM Systems", Proceedings of International Pipeline Conference, ASME, 1996

8 Geographic Information Systems

This chapter highlights some innovations introduced to the pipeline industry through Geographic Information Systems (GIS). GIS has shown to be effective in both the pipeline design as well as operations part of the lifecycle. We begin with the concept of Spatial Data Management, which serves as the cornerstone of any robust GIS by thoroughly guiding the planning, implementation, policies, standards and practices for the acquisition, storage, and retrieval of spatial information. The following section looks at some GIS enabled tools that have enabled pipeline engineers and other designers to automate their tasks or define new ways of working. This is followed by a discussion on how GIS supports the fulfillment of regulatory requirements. Finally, a summary is presented as a discussion on *The Central Database Paradigm*, the future path of data-centric engineering.

8.1 Introduction

The use of Geographic Information Systems (GIS) has been steadily growing in recent years as the amount of digital data continues to grow, becomes cheaper, and is easier to access and store. A GIS is essentially composed of hardware, software, data, standards, and processes to manage, analyze, and display all forms of geospatial (i.e. geographically referenced) information.

A map is an excellent metaphor to working in a GIS environment. In a GIS, a map is not a drawing, but rather a collection of features overlaid onto each other correlated by their geographic location. These layers are stored as separate data entities and can be used in any combination to create different maps, automatically generate alignment sheets, route a pipeline, model hydraulics, or assess pipeline risk factors. Rather than constantly updating drawings, we update the data on which our drawings are based.

Pipelines lend themselves particularly well to the geospatial world since they tend to cover long linear geographic distances. As a result, the unique innovation GIS has introduced to the pipeline world is the ability to manage large amounts of data based on its geographic location. Effectively managing data has enabled us to store data securely, manage change, control access, ensure integrity, and ultimately centralize it giving you *the single source of truth.*

Over the last decade, the pipeline industry has begun to adopt GIS for green field projects and pipeline operations and integrity management. In most organizations, this serves to provide basic GIS support (mapping, modeling, and spatial analysis, generally at the operational level) and to support engineers and managers through information transfer and communication. As the cost of infrastructure, hardware, and software has steadily declined and the financial benefits of using GIS realized, owner/operators have slowly accepted GIS into their engineering work practices and overall projects.

The more notable benefits of GIS in pipeline engineering include:

- Reduced engineering cost through use of more desk top exercises and by harnessing the immense computing power of the GIS;
- Greater safety in reducing field visits by enabling more desktop studies;
- A faster pace to the engineering process because of the availability of accurate, up-to-date information, and the ease with which it can be communicated;
- When data is made available early in the engineering process, increased accuracy and reliability in cost and schedule estimates are realized;
- The fundamental ability of a GIS to communicate information over vast distances to disparate groups;
- Ability to ensure all disciplines are on the same page by maintaining a single source of truth on project data;
- Reduced costs for certain tasks through automated report generation (e.g. Alignment Sheet Generators).

8.2 Spatial Data Management

Geospatial information is an essential asset of pipeline owners and operators. Hydrocarbon energy pipeline companies are increasingly generating and utilizing huge volumes of spatial data to support their business decision-making, whether assessing potential pipeline routes, determining integrity and maintenance programs, or predicting asset lifecycles. Furthermore, detailed records of the location of all pipeline and facility assets, as well as related land base, geographic features, environmental and socio-economic factors are all required to satisfy regulatory reporting requirements, to ensure efficient operation, and to facilitate design and construction cost savings. Ultimately however, it is not the quantity of data a company possesses that increases its competitive advantage but quality of that data, and its ready accessibility to those users who transform it into valuable corporate business intelligence.

Management of geospatial information assets is an evolving field of practice. Historically pipeline companies have collected this type of information from a range of sources, using a range of technologies, and storing it in various locations throughout the organization. Consequently, the evolution of geospatial data within hydrocarbon energy pipeline organizations has seen growth and strength in those areas where spatial information is created as a part of daily business activities. All the while information architecture to support spatial data has traditionally been application-centric. This, in combination with the department-level origin of information, has resulted in the emergence of silos of well-defined, yet disparate, spatial data sets. This isolated information environment is effective for those business units who have access to the data silos, however, this structure makes it virtually impossible to share information between departments and with other user-groups who might greatly benefit from access. As a result, the same data is being duplicated many times because it is stored in more than one location.

Pipeline owners and operators are now beginning to see that there is a huge increment of unrealized value lying dormant throughout their organizations, just waiting to be leveraged corporately through improved data management practices (1).

Our more detailed understanding of today's data management issues, and the enhanced information systems technologies available, make implementing a comprehensive spatial data center a fairly straightforward process. However, making the transition to a consistent, coordinated and centralized geospatial data center requires the development of a solid plan of action. This section provides a review of standards, infrastructure, data, and processes to enable effective spatial data management.

8.2.1 Standards

With the ubiquity of geospatial data and its growing use in pipeline projects and operations, geodetic standards are essential to maintaining reliable data. Robust standards ensure that common datums, projections, data types, and metadata establish a norm for working in a data intensive environment.

8.2.1.1 Datum Standards

Data should conform to datum standards which are widely accepted and recognized by government, industry, and the general public. Any data not belonging to one of the datum standards should undergo a transformation as specified by a given standard. Parametric data shifts should not be accepted for vertical or horizontal datum conversion. The key elements to datum standards are listed below.

In the US this could include National Geodetic Vertical Datum of 1988 (NGVD 88) and in Canada the Canadian Geodetic Vertical Datum 1928 (CGVD28). Vertical datums are always being refined, therefore, make sure one is chosen that adequately meets the needs of an organization.

The North American Datum 1983 (NAD83) shall be used as the horizontal datum. If data in the North American Datum 1927 (NAD27) is to be used, it must first undergo a datum transformation to NAD83. This transformation would be the Canadian National Transformation v2 (NTv2), or American NADCON (2.1).

8.2.1.2 Mapping Projections

The primary consideration when producing maps should be adhering to cartographic standards and guidelines. Projection parameters should suit the needs of the map and compliment the intended map use.

Universal Transverse Mercator (UTM) system is often the standard projection for data and mapping, however, the extensive geographic coverage of many pipeline systems (particularly east – west oriented) may require a custom projection such as Lambert Conformal Conic.

8.2.1.3 GIS Data Standards

The purpose of GIS data standards is to guide creating and maintaining GIS data. Effective data standards ensure data integrity and minimize confusion when new data is introduced or existing data is revised. These standards provide guidelines for data creation as well as the manner in which datasets could be validated. The main data types encountered in any system are vector, tabular, and raster.

1. Vector Datasets

Vector datasets represent geographic objects with discrete boundaries such as roads, rivers, pipelines and land owner boundaries. Vector features are represented as points, lines, polygons or annotation. Of primary interest with vector data are:

- Format – For example, vector data shall be stored and maintained as ESRI ArcGIS feature classes or feature datasets within a geodatabase;
- Content Standards – For example, vector datasets created using geographic coordinates shall be stored in latitude and longitude and expressed in decimal degreesain proper negative values for longitude;
- Attributes – For example, attributes for shape files shall be stored in a tabular (.dbf) format, and shall adhere to attribute guidelines.

2. Tabular Datasets

Tabular datasets represent the descriptive data that links to geographic map feature. The file format is critical to maintain continuity among users. For example, one may only want point feature class data stored as geographic coordinates in tabular format. All other spatial data shall be stored in the appropriate vector file format.

3. Raster Datasets

Rasters are most commonly used for the storage of aerial or satellite imagery. However, raster datasets also represent continuous layers such as elevation, slope and aspect, vegetation, and temperature. Raster format data is typically seen as data grids or imagery, each with unique requirements

Raster image files created through satellite, aerial photography, or scanning can be stored as uncompressed TIFF or GEOTIFF files and include a TIFF world file (.tfw). However, uncompressed formats require large amounts of disk storage and occupy the greatest proportion of disk space in an enterprise. As a result, lossless compression such as ECW or MrSID helps alleviate this problem.

4. LiDAR

LiDAR derived data is often assessed by the accuracy required to produce topographic maps and products that meet the National Standard for Spatial Data Accuracy (NSSDA).

A full-featured point cloud should be free of all data voids within the project boundary and have an average point spacing of at least 1 point per square metre. The point cloud should have a horizontal accuracy of better than 40cm RMS and a vertical accuracy of better than 15cm RMS in normal imaging conditions.

8.2.1.4 Metadata

Metadata is a cornerstone of the data management philosophy because it provides a record of change and history for any dataset compiled, stored, and maintained. Where possible, metadata should be created and maintained for all vector, tabular and raster datasets within any data structure. All metadata should adhere to the format and content standards outlined by a data manager.

Each metadata file is typically named identically to its corresponding spatial export file, and is delivered in Extensible Markup Language (.xml) format. Such a format is compatible with the output standards published by the Federal Geographic Data Committee (FGDC; www.fgdc.gov/metadata/geospatial-metadata-standards). Standards for the creation and transfer of metadata commonly follow the FGDC's profile of ISO 19115, as described in *Content Standard for Digital Geospatial Metadata*.

8.2.1.5 Summary

The adherence to established standards is critical to ensuring the accuracy, currency, and completeness of datasets in a centralized data environment and outputs generated from these datasets. Existing database standards, Geodetic standards, and GIS data standards should be reviewed to ensure they meet the needs of the project and will result in high quality datasets and project deliverable outputs.

While every effort is made to ensure new and legacy files adhere to standards, it is worth remembering that there may be some instances where the amount of effort required to create, amend, or conform to standards is extensive. Hence, any of the guiding principles outlined in standards documents should be fit for purpose and provide a framework, not a deterrent to ensure that data integrity is maintained.

8.2.2 The Database and Data Models

Given the vast amounts of data collected and stored for pipelines, such project data needs to be properly managed so as to ensure that it can be easily accessed, retrieved, and utilized within the GIS framework. This is most effectively achieved with an expert Spatial Data Management System (Figure 1), providing a database-centric configuration complimented with an industry-standard pipeline data model. Leveraging the database and data model is typically a customized front-end application suite, which is responsible for data management, analyses, and reporting activities, all of which help compose the larger GIS context.

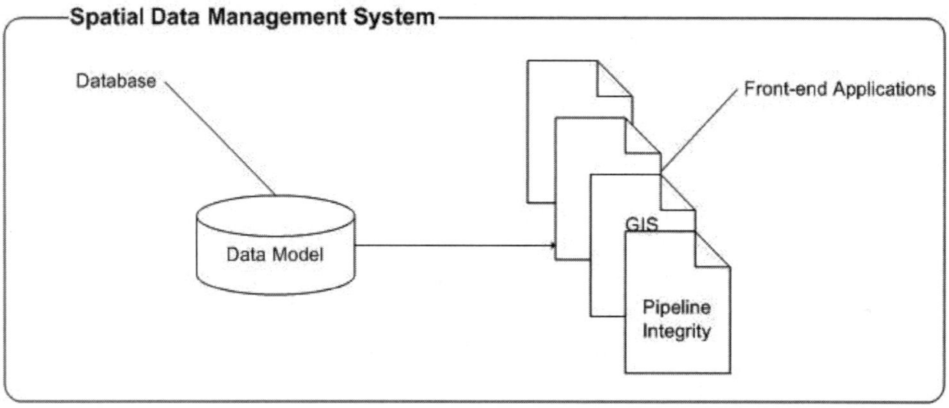

Figure 1 Spatial Data Management System

8.2.2.1 Database

The database is a major component of a Spatial Data Management System as it serves as a warehouse for project data. This means that the database can store both spatial and non-spatial data. Spatial data includes base maps, satellite imagery, aerial photographs, digital elevation models, pipeline centerline, cadastral layers, and any other data associated with geographic locations. Non-spatial data includes engineering reports and drawings, administrative documents, pipeline assets and operations information, stakeholder information, and other business information pertaining to the project. Since project data is being utilized by many disciplines within an organization, the problem of having data 'accidentally' modified from the original or removed from its intended storage location is inevitable. As a result, the integrity of the data is lost and time will be spent to investigate or rectify the problem unless a proper database is employed.

Since pipeline data continues to accumulate throughout the life of the pipeline, the underlying database of a Spatial Data Management System for any pipeline project should always be built with sufficient capability to house current project data inventory and support future project growth.

8.2.2.2 Data Models

Over the last decade, pipeline operators have formed associations to develop industry-standard data models for managing gas and liquid pipeline assets, as well as operations information. These pipeline data models have served to enhance pipeline companies' business operations in terms of cost, effectiveness, efficiency, and reliability. Benefits of these pipeline data models include:

- Elimination of the need to develop a data model from scratch;
- Reduction in integration and implementation time for new software applications;
- Adoption of proper data management techniques resulting in improvement of effective business processes; and
- Customization of existing data model to meet specific project needs.

There are several pipeline data models available in the pipeline industry; however, ISAT, PODS (Pipeline Open Data Standard), and APDM (ArcGIS Pipeline Data Model) are the most widely used.

1. ISAT Data Model

The ISAT data model was created as an open, publicly-available, standard data model for the pipeline industry. The model provides support for all the basic facilities routinely managed by a pipeline company along with operational data required by federal regulatory agencies. ISAT is database independent and supports the following activities: facility data maintenance, as-builting, alignment sheet generation, risk assessment, field data collection, pipe inspection, integrity management, web-based data, reporting and analysis; and, analyst tools including class location analysis, MAOP calculations and marketing. Many organizations have implemented a modified ISAT model to meet their unique business needs.

2. PODS

PODS (Pipeline Open Database Standard) is developed and managed by the PODS Association, which is a non-profit pipeline trade association. As a popular database model in pipeline data management, it is a completely open data standard. PODS provides a general model by which we can create a standard pipeline GIS database, and is easy to expand by adding different aspects as required.

The objectives of PODS include:

- A standard data model to minimize data design activities;
- Reduce database development risk;
- Easy to expand and maintain;
- Provide an integration environment with multi GIS software platforms;
- Exchange and share a standard database with partners;
- Formalize and optimize data relationships;
- Developed in RDBMS such as Oracle, Microsoft SQL Server, and MS Access.

PODS is widely used in the oil and gas pipeline industry, and many companies currently use this standard data model to manage their data. Furthermore, many commercial pipeline applications (such as automated alignment sheet generators) are using PODS as the backend data model for their operation.

3. Customized APDM

The ArcGIS Pipeline Data Model (APDM) was initially derived from both the ISAT and PODS data models and is expressly designed for implementation as an ESRI geodatabase for use with ESRI's ArcGIS and ArcSDE products. The APDM was designed as a template that could be customized by operators to account for their unique processes and assets, with the primary objective of incorporating tools for linear referencing.

8.2.2.3 Front-end Applications

The database and data models form the foundations for the Spatial Data Management System. However, data cannot be adequately leveraged without any form of interpretations. End-users are typically not interested in how the data is stored, but rather much more interested in what the data represents. For instance, a DEM stored in the database has little meaning if the user has no way of visualizing it to understand the terrain representation. Once the user has an understanding of the terrain, pipeline micro-routing or hydraulic studies can be electronically performed with the availability of the appropriate customized GIS tools and other project data residing in the database. In order to assess the integrity of the pipelines to detect potential issues and plan for remediation, inline inspection data needs to be analyzed and represented in the form of charts, reports, and geographical visualizations. It is the representation of the stored data that allows end-users to make informative decisions to better serve their business operations in terms of cost, effectiveness, efficiency, and reliability.

Since many pipeline companies have increasingly adopted the implementation of industry-standard pipeline data models, pipeline engineering software vendors have begun to use these as their underlying data models for application development. As a result, integration to a variety of readily available software is seamless when the foundations of the Spatial Data Management System are in place.

An expert Spatial Data Management System is one that leverages the technologies and data models to make available front-end applications for data management, analysis, and visualization activities.

8.2.3 GIS Infrastructure

A high-quality Spatial Data Management System requires a robust infrastructure to ensure data integrity and reliability in its use. A common configuration is shown in Figure 2. The main components of this arrangement are: database server, GIS server, web server, and workstations.

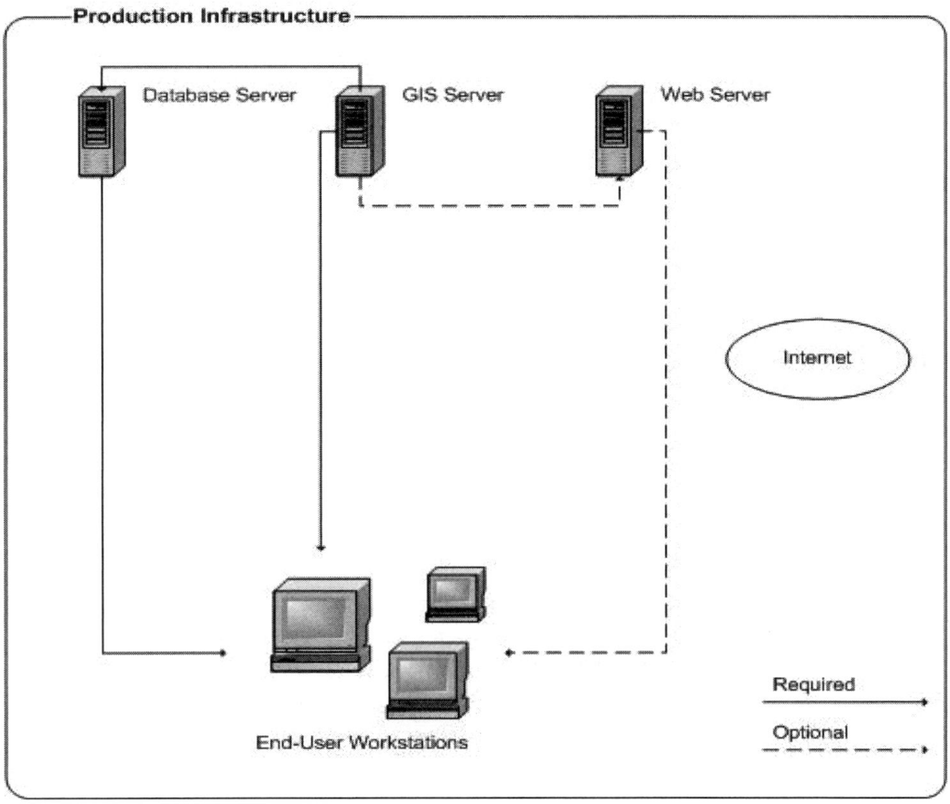

Figure 2 GIS Infrastructure

8.2.3.1 Database Server

The database server of a Spatial Data Management System has similar behaviors as any client-server architecture. This means that the database server stores data, processes and responds to data requests sent by any client machine running front-end applications.

Since the database server is a central system, it has the ability to enforce strict security and control access by allowing only authorized personnel to update the data. As a result, data integrity is always maintained. In addition, the client-server architecture is versatile such that changes or upgrades performed in one environment do not disrupt the services of any other. This is a great advantage, as computer system down time should be an avoidable expense for all business operations.

8.2.3.2 GIS Server

The GIS server is used to deliver GIS capabilities to client machines and it behaves in a similar manner as any client-server architecture. This means that

GIS requests are processed on the server side and results are delivered to the client machines in a very efficient and timely manner. This is an advantage, as the client machines are not bogged down with processing heavy-load GIS tasks. The GIS server can also deliver maps and GIS functionalities to the Internet.

8.2.3.3 Web Server

With the convenience of the World Wide Web, we are able to access almost any information in any format at the click of a mouse button from anywhere in the world. In the context of a Spatial Data Management System, the web server is responsible for rendering front-end web-based applications to web browsers on client machines.

In general, web applications are in the forms of HTML or ASP pages containing scripts, macros, and multimedia files. Web servers then serve these pages using HTTP (Hypertext Transfer Protocol), a protocol designed to send files to web browsers.

8.2.3.4 Development Workstation

A development workstation is a client machine for software developers to develop and test the Spatial Data Management System. This is where various software, development platforms, and software developer kits are installed to allow software developers to migrate data into the database and create specific programs or applications.

8.2.3.5 End-user Workstation

This is where front-end applications are installed and available for end users. If third-party tools are used in creating the front-end applications, then this client environment must have the necessary runtime objects or desktop applications of those tools to be able to access the desired functionalities. If the front-end applications are web-based, then the client machine must have a web browser, internet connection, and be configured to accept scripts, applets or other specific criteria that the web applications require.

8.2.4 Data Management Workflow

Many people within an organization or project will create, handle, or edit data at some time. In order to minimize data mix-ups, deletions, or unwanted changes, a data management workflow should be followed.

A central database is composed of data layers and tables held either in the *Staging* or *Production* storage areas. In the staging area, data is available for use in preliminary mapping and analysis, but is undergoing quality checks, editing or revisions. Data in the production area is considered the current version and most accurate given a specific point in time. It is production data that should be taken as the source of truth for deliverables, maps and information products to the document management system.

The main features of a standard data management workflow are (Figure 3); acquire/create datasets, staging, staging Quality Control (QC), production, production QC, rev-up data.

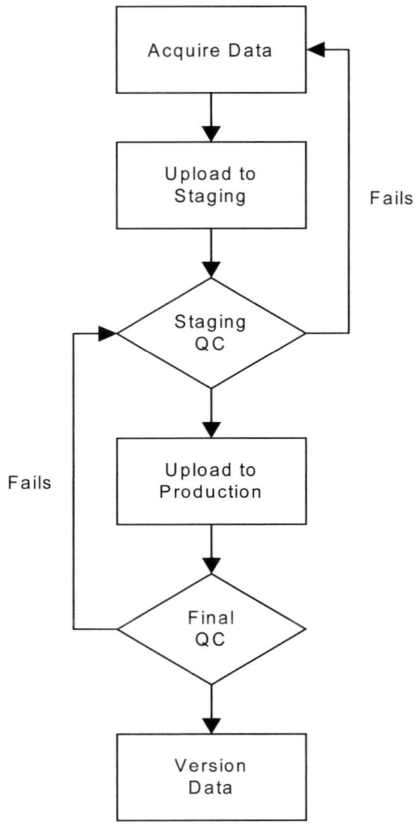

Figure 3 Data Management Workflow

8.2.4.1 Acquire/Create/Edit Datasets

New data is obtained either internally (through a variety of disciplines), or from an external source (third party contractor, government sources, etc.). The procedure to prepare data for uploading is the same for both internal and external data sources. Examples of datasets are discussed in the next section (8.2.5).

When an existing dataset needs to be modified, it typically will go through a Change Management process, which is outlined in section 8.2.7.

Once data has been acquired, created, or edited, metadata is compiled. This is an important step since metadata should be created by the data owners since they are most familiar with the data being submitted.

8.2.4.2 Staging Quality Control

When a dataset is submitted with current metadata, the Staging Administrator performs a QC check. This QC check is typically based on standards developed for the functioning GIS.

8.2.4.3 Pass or Fail

If the Staging Administrator identifies any issues with the spatial data, metadata, and/or attribute table, the dataset does not pass the initial QC and the GIS Operator, upon request, amends the dataset. Once the Staging Administrator is satisfied that the dataset is ready, the Production Administrator is notified of the datasets to be independently QC'ed and migrated to Production status.

8.2.4.4 Production Quality Control

The Staging Administrator then notifies the Production Administrator and the Production Administrator performs an independent QC of the specified datasets according to an established set of requirements. If any dataset fails the QC, comments are sent back to the Staging Administrator to be fixed. Once all the datasets pass the final quality check the datasets are migrated to the Production environment and privileges to view the datasets are granted.

8.2.5 Data Composition

Centralized pipeline data sets typically fall into at least one of the categories below. Table 1 outlines some datasets encountered during pipeline design and Table 2 shows data maintained during the operations and integrity part of a lifecycle. It is important to note that one of the greatest benefits of instituting a central database early in the pipeline lifecycle is that data collected is carried forward and leveraged in future operations.

8.2.5.1 Base Mapping

Base mapping is the collection of regional maps such as government map sheets. These are used as a foundation map from which other data sources can be overlaid. Because of the ubiquitous coverage available, these maps provide information in areas where there is no data collected as of yet for a specific project. They also provide reasonably good data that can be used as a substitute when expensive data collection would not yield enough value to justify the cost. Features could include contours, hydrography, land cover, land use, roads and access, political and legal boundaries, cities, and towns.

8.2.5.2 Imagery

This includes satellite imagery and ortho-photography. Satellites have the advantage of acquiring imagery over a very large area within a single image. In addition, orbiting satellites can capture imagery in foreign countries without the need to send out a flight crew, and use parts of the spectrum not visible to the human eye to provide additional information. Although satellite data has advantages, aerial images are still relied upon to collect very high resolution images with near-survey grade accuracy.

8.2.5.3 Terrain Models and DEMs

Of critical importance to pipeline projects are models of the Earth's topographic surface, which allow for more accurate modeling of hydraulics, slopes, routing, profiling, cut and fill, line-of-sight, grade, general mobility, and others. A digital elevation model (DEM) is a raster where each pixel is encoded with an elevation value above sea level. A DEM can represent the elevation of the surface of the bare earth, or it can reveal the elevations of all features on the surface of the earth, such as vegetation, electrical lines, buildings, etc.

DEMs used in pipeline applications are typically produced through photogrammetry or light detection and ranging (LiDAR). Traditional photogrammetry uses stereo air photos to compile the elevation of ground features like break-lines and mass points. LiDAR uses a laser mounted in an aircraft to rapidly pulse a laser beam down to the earth's surface and measure the return time to determine the distance to the earth's surface. After thousands of laser pulses are processed into a raster, a high resolution and accurate representation of elevation changes along the earth's surface can clearly be seen.

8.2.5.4 Engineering Data

Engineering data consists of data relating to specific engineering details, particularly any event which is near, on, or crosses a pipeline. Common types of engineering data can include items such as the centerline, pipe material, coatings, welds, crossings, profile, in-line inspection data, digs and repairs, cathodic protection, assemblies, right-of-way, and workspaces.

8.2.5.5 Environmental

Data in this category is related to environmental factors which are not engineering specific, and includes data regarding the use of environmental resources. Common types of environmental data include aquatics, fish and fish habitat, water quality, groundwater, vegetation, landforms, permafrost, air quality, wildlife, historic resources, land and resource use.

8.2.5.6 Administrative

This category contains data associated with government or survey activities such as municipal and provincial boundaries, private lands definition, registered plans, land owners, mineral rights etc.

8.2.5.7 Questions to Ask

When acquiring GIS data it is important to consider several issues.

- If the data desired is not readily available, how long will it take to produce?
- What is the quality of the data in terms of the positional accuracy (location), and attribute accuracy (ex if a polygon shows an area having a population density of X, how close is the actual population density to that value)?
- Are data sets being combined from different sources (ex two adjoining areas surveyed for soil properties by two different companies), how will they align/match up to one another?
- How current is the data? Is the data too old to be useful?
- What is the financial cost of obtaining the data?
- Will the data be delivered with a file size and in a file format that is usable and convenient?
- Not all GIS data can be freely shared or copied within an organization, or between companies. What are the licensing/legal agreements associated with the data?

Table 1 Typical Datasets for Pipeline Design

Data Descriptor	**Data Example**
Pipeline Facilities	Pipe information, external coating, internal coating, valve information, casing, launcher receiver information, pipe bend, flange, fabrication,
Geographic Features	Roads, access, foreign line crossing, railroad, land use, right-of-way, workspaces, hydrography, land owner, cadastral
Location	GPS point, monument, profile, centerline geometry
Raster/Imagery	Scanned mapsheets, aerial/satellite imagery, LiDAR, cross/longitudinal slope grids
Environment	Aquatics, fish habitat, water quality, groundwater, vegetation, landforms, geotechnical, air quality, wildlife, historic resources, land and resource use
Stationed Centerline	Line, route

Table 2 Typical Datasets for Pipeline Operations and Integrity

Data Descriptor	Data Example
Inline Inspections	Temperature, velocity, pipe length, wall thickness, wall loss, odometer, diameter, GPS coordinates, weld distance, anomaly type, pigging date, burst pressure, accuracies
Physical Inspections	Corrosion, cracking, mechanical damage, metal loss, material defect, excavation, soil, water, pipe condition, coating, relief valve, girth weld
SCC Potential	Soil potential, pipe susceptibility
CP Inspections	Anode reading, rectifier reading, casing reading, ground bed reading
Pipeline Facilities	Pipe information, external coating, internal coating, valve information, casing, launcher receiver information, pipe bend, flange, fabrication, repairs
Operating Measures	Odorant, temperature, pressure, operating history
Offline Events	Structure, populated area, HCA site
Regulatory Compliance	HCA, DOT class, Activity zone, test pressure, leak history
Geographic Features	Roads, access, foreign line crossing, railroad, land use, right-of-way, workspaces, hydrography, land owner, cadastral
Raster/Imagery	Scanned mapsheets, aerial/satellite imagery, LiDAR, cross/longitudinal slope grids
Location	GPS point, monument, profile, centerline geometry
Stationed Centerline	Line, route

8.2.6 Data Quality

An essential part of any GIS functioning as expected is quality data. With the amount of digital data available today, combined with new and traditional ways of developing digital data, it is critical to have a common means to identify and describe quality data. To assist the identification of quality data a set of GIS data standards should be prepared (section 8.2.1.3) as well as a Quality Control/Assurance procedure that will ensure that only high-level quality data is accepted and introduced into the GIS database.

When defining GIS data quality the following prime components of the GIS data should be reviewed and accuracy standards applied to them:

- Spatial Placement;
- Connectivity;
- Database Design Conformance;
- Database Attributes;
- Age of Data;
- Data Completeness.

8.2.6.1 Data Quality Validation Guidelines

1. When the spatial accuracy and connectivity of the data is validated the following should be considered:
    - o Use of correct spatial reference system;
    - o Data should be free of topological errors;
    - o Data format used should be consistent within the database.

    Geometric networks should be used to validate the GIS data connectivity, such as: object, database and device connectivity, assure the GIS database accurately models the real life network system.
2. For database design conformance and database attributes validation the following should be considered:
    - o Using a suitable Data model standard will provide easy assessment of existing data in the database and the ability to apply standard business rules for multiple datasets.
3. The age of data has a direct correlation to the source used to provide the data. In order to assure that the GIS system users make the appropriate decisions every effort must be made to use only the most recent data and properly document the time data was collected.
4. For the completeness of the GIS data it is critical that the database files capture all objects of the automated pipeline system, such as number of compressor stations, meter stations, valves, and pig traps. Also, generalization, omission, selection criteria and other rules used to define datasets must be properly explained and a clear definition of all possible error specifications has to be provided.
5. Creation of proper metadata and establishment of metadata standards or adoption of existing metadata standards is another important aspect of creating and maintaining quality GIS data.

8.2.6.2 Benefits of Using Quality GIS Data for Pipeline Applications

1. Quality integrated spatial data from GIS systems and real-time data from SCADA can be used as a decision making tool by relating the pipeline data collected at a specific location (meter station, pump station) to changes in the data over time.
2. Assures efficient emergency response planning, by providing the added functionality of a GIS such as use of spatial queries to assess and evaluate existing infrastructure, demographic data and available access routes relevant to specific pipeline infrastructure.
3. Accurate 2-D and 3-D graphic visualization and modeling of pipeline systems and operating devices.
4. Provides an efficient and cost-effective creation of digital data ready for integration with SCADA of detailed as-built plans for meter station, pump station, pipelines.
5. Effective pipeline integrity work documentation and maintenance planning (section 8.3.2 Integrity Tools).

8.2.7 Change Management

The main reason for spatial data management is essentially to manage change. This occurs by providing an information infrastructure for the life cycle of the pipeline and adds stability to an information management system. A certainty in this digital world is that the amount of data required and generated is immense and continues to grow.

Change Management means accountability. For an organization, change management means defining and implementing procedures and processes designed to deal with changes to any item that should fall under some form of version control.

Change management provides a paper trail giving an accurate history of any changes applied, the reason for the change and of course pre-approval and post approval signatures. Change Management must be a structured process resulting in the validation of the proposed changes as well as an accurate status reporting tool for all affected parties.

8.2.7.1 Controlled Change Management

The role of Change Management is to add organization to the potential for disorganization. Left long enough uncontrolled change can become a logistical nightmare. It plays an important role in any organization since the task of managing change is not an easy one. Successful change management should involve selling the process internally and creating an efficient infrastructure for change management.

Managing change requires a broad skill set of political, analytical and people skills. These skills are important during and after the initial implementation. The implementation of Change Management process demands a whole new way of working in an organization, and also involves looking differently at other processes that may be affected.

A Change Management system is generally tailored to suite the needs of a Project or Company, but should also be adaptable to any project within the organization.

It is important to consider the two user groups when dealing with controlled change, the *requestors* and the *implementation team*. Left uncontrolled the potential for multiple changes by multiple requestors can often lead to confusion on the implementation side and issues with the integrity of the changes. Through control and approvals only the requested operation should be implemented with little room for after the fact amendments.

8.2.7.2 Proactive not Reactive Change Control.

Identifying the need for a Change Management system should be a priority in the initial stages of a project, playing catch up at a later stage can introduce integrity and revision issues.

Implementing controlled change in the initial stages of a project will help reduce the resistance to change as most people have negative attitudes and perceptions towards change. If the process is implemented as part of the ramp up it will almost always be seen as part of the project procedure.

8.2.7.3 Guidelines to Implementing and Sustaining a Change Control Process

The potential for resistance must be taken into consideration, and why that resistance may occur. The potential for resistance means implementation of a Change Management process will involve political, analytical and people skills.

- Know your data structure and understand what data must be controlled.
- When possible assign ownership.
- Implement the process in steps, so as to minimize the overall impact on current structures and processes by identifying the Change requirements through data collection and analysis.
- Minimize the administrative traffic jam by making the process user friendly

Spatial data management provides organization and structure to the complex web of information demands for a project or organization. The structure it provides allows data users to conduct their activities in a stable environment despite the constant changes being made to the data. With a dependable information system backbone, data use and sharing become valuable additions to workflow.

8.3 GIS Tools to Support Pipeline Design and Operations

To take full advantage of the data you are securely maintaining within a centralized GIS environment, automating engineering processes through custom GIS tools proves very effective. If you are using data that is structured within a database, incorporating it into applications is efficient and automating tasks becomes easier. This section reviews tools and techniques that help automate engineering tasks during pipeline design and operations.

8.3.1 Automated Alignment Sheets

Alignment sheets play an important role in pipeline projects. Used as a tool or deliverable for every phase of a pipeline project, alignment sheets are a visible representation of where the pipeline exists in the real world. Over time, alignment sheets have become richer in content as larger amounts of digital data are captured to aid in design. The use of Computer Aided Drawing (CAD) has significantly aided in the creation of alignment sheets but with pipeline projects becoming larger and more complex, traditional methods of CAD drafting do not suffice in managing the increased digital data.

With the use of databases and constant improvements in computing hardware, developers have been able to create programs that handle the more tedious areas of alignment sheet generation, namely the data management process and sheet generation process. Automating the creation of alignment sheets does not result in products that are radically different from those produced manually, but rather the new technology results in time savings and reductions in overall effort.

8.3.1.1 Traditional Alignment Sheets

The traditional method for creating alignment sheets typically requires a large number of CAD operators, with each operator maintaining a specific number of sheets for a designated section of the pipeline. These sheets are updated and changed throughout the duration of the pipeline project and with each manual change, the integrity and quality of the data shown on the sheets is reduced. This reduction in quality is caused because changes made in one sheet are not automatically reflected in other sheets; the CAD operator must manually change them. Even with a proficient and effective group of CAD operators and engineers, the job of manually maintaining data on the alignment sheets ultimately fails as the volume of the data becomes too great or edits happen over time. The use of automated alignment sheets minimizes, if not eliminates, the pitfalls of manually drafted CAD alignment sheets.

8.3.1.2 Benefits of Automated Alignment Sheets

There are many benefits to using automated alignment sheets. The actual sheet production process takes only a fraction of time compared to manual drafting, and efficiencies are continually increasing with advancements in hardware and software development kits. A single person can easily manage the entire process of automated sheet generation.

In addition to reducing generation time, automated alignment sheets can be very flexible in terms of the data and content included. Specific sets can be created for engineers who need preliminary routing information, and a completely different set can be created for regulators who are interested in environmental aspects of the routing.

8.3.1.3 Implementation

Creation of automated alignment sheets can only occur if a pipeline project has implemented a centralized spatial database. This is important as the automated alignment sheets use the centralized spatial database as its single source of truth from which all data is extracted. On a basic level, automated alignment sheets are just a complex database report. Each set of alignment sheets is essentially a snap shot of the data within the centralized spatial database.

A shift in thinking is required to ensure the successful implementation of automated alignment sheets. Using the manual technique requires the majority of effort be spent maintaining the data on each individual sheet. If the data is found to be incorrect, the CAD operator must manually correct it. Using the automated approach, this effort must be shifted from sheet-by-sheet maintenance to maintaining the datasets within the centralized spatial database. When a data change is required on an automatically generated alignment sheet, the change must occur in the database and not on the sheet itself. By managing change within the database, all subsequent sheets generated will reflect the most current and accurate data. As a result of this mentality shift, the front-end effort required prior to sheet generation is very important, since each dataset should be verified for accuracy before the sheets are run.

8.3.1.4 Data Quality

Users planning to implement an automated alignment sheet solution must understand that the front-end effort of performing quality assessments and checks prior to sheet generation is vital to the quality of the output product. Automated alignment sheets can be considered as a complex database reporting tool that displays data exactly as it exists within the centralized spatial database. If the data within the database is incorrect, out of date, or incomplete at the time the sheets are run, these data deficiencies will be reflected in the output sheets.

The responsibility regarding data quality rests between the database administrators and the pipeline designers. These two groups must work together to ensure that the data created by the pipeline designers is properly represented in the centralized spatial database and thoroughly check against quality control measures by the database administrators. The channels of communication and responsibilities between these two groups must be well established and clearly defined in the change management process.

In most projects, a pipeline data model has already been accepted and all the input data is parsed into the corresponding tables of the data model. Whether or not a project has opted to use a pipeline data model, the data within the centralized spatial database must be organized in a manner where linear referencing can be applied.

8.3.1.5 Linear Referencing

Automated alignment sheets rely heavily on a method called linear referencing that is used to integrate the data in the centralized spatial database. Linear referencing is a straightforward method for associating attributes or events to locations or segments of a linear feature. Using linear referencing, events are easily located based on their distance from an established starting point, rather than from an exact x, y location using measurements of latitude and longitude. For pipeline projects, the pipeline route would be classified as the linear feature and all corresponding events and elements would be referenced to the route via the measure (e.g. Milepost). The value of linear referencing is that it *dynamically* adjusts event locations along the pipeline as the route changes, or as alternate design options are explored.

Linear referencing can easily be applied to features such as valve locations, road crossings, creek crossings and utility crossings because these features can only occur at a specific measure along the route. This also applies to features such as pipe protection and erosion control that are required to run along the lengths of the pipeline and are denoted by a start measure and end measure.

Features that are area-based (legal land boundaries, soil types and slope boundaries) can be linearly referenced to the pipeline by the intersection points where the pipeline enters and leaves the area. The success of linear referencing relies heavily on the quality and maintenance of the centralized spatial database.

8.3.1.6 Software Solutions

There are several software companies that provide automatic alignment sheet generation (ASG) programs, but the workflow dictating how sheets are generated is common to all at a high level. The ASG first queries the database and retrieves the corresponding spatial data; then translates the spatial data from real-world coordinates to paper-space coordinates; and, finally generates the sheets according to the layout template specified by the designer. There is very little human interaction between the first step of querying the database to the final step of sheet generation. As a result, the sets of alignment sheets generated are consistent as each sheet was created using the same method and data as the previous sheet. This eliminates the potential non-systematic errors that are typically encountered during the manual drafting of alignment sheets.

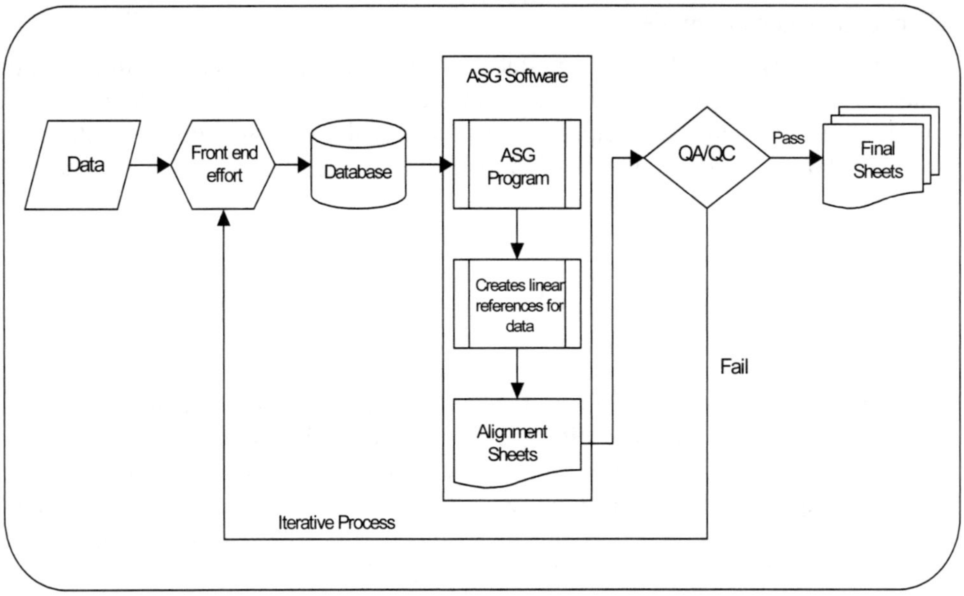

Figure 4 Automated Alignment Sheet Process Overview

8.3.1.7 Process Overview

It is important to note that despite the changes in implementation and mentality shifts, the process of generating automated alignment sheets remains iterative. A set of sheets will go through several revisions before reaching its final form. Feedback should always be solicited from the appropriate stakeholders, and a final quality assurance review should be the final stage of the process. Figure 4 illustrates the high level workflows required to generate automated alignment sheets.

The iterative process starts at the "Front-End Effort" and ends at "QA/QC". The process is complete when the sheets have met the specified requirements of the project. Though there are six steps listed in the iterative process, only three require human input. The remaining three steps represent areas where manual drafting would be required in the traditional sheet generation process.

8.3.2 Integrity Tools

Pipeline companies want to operate their lines safely and efficiently and at optimal capacity for as long as possible while incurring minimal maintenance costs. Faced with ever increasing costs and regulations, pipeline operators must find new ways to maximize output and extend the life of their pipelines.

Engineers design pipelines based on manufacturing and operating principles. From the outset, pipelines are planned to operate for a designated number of years of service, yet many pipelines continue to carry products many years beyond this designated limit. The proper maintenance schedule and continuous testing of defects has allowed such lines to continue to operate, and save companies the enormous expenditures associated with constructing a new pipeline.

Integrity tools use vast amounts of pipeline data, as well as data related to the surrounding population and environment, to map areas of risk and possible mitigation strategies. For example, if a pipeline is located in close proximity to a densely populated area, then integrity tools can be used to identify the zone where people may be endangered, and model mitigation options such as the use of thicker pipes, warning signage, and fences. Based on results from this analysis, decision makers can determine if additional safety measures are required.

8.3.2.1 Public Safety

Keeping people away from pipelines is good practice, though it's not always possible. In North America, pipelines are well protected and far away from urban centers, however in many parts of the world exposed pipelines next to high density residential areas are commonplace. When people reside, permanently or temporary, inside these corridors, then special measures must be taken to protect public safety. The US Department of Transportation (DOT) regulates protective measures based on the potential casualty to the population living or gathering at structures near pipelines. The DOT regulation dictates that pipeline operators must produce a classification of geographic locations based on the density of population and proximity to the pipeline.

The identification of pipeline proximity to people and structures is automated through spatial measurements within the integrity tool. Modern technology in GPS, satellite imagery, digital survey equipment and data collectors provide the data used for integrity analysis.

8.3.2.2 Environment

Protecting the environment is difficult to manage, in remote areas small leaks may go undetected for long periods. The management and regulatory methods are different for pipelines on land, offshore or at crossing locations of streams and rivers. Some regulations require sampling of water and inventory of wildlife within a certain distance from pipelines many times each season. There are many forms of data collections to study environmental conditions, from aerial reconnaissance with sophisticated measurements instruments to individuals performing site surveys. These data collection methods can generate large volumes of spatial data that can be managed in a GIS to produce the required reports.

8.3.2.3 High Consequence Areas

The severity of harm to people, property, and environment is a function of the pipe material characteristics, operation conditions, and type of product. People have the greatest degree of production by regulators.

A safety zone is defined by the regulators, dependant on occupancy of people living or assembling within a certain distance from the pipeline, and operators must mitigate possible risk within this safety zone. These safety zones are defined by regulators and commonly known as High Consequence Areas (HCA), the US DOT definition of HCA is a function of spatial distance of the pipeline to the number of people that are in the area on a permanent or occasional basis. Identification of the number of people residing in a structure is a function of the occupancy capacity of structure such as residential, church, sport stadium, playgrounds, etc.

It is not feasible to determine the number of tenants in all the buildings by site visits, therefore, regulators have predefined the number of occupants based on the structure type. A residential building of a certain size or height has a predefined number of people. Other structures have predefined number of occupants based structure guidelines of the regulator. Based on the population and structures definition provided by regulators, a map showing building structures and its proximity to the pipeline allows for creation of HCA.

Aerial and satellite imagery have been very effective datasets for establishing and updating structure identification and location (2). When there are changes to the pipeline or surroundings, new maps of structures and pipeline are inputted in the GIS to automatically recalculate new HCA.

8.3.2.4 Linear referencing system

Pipes are manufactured in sections and assembled together to create a large network of pipelines. The linear referencing system, also known as dynamic segmentation, is a core component of GIS for pipelines to handle any complexity of pipe network geometry in initial layout to future cutouts and repairs. Linear referencing system defines how a network of pipes is connected together.

Non-positional attribute data related to the pipe section such as material and operating data can now be linked to a location on the earth. With GPS coordinates of each section an entire pipeline network and its complexity of attribute data can be represented in three-dimensional view to show the exact location and attributes of the pipeline. The exact geographic representation the pipeline in a GIS database system enables automated methods to update maps and any reports that require a physical location of a pipe attribute. It is possible to map the entire pipeline with detailed related information stored in the database, even cutouts and decommissioned segments.

Complex spatial queries to retrieve data related to the entire pipeline network are easily accomplished. A few sample spatial queries are below.

- Find the length of the pipeline that …
- Show all pipe segments connected to compressor station A.
- A pipe segment has internal corrosion, query all other segments

connected to this segment to check if similar internal corrosion has common contributing factors.

- Pipes manufactured by company X reported a high rate of defects, find all section of pipe manufactured by company X.
- Map location of all pipelines of steel type X, and of age Y and manufactured by Z.
- Map location of warning signage where there are playgrounds
- Map location of pipelines carrying X liquid product and on a slope of Y degrees with buildings at Z distance from the pipeline.
- Map location of all pipelines that has been decommissioned and not purged with nitrogen within X years.

8.3.2.5 Benefits

In regions where safety and mapping standards are poor, pipelines could reside anywhere. It could be transporting poisonous gases and located underneath your house. Fortunately, the US and other governments have regulations requiring operators to identify all potential threats to each pipeline segment, with more stringent standards in high consequence areas (HCA), and requirement to conduct risk assessment that prioritizes individual segments for integrity management.

With widely available accurate aerial and satellite imagery, management of pipeline assets can be a regularly scheduled event to obtain an accurate inventory of all pipeline assets.

8.3.3 Maps

Traditional methods of designing, selecting, and maintaining a pipeline route have their roots in manual drafting; a very time and labor intensive process. Increasingly, GIS is being recognized as a valuable tool that can increase efficiencies, while also maintaining the high quality cartographic outputs typically associated with manual drafting. Not only can GIS be used to create and maintain spatial data, design and evaluate potential route locations, and perform analysis or modeling to determine potential impacts; GIS can also be used to produce high quality cartographic outputs in a variety of formats.

8.3.3.1 Using maps to tell a story

Maps help to tell a story, and are often more powerful and convincing than the written word. Whether it's a map of river crossing sites along the route, a map of pit development and facility locations, or a least cost path map for a new route, visualizing man-made, natural, or cultural features on a map allows us to better understand the relationships between objects in the real world. Maps help us to see the whole picture, and help us to identify potential impacts we may not have considered based on our practical knowledge of the project and the site.

To create an effective map, the GIS operator must start with a well defined question. What story does the map need to tell? For example, *"Where is the best location for a new pipeline route"*. In order to produce a map that illustrates the best route, compared to other route alternatives, the GIS operator must first determine what data layers will be required to answer the question and then develop a methodology for processing the data.

Supporting data may need to be compiled if it does not already exist. The GIS operator may need to digitize, scan, download or otherwise prepare data prior to producing a map. Datasets should be compatible and referenced to a common datum, so they are displayed in a common map space, allowing for the overlay of features. It is also important to ensure the attribute values contained within the spatial data are accurate as tables, graphs, and map labels are produced based on the values stored in the database. The GIS operator will then perform the necessary analysis or modeling to answer the question posed. Once a satisfactory answer is produced, for example a preferred route location is identified, the results must be effectively visualized and distributed to the appropriate stakeholders. Usually this is done with a combination of hard copy maps, tables or graphs, but the increasing performance of Internet technologies is also making it practical to share map products via the Web.

8.3.3.2 Hardcopy Map Products

Results produced from analysis and modeling can be displayed and examined in many different formats including tables, charts, graphs, reports and maps. Hard copy maps can be an excellent vehicle for facilitating group discussion, providing information to regulators, and keeping stakeholders and members of the public apprised of progress or proposed changes to the project. Maps are a valuable visualization tool that can illustrate the location of the pipeline route vis-à-vis other physical and cultural features, combined with traditional cartographic elements.

Cartographic elements typically include the map title, scale bar, north arrow, legend and grid or neatline. Map products may also be combined with other elements such as tables or graphs that relate to the map. For example, automated alignment sheets combine a map window with a variety of data frames that are used to illustrate the route profile and other phenomenon that occur along the pipeline route. The complexity and options related to output mapping are infinite, however user-defined templates can be constructed to automate the desired cartographic elements and layouts. Using map templates can maximize the efficiency of producing multiple output products that require the same look and feel.

8.3.3.3 Web-enabled Maps

The inherent nature of the Internet is to allow the sharing of data. Since a Web-based GIS can be made available to the Internet, it provides the facility for the sharing and publishing of maps. This proves extremely beneficial for the collaboration of groups not physically working together and communicating with people in the field. Web GIS applications are ideally suited for basic map viewing operations because they are lightweight, widely available and ensure the most up to date version of a map is being viewed. Web GIS is discussed in greater detail in section 8.3.6.

8.3.3.4 Advantages of using GIS-based maps

Maps produced in a GIS provide pipeline companies with a method for quickly and effectively visualizing the natural environment, social factors, potential pipeline routes, and as-built existing features in one cohesive map. The full pipeline route, from origin to terminus, can be observed using an overview map, supplemented with site specific detail maps.

Using GIS to generate maps, rather than traditional manual drafting methods, allows for data and design changes to be reflected in the maps instantly. Large series of multiple maps, such as automated alignment sheets or route sheets, can be generated using GIS, with significantly less time and effort than when produced manually. Output products can be tailored to meet specific business needs, for example, maps and graphs in a hardcopy document, large scale map posters for public consultation, or web-enabled maps for field programs or round-table business meeting discussions. The flexibility of map production functionality in a GIS environment makes it easy for users to quickly paint a concise picture of the story they wish to communicate.

8.3.4 Modeling Engineering Processes

A GIS is a tool that allows users to model or simulate the real world by introducing and manipulating controlled variables, and analyzing the output. The purpose of building and using GIS models is to leverage information from data by replicating or estimating a real world phenomenon. The resultant information can be analyzed and used to make better decisions during pipeline system design or operations.

Models within a GIS are essentially mathematical equations with real world spatial reference that attempt to predict real world situations. By modeling scenarios in a GIS, complicated process are automated. These processes can be repeatedly tested, expensive field work is minimized, cost estimates can be refined, and project scopes narrowed to increase work and cost efficiencies.

8.3.4.1 Requirements Gathering for Model Development

Proper requirements must be gathered to ensure a model successfully performs the desired tasks. Requirements gathering is a structured and iterative process where the needs of the engineer (or model user) are realized, captured, and prioritized.

Requirements gathering is beneficial for several reasons. Working together creates a partnership between engineers and developers and ensures a mutual understanding of what is required and how it will be achieved. A clear vision of user needs helps to prevent time and money from being wasted on developing unnecessary features and will also deter misunderstandings that may result in costly rework at various stages throughout the project. More importantly, good requirements help to ensure that a project maintains its scope and schedule.

8.3.4.2 Examples and Expectations of Models

Modeling can be used in all stages of pipeline system design from pre-field work planning, detailed design, construction, and operations. By modeling on the desktop, engineers and project managers can become familiar with the project area and plan for the terrain, project variables or other phenomena that may affect their pipeline projects. With the ability to model design options from the office, projects can be safer, costs can be reduced, and scopes refined. Some common GIS engineering models are shown in Table 3.

Table 3 Common GIS-Based Engineering Models

GIS Model	Applied Uses
Buoyancy Control	Using terrain type modeling to determine the amount and type of buoyancy control to use in areas where the pipeline traverses swampy terrain.
Cut / Fill / Volume	Using a DEM to calculate how much earth to move, remove or add.
Least Cost Path	Manipulating multiple variables to determine the best routing for a pipeline.
Terrain Analysis	Determining various geohazards, such as slope creep, frost heaves, or acid rock drainage.
Line of Site	Determining how visible a pipeline will be from a town or designated point of view.
Cross and Longitudinal Slopes	Calculating slope and its affect on pipeline routing (used with or without other variables).
Hydraulics	Using a DEM to calculate the hydraulic flow rates of a pipeline system and determining the optimal location for pump or compressor stations.
Hydro Networks	Compiling a stream network, to model a change in stream flows resulting from the influence of a pipeline and its effects on the local environment.

GIS Model	Applied Uses
Pipeline Crossings	Analyzing various ROWs to determine optimum pipeline crossing locations.
Stationing (Linear Referencing)	Using DEM's to calculate chainage of pipeline systems and features along the ROW.
Air Quality Monitoring	Determining general wind directions and their potential effects on the spread of exhaust from facilities.
Dynamic Segmentation	Transforming tabular data into segmented linear features using from and to measures.
Timber Volumes	Calculating the volume of trees cleared during pipeline construction.
Risk Models	Estimating external corrosion, internal corrosion, and third-party damage.

Models cannot provide all the answers for pipeline design and will not replace field work entirely, but they are valuable tools that aid the decision-making process. The value in using GIS models is that they empower engineers and designers to simulate countless scenarios and sensitivities, thus enabling more informed decisions.

8.3.4.3 Model Development Process

Modeling involves asking questions (or queries) of spatial data and allowing the GIS to perform a controlled process where the output provides information that can be used for decision-making. Asking the right questions with the right data is paramount to retrieving relevant information and desired results. As a result, rigorous development protocol will ensure proper model operation and results that are meaningful, reliable, and repeatable.

The general progression of model development follows: Conceptual Design, Preliminary Design, Final Design, Initial Development, Initial Quality Control, Final Development, Final QC, User Training, Deployment, and Product Support.

1. Conceptual Design

The purpose of conceptual design is to give the developer the opportunity to capture the problem and provide the client (i.e. engineer or other model user) with an overall solution to the problem, in order to optimize business benefits for the client. The conceptual design will serve as the basis for further discussions to finalize the details of the model. It will give both the client and the developer the opportunity to present questions of what to expect of each other.

2. Preliminary Design

The preliminary design stage will give the developer the opportunity to prepare design specifications for the project based on discussions after the conceptual design stage. Typically, the client will receive a preliminary design document, which will include project requirements, work breakdown structure, man-hour estimates, as well as a schedule. The client will have the chance to review and amend the project specifications set out in the preliminary design document.

3. Final Design

At this stage, both the client and developer have exchanged adequate information and agreements to finalize the details for the project. This is when final engineering decisions, design prototypes, and detailed plans are made in preparation for the next stage of the development cycle. Depending on the project scope, this stage can be combined with the Preliminary Design.

4. Initial Development

With the final design plan in place, the developer (team) proceeds. As there should be no design changes taking place at this stage, the developer team can focus their complete attention to developing a high quality model.

5. Initial Quality Control (QC)

This stage is where testing of the initial development takes place. At this stage, any software issue arising from testing will be flagged and brought to the developer team's attention.

6. Final Development

Issues from initial QC are addressed at this stage. Again, there should be no design changes at this point so to allow the developer team to focus on rectifying the problems identified from the initial QC.

7. Final QC

The QC group completes a thorough testing of the software at this stage. Any obvious software issues should have already been addressed in the Final Development stage.

8. User Training

Different levels of training can be discussed with the client to tailor the required training sessions to accommodate all individuals. User manuals can be compiled to assist with software usage during the training and after deployment.

9. Deployment

This is where the model is installed and tested. Technical staff are available to answer any questions that the client may have.

10. Product Support

After the product has been deployed, technical support can be provided to the client as required. Although this entire development process may seem too rigorous and perhaps cumbersome for small applications, the general process should still be followed to ensure that development is done efficiently and only once. It is when developers and engineers try to take a short cut that development projects grow out of control.

8.3.4.4 Benefits of Modeling

The use of GIS models to automate engineering processes has many benefits from increasing cost and time efficiencies, to refining and optimizing work scope and schedule.

Modeling on the desktop from the beginning of a project, can empower engineers to make better decisions about routing, which in turn can save time and money. Models can be used to reduce material costs if the routing is optimized during the planning stages. For example, if a route is optimized to avoid a large wetland area, incremental screw anchor costs can be minimized. Beyond material cost savings there are savings to be made in minimizing redundancy or ineffective fieldwork.

Desktop modeling can result in fewer field visits and reduced reliance on survey work from the outset of projects. During the initial planning stages, models can significantly reduce the need for survey work. For example, if engineers know from the outset that a certain routing option is not viable based on a slope analysis model, efforts can then be focused on the more viable options, therefore optimizing fieldwork budgets and schedules.

During or after construction, models can be used to build efficiencies into workflow processes. Flow scenarios can be modeled and dynamic segmentation can be used to provide detailed and localized information at any given point along a route. If a route needs to be moved or recommissioned, previous models can be modified to provide insight into optimal relocation routes.

Automation should be inherently repeatable by design. Once established as being a useful tool, there is value in its reusability. Repeatability allows engineers to apply tools that work from one project to the next, with dependable and reliable results. By using well thought out and proven models, repeatable results are achievable on multiple projects, saving both time and money.

In summary, automating complex pipeline processes is a continually evolving craft. Improvements in hardware and software allow for infinitely more detailed and complex model design. These advances in technology will continue to improve modeling efficiency and in turn create better pipeline designs in the future.

8.3.5 Visualization

Visualization techniques are used to enable 3D visualization by overlaying regular 2D vector files with enhanced rasters to simulate a 3D environment. With the introduction of Google Earth, people have quickly come to understand how visualization can be an effective communication tool. This section outlines different forms of visualization, starting from the simple pictures from space or an aircraft, to complex 3-D renderings.

8.3.5.1 Examples of Visualization

Rasters in a 3D environment provide the user with a realistic view of the terrain's characteristics. These characteristics are captured not only by photography but also in digital elevation modeling. In this section, several raster examples are provided that are commonly used in a 3D environment.

1. Imagery

Imagery provides the scenery of the terrain that is captured from a higher altitude above the area of interest. The two most common sources of imagery are aerial photographs and satellite imagery. Aerial photos are collected by a camera in an aircraft, whereas satellite images are collected by satellites orbiting around the earth.

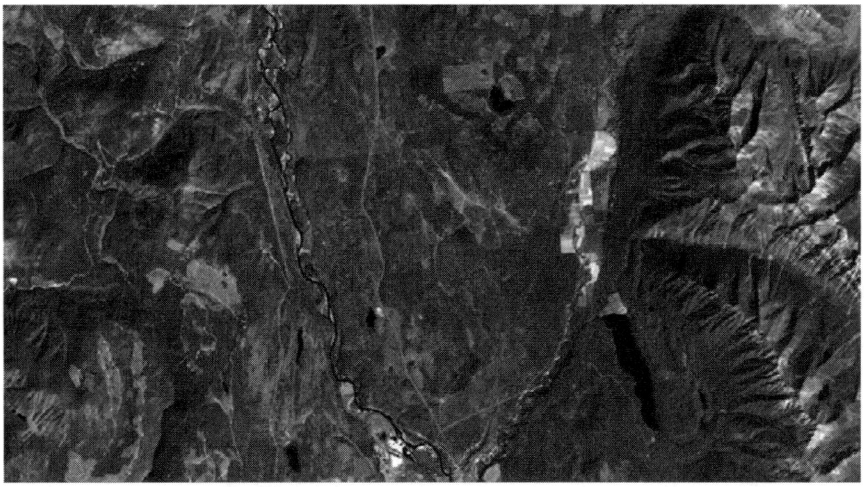

Figure 5 Example of Satellite Imagery

2. Pseudo-Color Digital Elevation Model (DEM)

Pseudo-color DEM represents the relative elevation value with a color pixel in a 24-bit raster. Pseudo-color DEM can be created in two ways, either as a discrete Pseudo-color DEM can be created in two ways, either as a discrete color ramp or as a continuous color ramp.

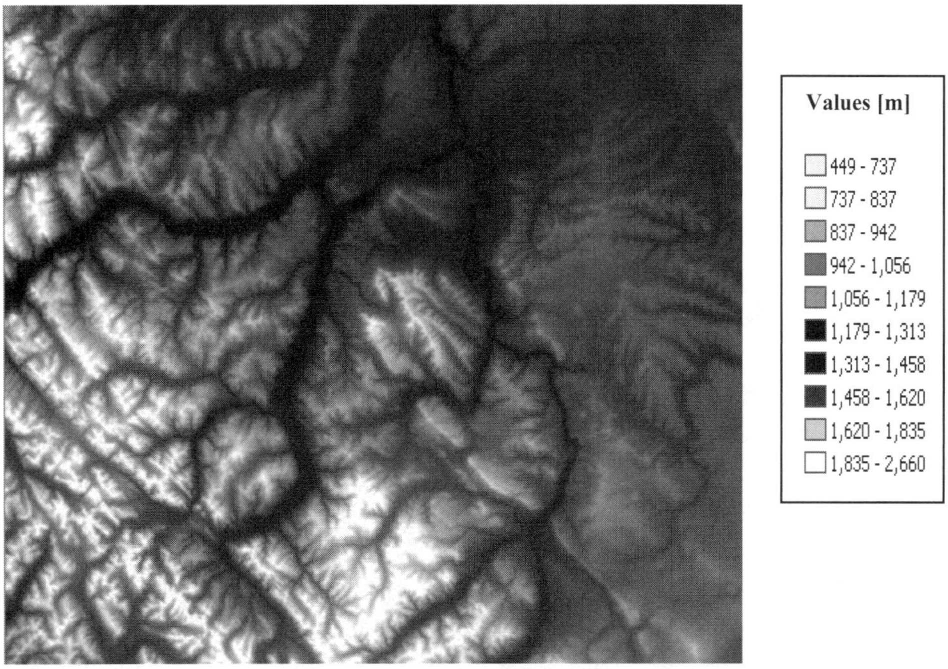

Figure 6 Discrete Pseudo-Color DEM

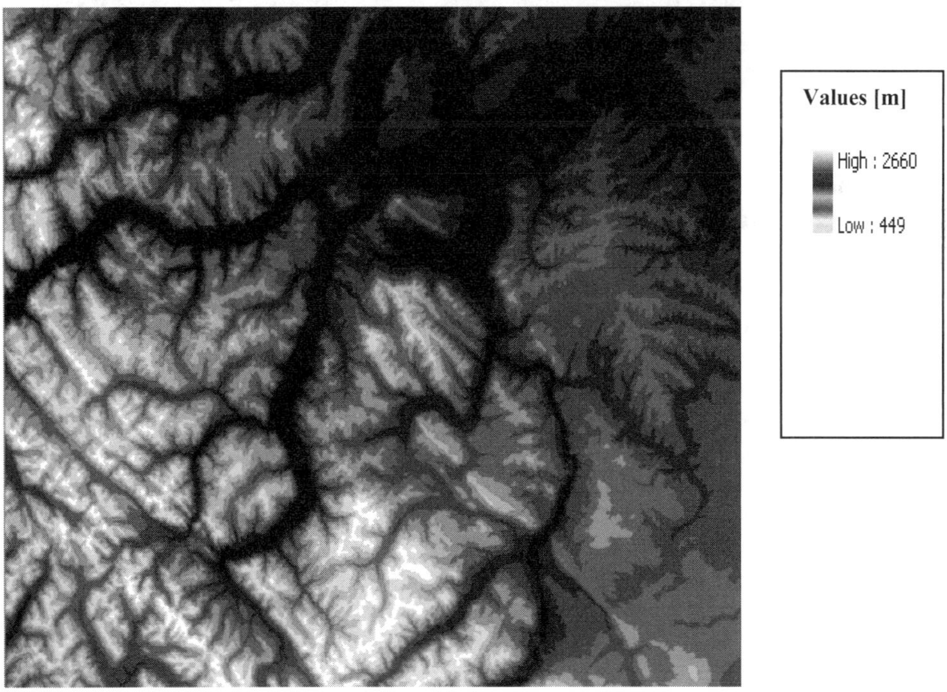

Figure 7 Continuous Pseudo-Color DEM

3. Hillshade Relief Map

A hillshade relief map employs an algorithm to represent terrain features with artificial shadows. This algorithm is used to convert the elevation values in the DEM to fit between 0 and 255 for an 8-bit raster. It also allows the user to set both the sun direction and the sun's angle above the horizon when creating the hillshade.

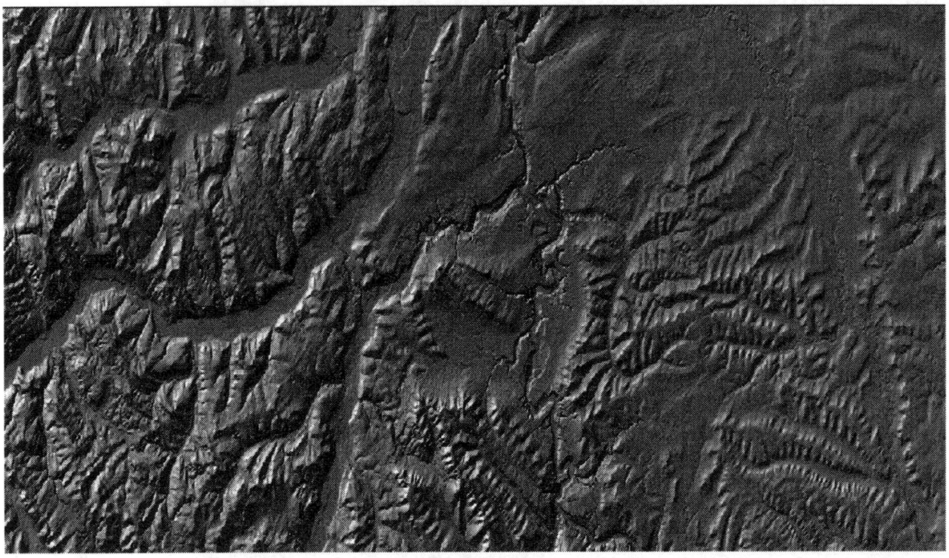

Figure 8 Example of a Hillshade

4. Fusion Mapping

Fusion mapping is a technique of combining two or more rasters into one 24-bit raster. Imagery, pseudo-color DEM and hillshade are generally used as input rasters for fusion mapping. The figures below display examples of fusion mapping.

Figure 9 Example of Fusion Pseudo-Color DEM and Hillshade Mapping

5. Vector

There are three types of vector files: point, line and polygon. All features can be represented by one of these vector types. In a 3D environment, vectors have elevation attributes and can be extruded to provide a 3D visual representation. The following sections provide two examples that are commonly used in pipeline projects.

6. 3D Contours

3D contours are used for pre-planning stages of facility and pipeline design. 3D contours are created by extruding 2D contours with elevation values. This is generally done with a DEM.

7. Proposed Facilities

Above ground pipelines, drill rigs, and barge landings are examples of proposed facilities. These facilities have 2D coordinate information, and when combined with the DEM, a realistic 3D view of the proposed facilities can be generated.

8.3.5.2 3D Visualization with Raster and Vectors

Software packages intended for a 3D environment allow the end user to adjust zoom scale, position, and rotate the viewer's perspective. Draping raster over DEM can be done in such programs to provide 3D visualization of the terrain characteristics. Vectors can be added to aid with the pre-planning stage of proposed facility and pipeline. Figure 10 is an example of 3D visualization with raster and vectors.

Figure 10 3D Visualization with Raster (Fusion of Pseudo-Color DEM and Imagery) and Vectors

8.3.5.3 Using and Interpreting Visualization Results

Visualization products can be used in a number of different ways and for different objectives. One such use is for business development purposes to gain client interest or to plan for a project. For example, a simple but colorful 3D map of a proposed pipeline route can help high-level decision makers envision the size and scope of a proposed project, especially with true-life scaled 3D imagery and models. Comparisons can be made of the "before" look versus the "after" look, to help assess project size, costs, area, etc. In addition, 3D environments can be explored via a 3D fly-through movie, an exciting and eye-catching promotional tool. As they often say, a picture is worth a thousand words, which is especially true when slide after slide of charts, figures, and numbers can be succinctly summarized in a proposed picture, map or movie.

Quite often 3D models and maps, using various visualization products, are used in open-house information sessions. When proposed pipeline projects are presented to the public, it is very important that the technical detailed drawings not complicate or overwhelm the public, creating unnecessary questions and concerns. The use of mapping products can help the general layperson grasp a better understanding of how the project will affect them, their property and the community. Thus, public awareness can be greatly aided with easily interpreted and recognizable mapping products, such as DEM draped with aerial imagery and identifiable property ownership zones.

Micro-routing is a common visualization product that can be used for engineering design. These visuals can serve as tangible reference maps, without having to go through expensive and time-consuming field visits, especially when pipelines can be thousands of kilometers long. Micro-routing can help assist with maintenance through the identification of intersecting cross sections of non-visible boundaries (i.e. property ownership), and is useful for envisioning the pipeline project area in times where poor weather affects visibility, travel and work ability.

Figure 11 Imagery Draped DEM of Rocky Mountainous Terrain

Visualization can also help analyze terrain structure. For example, draping imagery over a DEM allows engineers to quickly assess terrain dangers such as steep slopes and failures. This helps engineers quickly identify and avoid dangerous areas where construction is not recommended. Geologists can determine the overall geological structure, using a combination of imagery showing photographed forestry growth and DEMs with texture enhancements to identify concerns such as fault lineaments, water and soil erosion lines, or water crossing features. These could indicate unstable soil or bedrock, which can affect construction costs and feasibilities. Also, engineers can quickly indicate areas where further investigative detail is needed, such as elevation profiles or environmental risk models, or where higher resolution data needs to be purchased. This is especially relevant over rocky and mountainous terrain. Overall, visualization techniques can be used to automate the design procedure processes before pipelines have even been created.

In addition to viewing basic pipeline routes, the design and placement of related facilities can also be greatly assisted with visualization products. Using techniques such as hillshades, engineers can evaluate the flat areas where pump stations, staging, waste and camp areas can be optimally located with minimal 'cut and fill' requirements and thus minimal construction and logistic costs. Geological concerns, such as landslide or avalanche areas, rock deformations, water drainages or land instability can be easily identified, and considered for optimal and cost-effective pipeline construction methods.

Overall, the advantages of visualization are easily understood since they can be seen. One can quickly envision a 3D representation of a project, without the costs, time, and safety risks incurred by field visits. A desktop study can be shared with many others in various formats such as maps, interactive on-the-fly displays in a conference room, or 3D videos or screen captures. Although the upfront costs of data acquisition and software/hardware needs deter some users, the benefits of visualization typically outweigh the initial costs. This builds up the GIS infrastructure and further leverages the information assets help within the data management system.

8.3.6 Web GIS

As the use of GIS continues to increase within the industry, so too has the adoption of Web GIS. At its most basic level, a Web GIS refers to an implementation of a GIS over a TCP/IP computer network, which could be a corporate intranet or even the Internet. At its most complex, a Web GIS could represent a highly complex, distributed geographical information system that can be made easily accessible to every involved party of a pipeline project.

8.3.6.1 Web GIS Architecture

The architecture of a Web GIS is an extension of a traditional client/server computer system. Much like a client/server environment, it is necessary to have a server, that is, a powerful computer that will encapsulate the functionality and data of a GIS, and one or more clients, being computer applications that access and use the data and functionality contained on the server. Where a Web GIS differs however is its ability to completely centralize the data and functionality of the system on the server, thereby requiring the clients to be no more than a basic web browser application such as Internet Explorer.

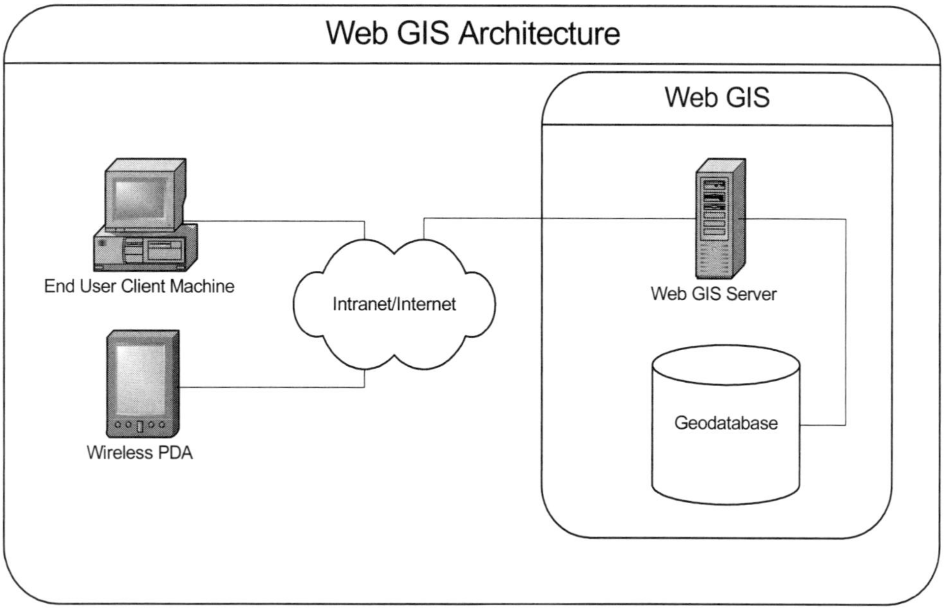

Figure 12 WEB GIS Architecture

8.3.6.2 Advantages of Using a Web GIS

Since the architecture of a Web GIS follows a heavily centralized model, it offers the following advantages:

1. Rapid System Deployment

As a project progresses and evolves, so too must the GIS that supports it. By using a Web GIS model, the system can be modified in one place (i.e. the server), and subsequently, all users will always be accessing the latest version of a system, without having to perform software upgrades/installations.

2. Centralized Management

The Web GIS model allows for the management of data, functionality and security from the central server location. This greatly reduces time spent performing computer administration tasks, and ensures data security.

3. Unrestricted Deliverability

Traditional computer systems are often difficult to make accessible to users outside of a particular area, whether it be a geographical area (i.e. must be in the same building, same city) or computer network. But since a Web GIS uses standard communication protocols (TCP/IP and HTTP), it can be made readily available to users across different areas and computer networks.

4. Support for Different Clients

As stated before, the Web GIS model requires that the end user have only a basic web browser to use it. This makes it possible for users of different operating systems and hardware platforms to easily access a Web GIS. So whether the user is sitting at a desktop computer, or in the field with a wireless PDA, the functionality of the Web GIS will still be available.

5. Data Sharing

The inherent nature of the Internet is to allow the sharing of data. Since a Web GIS can be made available to the Internet, it provides the facility for the sharing and publishing of data. This proves extremely beneficial for the collaboration of an alliance of companies, or even for the regulatory and public consultation process of a project.

6. Systems Integration

A Web GIS is in itself part of a Service Oriented Architecture (SOA). That is, it provides a service, being GIS functionality, to a variety of clients, including other systems. Given this, it lends itself to being easily integrated with other server-based services, for example, a SCADA system could provide real-time data that could integrated with the Web GIS and displayed on a map.

8.3.6.3 Applications of a Web GIS

While it is quite possible for a Web GIS to provide extremely complex geoprocessing capabilities, they are ideally suited for lightweight applications including:

1. Map Viewing

Web GIS applications are ideally suited for basic map viewing operations because they are lightweight, widely available and ensure the most up to date version of a map is being viewed. This is especially useful when viewing pipeline routing maps.

2. Spatial Searches

Web GIS applications are effective when there is a need to provide the ability to perform a search on a map because the server handles the storage of the data and the processing of the search. For example, a user may wish to search for all pipeline water crossings within a certain range of a given location.

3. Regulatory and Public Consultation

Since a Web GIS can be easily published to the Internet, any relevant data/maps that need to be viewed by outside parties can be made available to them in their most current state.

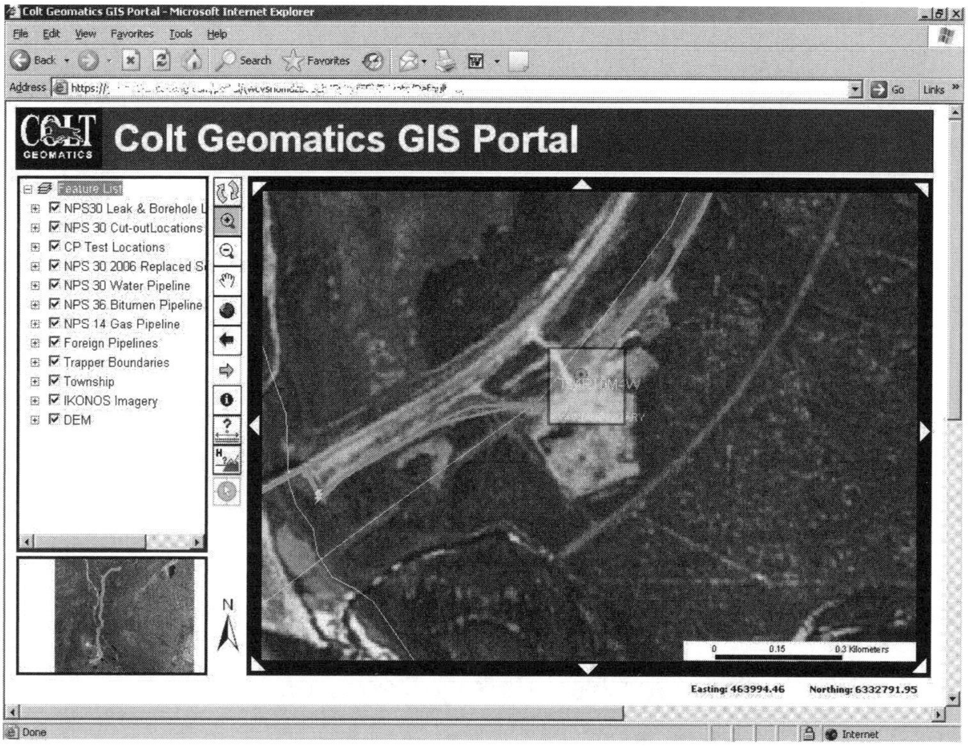

Figure 13 WEB GIS Map Viewer Application

4. Field Programs Support

As Web GIS applications can be accessed by a variety of clients, including wireless PDA's and other mobile devices, they lend themselves to being able to support various field programs. A Web GIS application could be written to not only provide the field program worker with the latest data, but also allow them to make changes from the field.

8.3.6.4 Web Services

Over the last few years there have been major improvements in the area of systems interoperability. Of particular note is the development of Web Services. Essentially, a Web Service is a Web-based software application that is used by another software application. That is, it is a self-contained computer system whose functionality and data can be leveraged by another computer system, even over great geographical and corporate boundaries, and across heterogeneous computer operating systems, development environments and networks.

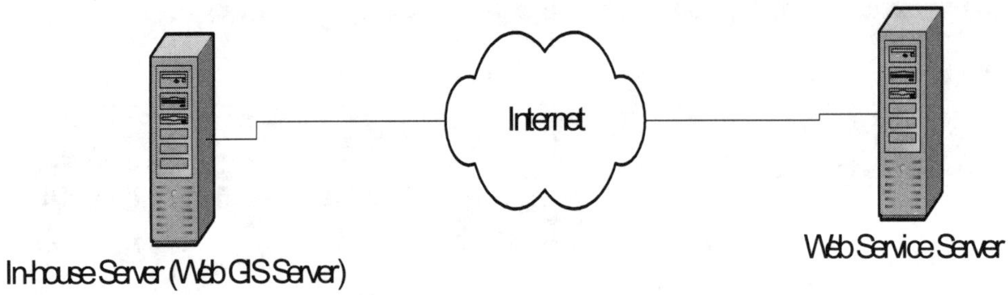

Figure 14 Web Service Illustration

The implications of Web Services within the context of Web GIS are vast. By being able to access the functionality and data of various systems, the need to implement said functionality or acquire said data is removed. For example, a government agency may expose vital geographical transportation infrastructure or vegetation data via a Web Service. It is then possible to create a system that can access this Web Service in-house, even if the systems are running on two different operating systems, and developed in two different development environments.

8.4 GIS Support for Regulatory Requirements

Issues of safety and environmental health are of primary concern to pipeline companies, regulatory bodies, stakeholder groups and the general public alike. As part of day-to-day operations, pipeline owners and operators are required to document a growing number of statistics to demonstrate their compliance with engineering standards, safety and environmental regulations, and record management requirements set out by various regulatory bodies. In this section we will examine a variety of ways in which GIS technology can be used to automate business processes that help pipeline companies meet increasing regulatory requirements.

8.4.1 United States Regulatory Bodies and Legislation

In the United States there are several governmental and independent regulating bodies whose mandates focus on ensuring national pipelines are designed, constructed, operated and decommissioned with minimal risk and impact on humans and the environment. Here we examine the major regulating agencies within the United States and some of the most common regulatory requirements imposed on pipeline companies.

8.4.1.1 Federal Energy Regulation Commission

The Federal Energy Regulatory Commission (FERC; http://www.ferc.gov/) is an independent agency that regulates the interstate transmission of electricity, natural gas, and oil. One of FERC's top priorities is to ensure environmentally safe infrastructure. Under this mandate FERC is responsible for: regulating the transmission of natural gas and oil by pipeline in interstate commerce; approving the siting of and abandonment of interstate pipeline, storage and liquefied natural gas facilities; and, using civil penalties against energy organizations and individuals to ensure compliance with FERC regulations.

FERC performs operational inspections and audits of industry participants to ensure compliance with rules, orders, regulations, and statutes. Inspections and audits of electric power, natural gas and oil pipeline industries concentrate on materially relevant issues, and enforce penalties for non-compliance.

8.4.1.2 Office of Pipeline Safety

The U.S. Department of Transportation, Research and Special Programs Administration, Office of Pipeline Safety (OPS) is responsible for ensuring the safe, reliable, and environmentally sound operation of the U.S. pipeline transportation system through mandated regulatory and enforcement activities. This mandate is enforced through a number of initiatives including: Compliance Safety; the National Pipeline Mapping System; the Integrity Management Program; Pipeline Safety Data Analysis; Regulatory Development; and, the identification of Unusually Sensitive Areas.

These federal pipeline safety regulations:

- Assure safety in design, construction, inspection, testing, operation, and maintenance of pipeline facilities;
- Set out parameters for administering the pipeline safety program;
- Incorporate processes and rule-making for integrity management; and,
- Delineate requirements for onshore oil pipeline emergency response plans.

OPS has regulatory oversight of approximately 330,000 miles of gas transmission pipeline and 160,000 miles of hazardous liquid pipeline operating in onshore and offshore territories of the US (3). The Pipeline Safety Act, adopted by congress in 1992, directs that OPS must require pipeline operators to identify facilities located in environmentally unusually sensitive areas, to maintain maps and records detailing that information, and to provide those maps and records to federal and state agencies upon request. To store and manage this location information, OPS implemented the National Pipeline Mapping System (NPMS).

The NPMS is a GIS database that tracks and visualizes the location of gas transmission and hazardous liquid pipelines, liquefied natural gas (LNG) facilities and breakout tanks within the command of OPS. First developed as a national repository, the NPMS now serves as a decision-support tool for inspection planning, community access and risk assessment identifying where additional precautions are required to guard against potential pipeline releases.

In 2001, after 9/11, the NPMS was removed from the public domain. This was done to protect the security of the pipeline infrastructure.

In 2002, participation in the NPMS was no longer voluntary for pipeline operators; congress mandated participation through an amendment to the Pipeline Safety Act. Section 15 of the Act details new requirements for pipeline operators, with specific regard to spatial information. The Act dictates that pipeline operators are now required to submit "geospatial data appropriate for use in the NPMS or data in a format that can be readily converted to geospatial data." The revised Act also requires that attribute data and metadata for all pipeline operation systems be submitted for inclusion in the NPMS.

8.4.1.3 Pipeline and Hazardous Material Safety Administration

Also under the jurisdiction of the U.S. Department of Transportation, the Pipeline and Hazardous Material Safety Administration (PHMSA; www.phmsa.dot.gov) administers the national regulatory program for ensuring the safe transportation of natural gas, petroleum, and other hazardous materials by pipeline to industry and consumers. PHMSA oversees the nation's pipeline infrastructure, which accounts for 64 percent of the energy commodities consumed in the United States. PHMSA conducts inspections and audits of pipelines to ensure compliance with safety and training requirements. The PHMSA audits include a review of the accuracy of mapping and survey information for the purpose of identifying the precise location of High Consequence Areas (HCA's). Pipeline operators who fail to provide accurate and sufficient documentation to demonstrate their compliance may be subject to significant fines and other disciplinary action.

8.4.2 Canadian Regulatory Bodies and Legislation

Similar to the United States, Canada also has a number of regulating bodies that govern the design, construction, operation and decommissioning of pipelines.

8.4.2.1 National Energy Board

The National Energy Board (NEB; www.neb.gc.ca) is an independent federal regulatory agency that regulates aspects of Canada's energy industry including: the construction and operation of inter-provincial and international pipelines; the export and import of natural gas, oil, and electricity; and, frontier oil and gas activities. The NEB promotes safety and security, environmental protection and efficient energy infrastructure.

In its role of regulatory oversight, the NEB controls the operation and maintenance of pipelines under the National Energy Board Act (2005). The Requirements and Guidance Notes define two requirements that can easily be supported by GIS: Conducting effective public engagement related to operations and maintenance; and, maintaining documentation for operations and maintenance activities.

Under the first of these requirements, pipeline operators are required to share information with members of the public, who may be affected by planned operations and maintenance activities, and to identify and resolve issues or concerns related to these activities. Examples of potentially affected members of the public include landowners, tenants, residents, Aboriginal communities, government agencies, non-governmental organizations, trappers, guides, outfitters, recreational users, other land or resources users, and commercial third parties. The Act requires not only that the public be engaged, but also that sufficient records be maintained and provided upon request.

In addition to requiring documentation on public engagement, the NEB requires that operating companies prepare and manage records related to pipeline design, construction, operation, and maintenance that are needed for performing pipeline integrity management activities. Requirements set out in the national standard for Oil and Gas Pipeline Systems (CSA Z622-03) dictate that such records be kept current and readily accessible to the operations and maintenance personnel requiring them.

8.4.2.2 Alberta Energy and Utilities Board

Alberta is one of Canada's most energy-rich provinces, with a large percentage of the nation's pipelines originating in, or traveling though the province. The Alberta Energy and Utilities Board (EUB; www.eub.ca) is an independent, quasi-judicial provincial agency that regulates the safe, responsible and efficient development of Alberta's energy resources, and the pipelines and transmission lines that move the resources to market. The goal of the EUB is to ensure compliance with regulations through inspections, surveillance, intervention, and education.

8.4.3 Using GIS to Support Regulatory Compliance

The very nature of Geospatial Information Systems (GIS) lends itself to supporting many types of regulatory requirements. GIS technology enables users to tie a myriad of information to any given location. This ability allows pipeline operators to maintain detailed records, in the form of automated maps and related databases, which store the location and attributes of their assets.

The value of GIS to pipeline automation is twofold. First, GIS provides an automated environment for performing analysis functions such as risk assessment or defining High Consequence Areas (HCAs). Second, GIS offers the ability to efficiently manage spatial data so that it can be easily recalled at any time, in response to regulatory requirements, information requests, and inspection or audit requisites. In this section we will examine in greater depth some of the most common ways in which GIS can be applied to support regulatory compliance and reporting.

8.4.3.1 Permit Applications

Before a pipeline is ever approved for construction and operation, pipeline companies may be required to submit countless permit applications for areas of the pipeline that cross watercourses or transportation routes, or for any facility that is proposed along the pipeline route. Maps form a key component of the permit application, and are required to identify the exact location of the proposed infrastructure vis-à-vis the surrounding environment and geological features. GIS is a necessary tool in developing permit application maps for regulatory approval. GIS technology allows for a variety of 'what-if' scenarios to be tested and visualized in a matter of minutes, allowing for various design options to be considered and evaluated quickly.

8.4.3.2 Risk Analysis and Geohazard Identification

As part of the approvals process, most regulatory bodies require extensive documentation reporting on risk analysis findings and the identification of potential geohazards. Geohazard identification involves the location, identification, and comparative assessment of historical failure incidents caused by geohazards. A growing movement towards proactive management has resulted in emphasis on terrain analysis using stereo aerial photographs and satellite imagery to map areas of potential influence surrounding the proposed pipeline, beyond the right-of-way and including surrounding watersheds. GIS tools are used to support: frequency analysis; consequence analysis; and, risk estimation, evaluation, control, and monitoring activities. Pipeline companies are increasingly being required to provide detailed risk assessment reports to regulatory bodies, to ensure they are compliant with safety protocols.

8.4.3.3 Responding to Information Requests

Regulatory agencies may require, at any time, that pipeline operators provide information regarding any aspect of their proposed or operating pipeline and/or facilities. Spatial data management, facilitated through a GIS environment, ensures that information on assets, potential environmental impacts, and safety concerns can be immediately collected for any spatial area. This eases the task of responding to anticipated and unexpected regulatory information requests in a timely fashion.

8.4.3.4 Pipeline Integrity and HCA Identification

As outlined above, the OPS enforces regulations related to pipeline integrity. OPS uses GIS technology and the NPMS to characterize and define High Consequence Areas (HCAs) for hazardous liquid pipelines. Under the requirements to develop an Integrity Management Program (IMP), operators must identify all pipeline segments that fall within, or could affect, a HCA. While OPS defines HCAs for hazardous liquid pipelines operators, integrity management for the natural gas industry requires that natural gas pipeline owners identify their own HCAs. This definition is based on stringent guidelines related to the presence of housing or other structures on or near the right-of-way.

Geoprocessing functionalities of GIS allows for the simple overlay of data layers, and proximity/buffer calculations to identify potentially hazardous parts of the pipeline network. Complex algorithms can be built to model the pipeline, surrounding infrastructure and natural features, and potential impacts on pipeline integrity.

8.4.3.5 Public Awareness Requirements

Public awareness and education programs are critical to minimizing threats to the integrity of a pipeline and ensuring a high level of public well-being. Safety regulations are increasingly incorporating requirements for public awareness programs, including the identification of target audiences, specific messages to be delivered, and the frequency and methods of delivery (4). The American Petroleum Institute's (API) Recommended Practice 1162 provides clarity on how to define target audiences for public awareness programs, which include: the affected public, emergency officials, and local public officials and excavators. While public information resources make it relatively easy to identify emergency officials, public officials and excavators, identifying 'the affected public' can be more difficult.

GIS technology can be used to visualize the pipeline right-of-way with surrounding landowner information, and through proximity analysis can easily identify those properties within the affected corridor. Leveraging the database capabilities of GIS, mailing lists can be automatically generated to support mailing programs that provide homeowners, tenants, and businesses with public awareness materials. GIS also offers the ability for pipeline operators to leverage their normal business practices to demonstrate their compliance with public awareness and education requirements. GIS can be used to capture and visualize to auditors the scope of the identified affected public, in addition to the type and frequency of the notification provided.

8.4.3.6 Compliance Reporting and Records Maintenance

This section has reviewed many of the major regulatory bodies that control the design, construction, operation and abandonment of pipelines. Regardless of the governing agency or jurisdiction, pipeline operators will continue to face increasing reporting requirements. Though the scope of reporting obligations is vast, documentation is typically required to demonstrate compliance with engineering standards, safety practices, and environmental rules. With the demands for archival and historic tracking of assets for compliance purposes increasing, the volume of data required to operate and manage a pipeline will continue to grow into the future (5). GIS technology and spatial data management techniques provide a systematic and disciplined approach to ensuring records are well-maintained and readily available to any requesting agency.

8.5 Summary: The Central Database Paradigm Shift

The material presented throughout this chapter describes the components of geographic information systems as they may be applied to managing pipeline data and automating pipeline engineering processes. A fitting summary and overarching framework to pipeline GIS is the concept of the *Central Database*. Since most forms of automation, modeling, and analysis require an input of data, the central database provides a robust support to these operational realities. In fact, the central database is as much a philosophy as it is computer infrastructure, software, and processes. By subscribing to the philosophy of *a single central-source of truth*, pipeline projects and owner/operators begin to treat data as a valuable asset, keep data organized, secure, accessible, and effectively managed through continuous changes, updates, and applications. Once data is respected, decisions founded on that data gain value. The term *Paradigm shift* is commonly used to denote a punctuated change in a discipline's thinking. Thomas Kuhn (6) brought this term into its modern usage by referring to scientific revolutions as paradigm shifts "Successive transition from one paradigm to another via revolution is the usual developmental pattern of mature science."

The central database philosophy fits into this model since it is causing the engineering disciplines to approach problems in a *data-centric* manner, one that is fundamentally new to the discipline. Furthermore, the central database approach has a sense of inevitability since the amount of digital data accessible and the volume subsequently generated will only keep growing. As we have seen throughout this chapter, data is the basis of many automated processes and the central database philosophy is the foundation to fully leveraging data's use to support pipeline engineering.

8.5.1 Why We Use a Pipeline Central Database

The main reason for having a central database is to manage change. It does so by providing an information infrastructure for a pipeline's lifecycle and adds stability to an information management system. The structure it provides allows data users to conduct their activities in a stable environment despite the constant changes being made to the data. With a dependable information system backbone, data use and sharing become valuable additions to a project (7). A central database provides this stability through four supports to the information management system. These are:

- Direct change management.
- Data security;
- Data integrity;
- Metadata (the data storyboard and legacy)

These four functions of a central database allow for the centralized control and quality assurance of all data. As a result, an organization will have a single source of truth despite the variety of disciplines and contractors contributing data and generating information. Through planning and construction, operations and integrity maintenance, and finally to decommissioning, the central database's contents and physical infrastructure may change, but the four functions will continue, thus providing the stability for the life of the project or enterprise.

8.5.2 Benefits of a Central Database

There are several significant benefits that have been realized on many pipeline projects as a result of subscribing to the central database philosophy. These include engineering tools and automation, web GIS, and information access and control. There are certainly other benefits, however, these have shown to be ubiquitous.

8.5.2.1 Engineering Tools and Automation

Perhaps one of the most significant applications of a central database in the pipeline world is seen in automated alignment sheet generation. The revolution that has occurred is that the alignment sheet has gone from a drawing to a report. It is considered a report simply because features that typically appear on an alignment sheet are stored independently as data within a database (8). A sheet is generated by querying the database and extracting the data layers which correspond to a sheet window. The efficiency introduced by this process is that the data is independent of the alignment sheet. Therefore, the most current and approved version of the data is used as it is in numerous other applications. As a result, pipeline projects and operators have found significant time and cost savings to automatically generating alignment sheets.

Other engineering tools are easily implemented. The only barrier is the quality of data on which the output relies. For example, a slope tool can easily be created for computing cross and long slopes. However, the output strongly depends on the resolution and accuracy of the input topographic data. In any case, these tools use existing data and in turn create more data which is secured in the database.

8.5.2.2 Web GIS

The inherent nature of the Internet is to allow the sharing of data. Since a Web GIS can be made available through the Internet, it provides the facility for the sharing and publishing of data. This proves extremely beneficial for the collaboration of groups not physically working together and communicating with people in the field. Since the architecture of a Web GIS follows a heavily centralized model, it offers many advantages:

- Rapid System Deployment. By using a Web GIS model, a system can be modified in one place (i.e. the server), and subsequently, all users will always be accessing the latest version of a system, without having to perform software upgrades/installations.

- Unrestricted Deliverability. Traditional computer systems are often difficult to make accessible to users outside of a particular area, whether it be a geographical area (i.e. must be in the same building, same city) or computer network.
- Support for Different Clients. Web GIS requires that the end user have only a basic web browser to use it. This makes it possible for users of different operating systems and hardware platforms to easily access a Web GIS.
- Data Sharing. The inherent nature of the Internet is to allow the sharing of data. Since a Web GIS can be made available to the Internet, it provides the facility for the sharing and publishing of data.
- Systems Integration. A Web GIS is in itself part of a Service Oriented Architecture (SOA). That is, it provides a service, being GIS functionality, to a variety of clients, including other systems. Given this, it lends itself to being easily integrated with other server-based services, for example, a SCADA system could provide real-time data that could be integrated with the Web GIS and displayed on a map.

8.5.2.3 Information access, release, and control

The bar has been raised in regulatory situations whereby regulators routinely require maps and other visual output to support applications. The production of these products is not difficult, however, maintaining coherent control on the data being reported can be, particularly on large projects. A central database controls what is being reported, who has access to change it, and when it was released. This security is comforting when volumes of output are being generated from massive amounts of project data.

This security and control certainly extends beyond regulatory needs and can be seen as a safety net for other activities ranging from preliminary engineering to pipeline integrity and risk.

8.5.3 Final Words

Pipeline systems are growing in complexity and many are aging, which require a constant accounting of their location, integrity, and potential impact on their surroundings. To help manage the volumes of data generated and maintained during this accounting, geographic information systems have provided many efficiencies. These systems keep data organized, secure, accessible, and effectively managed through continuous changes and updates. They also enable the automation of many engineering and integrity processes, thus offering innovation to pipeline projects and operations. However useful geospatial techniques have become, like many innovations, their introduction has been disruptive at times. With more ubiquitous software, cheaper computer hardware, and faster network and internet connections, adoption of GIS is becoming less disruptive and easier to understand, to the growing benefit of pipeline projects and owner/operators.

References

(1) Adam, S. and J.T. Hlady, "Data is an Asset that must be managed," *Proceedings of IPC: International Pipeline Conference, Calgary, Canada.* 2006.

(2) Adam, S., M. Farrell, "Earth Imagery Monitors Pipeline Integrity" *Imaging Notes*, 16(2): 18-19. 2001.

(3) Hall, S., "The National Pipeline Mapping System – A Review." *Business Briefing Exploration & Production: The Oil & Gas Review*, Vol. 2, pp. 84. 2003

(4) Johnson, D., "The Rules Covering Public Awareness and GIS: A Pipeline Safety/Compliance Perspective." *Proceedings of GITA's 15th Annual GIS for Oil & Gas Conference and Exhibition.* Calgary, Canada. 2006.

(5) Veenstra, P., "Meeting Future Challenges of Pipeline Data Management." *Proceedings of GITA's 15th Annual GIS for Oil & Gas Conference and Exhibition*, Calgary, Canada. 2006

(6) Kuhn, T.S., *The Structure of Scientific Revolutions*, 2nd. ed., Chicago: Univ. of Chicago Press, p. 206, 1970.

(7) Pinto, J.K. and H.J. Onsrud, "Sharing Geographic Information Across Organizational Boundaries: A Research Framework." In Onsrud, H.J. and G. Rushton (Eds.), *Sharing Geographic Information*, (New Brunswick, NJ: Center for Urban Policy Research, Rutgers), 44-64. 1995,

(8) Jones, B.A., "Using Geographic Information Systems for Pipeline Integrity, Analysis, and Automated Alignment Sheet Generation," *Exploration & Production: The Oil and Gas Review,* p. 52-54. 2003.

Appendices

1. Pipeline Flow Equations and Solutions

1.1 Pipeline Flow Equations

The mathematical models used for pipelines are based on equations derived from the fundamental principles of fluid flow and thermodynamics. Four equations are required to relate the four independent variables: pressure, temperature, flow rate and density. Flow through a pipeline is described by the momentum, mass and energy conservation equations together with the equations of state appropriate to the fluids in the pipeline. These three conservation laws can be expressed in partial differential equations: momentum equation, continuity equation and energy equation. The one-dimensional form of the conservation equation is adequate for pipeline flow simulation.

1. Momentum Equation

The momentum equation describes the motion of the fluid in the pipeline, requiring fluid density and viscosity in addition to the pressures and flows. Including the Darcy-Weissbach frictional force, it is expressed as

$$\rho \frac{\partial v}{\partial t} + \rho v \frac{\partial v}{\partial x} + \frac{\partial P}{\partial x} + \rho g \frac{\partial h}{\partial x} + \frac{f \rho v \mid v \mid}{2D} = 0$$

where

ρ = Density of the fluid
v = Velocity of the fluid
P = Pressure on the fluid
h = Elevation of the pipe
g = Gravitational constant
f = Darcy-Weissbach friction factor
D = Inside diameter of the pipe
x = Distance along the pipe
t = Time

The Darcy-Weissbach friction factor is determined empirically and represented by the following equations:

$$f = \frac{64}{\text{Re}} \; for \; \text{Re} \le 2400$$

and

$$\frac{1}{\sqrt{f}} = -2\log\left(\frac{e}{3.7D} + \frac{2.51}{\text{Re}\sqrt{f}}\right) for \; \text{Re} \ge 2400$$

where Re is the Reynold's number defined by

$$\text{Re} = \frac{|v|\rho D}{\mu} = \frac{|v|D}{\nu}$$

and μ is the dynamic viscosity and ν the kinematic viscosity.

2. Continuity or Mass Conservation Equation

The mass conservation equation accounts for mass conservation in the pipeline, requiring the density and compressibility of the fluid in the pipeline together with flows, pressures and temperatures.

$$\frac{\partial(\rho A)}{\partial t} + \frac{\partial(\rho v A)}{\partial x} = 0$$

where

A = Cross sectional area of the pipe

The cross sectional area can change due to the changes in pressure and temperature:

$$A = A_0\left[1 + c_P(P - P_0) + c_T(T - T_0)\right]$$

where the subscript zero refers to standard conditions. c_T is the coefficient for thermal expansion of the pipe material and c_P is defined as

$$c_P = \frac{1}{E}\left(\frac{D}{w}\right)\left(1 - \mu^2\right)$$

where

E = Young's modulus of elasticity of the pipe

w = Pipe wall thickness

μ = Poisson's ratio

3. Energy Equation

The energy equation accounts for the total energy of the fluid in and around the pipeline, requiring the flows, pressures and fluid temperatures together with fluid properties and environmental variables such as conductivity and ground temperature.

$$\left(\rho C_v + \frac{4 w \rho_p C_p}{D} \right) \frac{\partial T}{\partial t} + \rho v C_v \frac{\partial T}{\partial x} + T \frac{\partial \rho}{\partial T} \left(\frac{\partial v}{\partial x} + \frac{v}{A} \frac{\partial A}{\partial x} \right) +$$

$$+ \frac{f \rho v^2 \mid v \mid}{2D} - \frac{4k}{D} \left(\frac{dT}{dz} \right) = 0$$

where

Cv = Specific heat of the fluid at constant volume

T = Temperature of the fluid

ρ_p = Density of the pipe material

C_p = Heat capacity of the pipe material

k = Heat transfer coefficient

z = Distance from the pipe to its surroundings

4. Bulk Equation of State

In addition to these three conservation equations, an equation of state is needed to define the relationship between product density or specific volume, pressure and temperature. The simple equation of state given below is adequate for crude oils and heavy hydrocarbons:

$$\rho = \rho_0 \left[1 + \left(\frac{P - P_0}{B} \right) - \gamma (T - T_0) \right]$$

where

B = Bulk modulus of the fluid

γ = Thermal expansion coefficient of the fluid

The API equation of state is a variation of this equation of state (1). The bulk equation is used for custody transfer and widely accepted by liquid pipeline industry as a standard equation of state for hydrocarbon liquids such as gasoline or crude, while BWRS or SRK equation of state is suitable for light hydrocarbon fluid such as natural gas. There are various equations of state for other hydrocarbon and non-hydrocarbon fluids, which are extensively compiled by NIST (2).

1.2 Solution Methods

These four equations are solved for four primary variables: flow or velocity, pressure, temperature, and density. In real-time modeling, these equations are solved each scan using boundary conditions received from the host SCADA system. It is assumed that no chemical reaction takes place in the pipeline system and that the fluid remains in a single phase. Multiphase modeling and applications are not addressed in this book.

These partial differential equations are coupled non-linear equations. Since these equations cannot be solved by an analytical method, they are solved by numerical methods instead. In any numerical method, certain approximations are required such as replacing derivatives in the differential equations with finite differences using averages calculated over distance and time intervals and truncating certain terms in the differential equations.

The partial differential equations are expressed in terms of distance or pipe length and time variables. The solution requires initial conditions for the time variable in order to establish initial pipeline state and boundary conditions to provide boundary values for pipe length. A pipeline state is expressed in terms of four primary variables: flow, pressure, temperature and density. A typical real-time modeling procedure will follow the sequence described below:

- Establish initial pipeline state in terms of flow, pressure, temperature and density profiles along the entire pipeline. The initial pipeline state can be obtained by a steady state solution if there is no known pipeline state or by the previous pipeline state if it is available. When the pipeline model is first started, no pipeline state is known and thus the initial state is approximated by a steady state solution.
- At the end of a time interval, the current pipeline state is calculated from the four equations using the initial state determined in the previous time and by applying the boundary values received from the host SCADA system.

Depending on the method of handling boundary conditions, two real-time model architectures are available:

- Independent leg: In the independent leg architecture, the four governing equations are solved independently for each segment of the pipeline between two boundaries. Measured pressures and/or flows are used as boundary conditions, and a segment includes pipes, compressor or pump station, a regulator station, and/or valves. The equations for the pipeline system may include compressor or pump station and valves. The selection of segment boundaries and boundary conditions is flexible.

- Global model: In the global model architecture, the four equations are solved for the entire pipeline system. This can maintain conservation laws over the entire pipeline system and provide more accurate state estimation. However, the independent leg approach is frequently used in practice for an RTTM implementation, because measurement errors can be easily isolated and tuning efforts can be simplified.

There are many different ways to solve the difference equations representing the partial differential equations. Three popular solution techniques for pipeline flow simulation are briefly described below. The description includes only the aspects relevant to the real-time transient model. For more detail refer to specialized books for solving partial differential equations (3).

1. Method of Characteristics

Streeter and Wylie (4) applied the method of characteristics extensively in solving various pipeline related problems. The method of characteristics changes pipe length and time coordinates to a new coordinate system in which the partial differential equation becomes an ordinary differential equation along certain curves. Such curves are called characteristic curves or simply the characteristics.

This method is elegant and produces an accurate solution if the solution stability condition is satisfied. The stability condition, called the Courant-Levy condition, requires that the ratio of the discretized pipe length to time increment must be smaller than the acoustic speed of the fluid in the pipeline. In other words, time increment is limited by the disretized pipe length and the fluid acoustic speed. This is not necessarily a limitation for real-time applications where time increment is short. However, it can be a severe limitation if applications such as a training simulator require flexible time steps.

The method of characteristics is easy to program and can produce a very accurate solution, and also doesn't require large computer computational capability.

2. Explicit Methods

In explicit methods, the finite difference equations are formulated in such a way that the values at the current time step can be solved explicitly in terms of the known values at the previous time step (5). There are several different ways of formulating the equations, depending on discretization schemes and what variables are explicitly expressed.

The explicit methods are restricted to a small time step in relation to pipe length in order to keep the solution stable. Just like the method of characteristics, this is not an issue for real-time applications but a severe limitation for applications requiring flexible time steps. For applications extending over a long time, an explicit method could result in excessive amounts of computation.

Explicit methods are very simple for computer programming and can produce an accurate solution. The computer computational capability requirements are relatively light.

3. Implicit Methods

In implicit solution methods (6), the partial differentials with respect to pipe length are linearized and then expressed by finite difference form at the current time step, instead of the previous time as in the explicit method. The values at the current time step are arranged in a matrix, so the solution requires the use of matrix inversion techniques. There are several ways to arrange the numerical expressions, depending on discretization schemes and where values are expressed during or at the end of the time interval. Initially, an approximated solution is guessed at and then changes to the approximated solution are tried iteratively until change doesn't occur within a specified tolerance.

The implicit methods produce unconditionally stable solution no matter what size the time step or pipe length is. Unconditional stability doesn't mean the solution is accurate. Other errors may make the solution inaccurate or useless. The methods can generate accurate results if the pipe length and time step are short and the specified tolerance is tight. Therefore, they can be used not only for real-time model but also for applications requiring flexible time steps.

The disadvantages are that the methods require matrix inversion software, the computer programming is complex, and the computer computational capability requirement is comparatively high, especially for a simple pipeline system. However, the absence of a restriction on the size of time step generally outweighs the increase in the extra requirements, particularly for large pipeline systems.

There are other solution techniques such as variational methods (7) and succession of steady states. These are not discussed here, but interested readers are encouraged to refer to books on solution techniques of partial differential equations.

References

(1) "Petroleum Liquid Volume Correction", API Publication 1101, American Petroleum Institute, 1984

(2) NIST Standard Reference Database 4, Supertrapp version3, National Institute of Standards and Technology, Gaithersburg, MD 20899, U.S.A.

(3) Finlayson, B. A., Nonlinear Analysis in Chemical Engineering, McGraw-Hill, New York, N.Y, 1980

(4) Wylie, E.B. and Streeter, V.L., Fluid Transients, FEB Press, Ann Arbor, MI,

1983

(5) Carnahan, B., Luther, H.A. and Wilkes, J.O., Applied Numerical Methods, John Wiley & Sons, Inc. New York, N. Y., 1969

(6) Wylie, E.B., Stoner, M.A., Streeter, V.L., "Network System Transient Calculations by Implicit Model", Soc. Pet. Eng. J., 1971

(7) Rachford, H.H. and Dupont, T., "A Fast, Highly Accurate Means of Modeling Transient Flow in Gas Pipeline Systems by Variational Methods", Soc. Pet. Eng. J., 1974

2. Steady State Model vs. Transient Model

A steady state model calculates steady state flow, pressure and temperature profiles by ignoring the time dependent terms in the flow equations discussed in Appendix 1. The results of a steady state model are valid if the steady state assumptions are validated during pipeline operation. For a feasible flow, a steady state model can generate pressure, flow, temperature and density profiles along with listing of station suction and discharge pressures, and also:

- Determine pipeline capacity.
- Determine an efficient operating mode by selecting appropriate units if the line pack changes or transients in the pipeline network are relatively small compared to the system line pack.
- Calculate power or fuel usage and pump or compressor efficiency.
- Identify pipeline operations and an alternate configuration.

If the flow is not feasible, the model may provide the detailed information on operating constraints violated at each applicable location and allow a search for the maximum flow possible under specified conditions.

In general, the steady state model is suitable for the design and the following types of operation applications:

- Identifying what the system constraints are
- Determining an efficient operating mode, because the model can provide unit selection capability by showing the throughput vs. power requirements for various combinations of stations and unit line-ups
- Identifying stations that are most critical for continuous system operation
- Determining the maximum throughput under given conditions.

A steady state model is used extensively for pipeline system design, because it satisfies most of the system design requirements:

- The execution time should be fast to allow for large number of simulation runs.
- Most design tasks don't require short-term time-dependent pipeline behaviors, and extensive data is not required.

A transient model calculates time dependent flow, pressure, temperature and density behaviors by solving the time dependent flow equations discussed in Appendix 1. Therefore, a transient model generates hydraulically more realistic results than a steady state model, and theoretically the model is capable of performing not only all the above time independent functions performed by the steady state model but also time dependent functions such as line pack movement, effect of changes in injection or delivery, system response to changes in operation.

A transient model is an indispensable tool for studying pipeline and system operations. Although a steady state model is sufficient for most design tasks, a transient model is required to simulate transient responses to various operations before the design is finalized. The transient model is used for the following types of applications:

- Study normal pipeline operations – Pipeline operation changes are simulated to find a cost effective way of operating the pipeline system. The transient model allows the operation staff to determine efficient control strategy for operating the pipeline system and analyzing operational stability.
- Analyze startup or shutdown procedures – Different combinations of startup or shutdown procedures are simulated to determine how they accomplish operation objectives. The transient model can model a station, including the pump or compressor unit and associated equipment.
- Determine delivery rate schedules – The transient model can be used to determine delivery rate schedules that maintain critical system requirements for normal operations or even upset conditions.
- Predictive modeling – Starting with current or initial pipeline states, future pipeline states can be determined by changing one or more boundary conditions.
- Study system response after upsets – A pipeline system can be upset by equipment failure, pipe rupture, or supply stoppage. The transient model is used to evaluate corrective strategies by modeling various upset responses.
- Study blow-down or pipe rupture – The transient model allows the operation engineers to study the effects of blow-down on a compressor station and piping or to develop a corrective action when a leak or rupture occurs.

In general, a transient model is more complex to use and execution time is slower than a steady state model. It requires extensive data, particularly equipment and control data, which are often unavailable. However, a transient model is essential for the efficient operation of the pipeline.

3. Inspection and Continuous Sensing Methods

In addition to the CPM techniques, there are two other methods of detecting pipeline leaks: Inspection methods and continuous sensing devices. This appendix introduces these methods briefly.

3.1 Inspection Methods

This section describes several pipeline inspection techniques and discusses their operating principles, applications and advantages/disadvantages. The inspection methods described in this section include:

- Hydrostatic test
- Ultrasonic technique
- Magnetic flux technique
- Visual inspection
- Hydrocarbon detectors

3.1.1 Hydrostatic Test

Hydrostatic testing must be performed on new pipelines, as specified in ASME B31.4 and other standards, prior to in-service use. Hydrostatic testing is also used on operating pipelines to assess their structural integrity. It is the only reliable method of identifying stress corrosion cracking (SCC) problems.

For testing a new pipeline, a pipe segment is filled with water which is maintained at a high pressure for about a day. During this period, the pressure is monitored with accurate pressure gauges or dead weight testers. If the pressure holds, it is assumed that the line is free of defects. Because the line pressure varies with temperature, the temperature effect is accurately compensated for during the test period. The testing methods and procedures are discussed in several books (1), and therefore are not discussed here in detail.

When an operating pipeline is tested at a pressure above normal operating pressure using the fluid normally transported in the pipeline, it is called a dynamic pressure test. The purpose of this test is not to accommodate the increase in operating pressure level, but to confirm the pressure capability of the pipeline system. This test is known to be sensitive in detecting small leaks and is used for integrity assessment.

The Office of Pipeline Safety (OPS) of the U. S. Department of Transportation mandates that the requirements for integrity verification of operating pipelines be verified in high consequence areas. Hydrostatic testing is one of the primary methods for conducting pipeline integrity assessments for operating pipelines.

To evaluate the need for the testing of existing pipelines, the following factors need to be taken into consideration:

- Requirements for the line service
- Age of the line
- Leak history
- Exposure to environment

The main advantage of the hydrostatic test is that it can detect not only incipient failure reliably but also existing small pin hole size leaks. However, non-critical cracks may not be found and crack growth can accelerate due to pressurizing at the time of testing. For operating pipelines, it is costly and inconvenient to conduct a test using the resident transport fluid because the pipeline under test has to be removed from service. Furthermore, it is difficult to test because the temperature has to be stabilized for a long time. Finally, a test can be destructive if a line break occurs during the test.

Pneumatic testing with air can be an approved test method in areas where water is not readily available and where weather conditions are very severe (as in the high Arctic). Selection of test media and methods in the Arctic depends on environmental impact, availability and disposal of the test medium, accessibility of the right-of-way, and terrain types and thaw depths.

The pneumatic air test is the least costly and is the least environmentally risky. The pneumatic air test is not subject to freezing of the fill medium so it can be performed anytime. However, this test method does have several disadvantages:

- The test requires a long time to fill and perform in the test section.
- It has higher potential risks of pipe damage and personnel injury if a line break occurs.
- Neither test data nor actual experience in the high Arctic is available, so its effectiveness as a pressure test has not been fully proven in practice.

3.1.2 Ultrasonic Inspection Technique

Ultrasonic properties can be used to find pipe defects. Tools with ultrasonic properties are used to inspect internal and external defects and pipe welds on manufactured pipes and operational pipelines. The technique uses high frequency mechanical vibrations whose frequencies range from 1 MHz to 25 MHz. These ultrasonic waves propagate both transversely and longitudinally. The transverse wave is a shear wave, used primarily for detecting cracks in a pipe. A longitudinal wave is a compression wave and is used mainly for measuring pipe thickness.

The ultrasonic wave transmitter and receiver is made of piezoelectric crystal. The transmitter crystal generates ultrasonic waves when an electric current is applied to it. When a test pipe is in contact with the transmitter, the ultrasonic wave will be partly transmitted through the coupler and pipe wall, and partly reflected from the other side of the pipe wall. These reflected waves vibrate the receiver crystal, causing it to produce an electric current in response to a property of the piezoelectric crystal. Figure 1 shows the ultrasonic wave transmitter and receiver action (2).

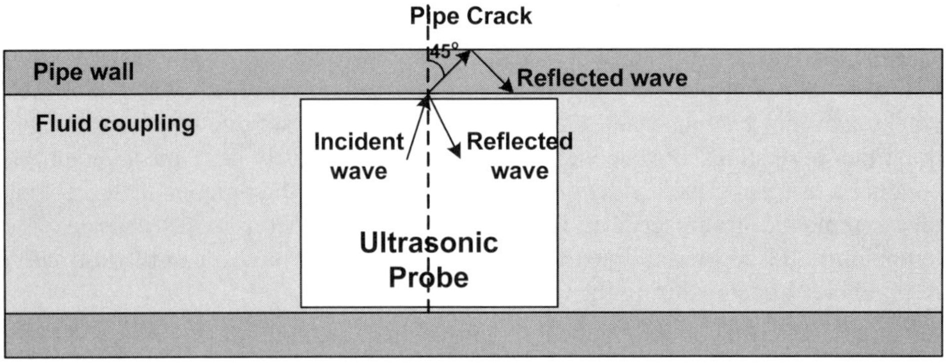

Figure 1 Ultrasonic Wave Action

To find cracks and defects from the outside, ultrasonic probes are put in direct contact with the clean surface of the exposed pipeline at intervals of about 100 – 200 m. To inspect a pipeline internally, an ultrasonic instrument is mounted on a smart pig and the instrumented pig is run with the fluid in the pipeline to measure the pipe's wall thickness and to detect cracks.

The sensitivity of an ultrasonic instrument is dependent on surface conditions, structural properties of the pipe, and coupling between the transducer and pipe surface. An ultrasonic inspection tool can detect small defects accurately under clean conditions assuming that it is well coupled with the pipe surface. Ultrasonic inspection techniques do not interfere with normal pipeline operations nor adversely affect the pipeline system safety. However, it is sometimes difficult to maintain good coupling between the transducer and pipe wall, particularly for gas pipelines.

At present, various ultrasonic inspection tools are commercially available. The latest development in signal processing techniques and computer technology enables ultrasonic inspection techniques to be practical, dependable and accurate.

3.1.3 Magnetic Flux Technique

The magnetic flux technique uses the magnetic properties of the material to be examined. When a strong magnetic field is applied to steel pipe, magnetic flux is formed in the pipe. If the pipe is uniform, so is the resulting magnetic flux. If the magnetic flux is distorted, the magnetized pipe may contain defects. Since changes in magnetic flux induce electric current, transducers measure the induced current. Magnetic flux is visually represented by magnetic lines as shown in Figure 2, which shows distorted flux lines around a defect (3).

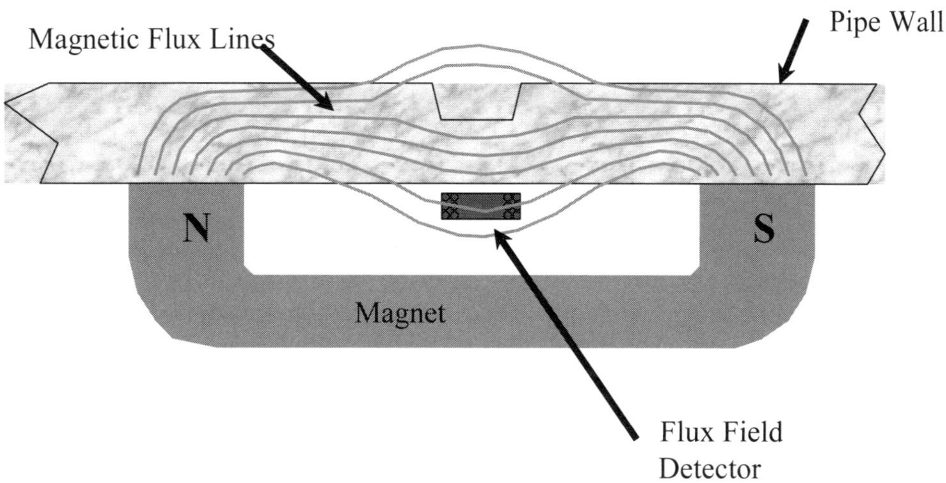

Figure 2 Magnetic Flux Lines

A strong magnetic field applied to a uniform pipe wall produces uniform magnetic flux lines. However, if the pipe wall contains a defect on either side of the pipe surface, the magnetic flux lines around the defect are distorted. If the pipe wall is reduced internally due to corrosion or gouging, the magnetic flux is leaked through the reduced pipe wall and the magnetic flux leakage can be detected by a transducer. The severity of the flux distortion is directly related to the severity of defects in the pipe wall, and thus the signals indicate the severity of the defects.

A magnetic inspection pig consists of three parts: a drive section, a magnetic flux detection section and a distance measuring section. The drive section is located in the front part of the pig. The magnetic detection section is in the middle and includes a strong magnet, a battery to power the magnet, transducers, electronics and a computer with a recording device. The distance measuring section is at the end of the pig. It contains an odometer to measure the pig travel distance.

The drive section allows the pig to be pushed by the transport medium and the distance measuring section measures the distance the pig travels from a reference point. The magnetic detection section performs the main functions. The magnet magnetizes the pipe, transducers measure the induced current generated by magnetic flux changes, the onboard electronics amplify the signals, and the computer processes and records the signals. The recorded signals are analyzed for defect assessment after the inspection pig is retrieved from the pipeline.

A magnetic inspection pig can detect pipe defects reliably and locate them accurately. It can run without interrupting normal pipeline operations. In general, it can produce a wealth of information for detailed defect assessment and future reference. However, a magnetic inspection pig tends to miss longitudinal defects and cracks, and is expensive to purchase or operate.

3.1.4 Visual Inspection Methods

Visual inspection is not only a form of leak detection but is also the final leak confirmation. This method has been used since the early days of the pipeline industry and is still popular.

Current visual inspection methods rely on detecting hydrocarbons along the pipeline right of way either visually or by using an instrument. Inspection crews walk, drive or fly the pipeline right of way searching for evidence of hydrocarbon leaks. Spillage evidence includes spilled hydrocarbons, vegetation changes caused by hydrocarbons, odor released from the pipeline, or noise generated by product escaping from a pipeline hole.

To reduce human errors, inspection crews often use special equipment such as infrared devices, flame ionization sniffing devices, or even trained dogs. Trained dogs are very sensitive to the odor of hydrocarbons released from a leak and have been successfully used for detecting them.

For inspecting transmission lines, pipeline companies often use an inspection airplane equipped with hydrocarbon detection sensors and cameras. It normally monitors vegetation changes and the amount of hydrocarbon vapor in the air above the ground of the pipeline right of way.

Visual inspection is simple and particularly useful for locating and ultimately confirming a leak. It is subject to human error and is not always reliable. Leaked hydrocarbons have to surface to the ground before they are detected. It can be costly and is dependent on external conditions including weather and pipeline accessibility.

Description	Air Surveillance	Ground Surveillance	Pipeline Employee	Third Party	Others
Leaks detected	10	6	178	77	15
False alarms	138	20	0	76	53
Annual Operating cost ($/yr-km)	$123	$39	$623	N/A	N/A
Surveillance length	27,431	19,940	27,067	N/A	N/A

The Canadian Petroleum Association (4) conducted a cost and effectiveness survey of visual leak detection methods using data collected over a five year period from 1985 to 1989. As shown in the above table, air and ground surveillance is relatively inexpensive, but not necessarily efficient for leak detection. Pipeline employee and third party detections were most efficient in this survey.

3.1.5 Hydrocarbon Detectors

When hydrocarbon products escape from a pipeline, hydrocarbon vapors are formed around the pipeline and eventually surface on the ground. Hydrocarbon detectors sniff the hydrocarbon vapor and alert the inspectors. One of the major problems with these detectors is that they detect methane gas (generated in the ground by such things as rotting vegetation or animals), giving a false positive.

There are many hydrocarbon detection devices and a few of them are discussed below for reference. Included are their operating principles and applications. Interested readers are encouraged to consult product vendors for further details.

Flame Ionization Device

A flame ionization device works like a home fire detector, utilizing some materials which are sensitive to hydrocarbon vapors. It is one of the most popular leak detection methods, particularly for detecting a leak from gas distribution pipelines. Visual inspection using such devices can be an effective method of leak detection.

A flame ionization device detects very small quantities of gas vapor, and is known to be sensitive and reliable. In addition, it can locate leak sites accurately. Operating personnel must keep the device directly above the pipeline at all times to maintain the devices sensitive and get reliable results. The device tends to be less reliable if a strong wind blows or the ground is wet.

Infrared Device

Most gases absorb infrared energy. Infrared spectra can reveal the wavelengths absorbed by a particular product. An infrared transmitter produces infrared light at the wavelength absorbed by the vapor to be detected and a signal receiver measures the transmitted light. Some devices use laser beam instead of infrared and the device works in a similar way. These devices can reduce dependence on visual inspection, thus minimizing human errors. These devices can be stationary or mounted on an aircraft.

This device is sensitive to hydrocarbon detection, making it particularly useful for offshore spill detection. These devices are simple to operate and cheap, and work at night and in fog. A stationary device can work even in bad weather conditions. However, they can only monitor a small area (in the order of one square foot).

Radar Device

A device using a radar beam can probe beneath the ground. This radar device sends signals generated from a radar transmitter into the ground and receives reflected signals from the buried pipeline and surrounding soil. By enhancing the received signals with a computer imaging technique, the leaking substance can be located and identified.

3.2 Continuous Sensing Devices

Continuous sensing methods use devices that continuously sense one or more of the leak phenomena listed in Chapter 7. This section discusses several continuous sensing devices. These devices provide very sensitive leak detection and accurate location capability. Some of the leak detection techniques that use continuous sensing devices are as follows:

- Acoustic Sensing Devices
- Optical Fiber Sensor Cable
- Vapor Sensor Cable

3.2.1 Acoustic Sensing Device

A leak continuously generates a sound wave which propagates at acoustic velocity in the upstream and downstream directions from the leak hole. A sound wave attenuates over distance as it propagates from the source. The magnitude of attenuation depends on the fluid in the pipeline, sound frequency of the leak, and the state of pipeline operations at the time of the leak.

The principle used in the acoustic leak detection method relies on the fact that when a fluid passes through a hole under high pressure, the resulting turbulence creates acoustic pressure waves that travel through the fluid. The turbulence also creates acoustic pressure waves in the pipe around the hole, which then travel through the metal structure. Acoustic sensors are placed on the pipe to detect these acoustic waves. The sound waves contain frequency components over a wide spectrum.

An acoustic leak detection system continuously monitors the pipeline for the sound characteristic of a leak. An acoustic system must be able to take into account background noise and pipeline operating characteristics to differentiate the leak signal from noises generated from other sources. Modern signal processing techniques allow the reduction of some such extraneous noises. The signals, after the background noise including operation characteristics are filtered, are compared to the appropriate thresholds to confirm or reject a leak. The acoustic leak detection system can also determine the leak location by correlating the sensor spacing, velocity of sound, and propagation time difference.

This leak detection system uses intrusive acoustic sensors such as hydrophones and/or special pressure sensors. A hydrophone is a passive listening device, converting acoustic energy into electrical energy. Hydrophones are installed in direct contact with the fluid like a normal pressure sensor and thus directly measure the sound in the fluid. Since sound is a pressure wave, pressure sensors can be used to detect the sound wave. Pressure sensors sense pressure wave signals and are installed outside the pipe or directly in contact with the fluid. The clamp-on type sensors measure the pipe deformation caused by pipe pressure waves and the direct contact type pressure sensors sense the pressure wave signals in the fluid. The clamp-on type sensors can be installed easily at any point along a pipeline while the pipeline is operating.

The acoustic leak detection system normally consists of several acoustic sensors, regularly spaced along the pipeline. The sensor spacing depends on the sensitivity of the sensor, the sound attenuation property of the fluid, required leak detection sensitivity and location accuracy, the power available, and cost. Typically sensor spacing is up to 15 km for gas pipelines and up to 30 km for liquid pipelines. Extra sensor installations are required near the pipe ends and the areas around pumps, compressors and valves in order to avoid spurious alarms.

Hydrophones used for this purpose are very sensitive. As a result, an acoustic leak detection system can detect a small leak (in the order of 0.5% of pipeline flow). In general, pressure wave sensors are less sensitive than hydrophones, and provide less sensitive leak detection performance. If pipeline pressure is lower than about 250 kPa, the leak sound is so weak that special hydrophones or pressure sensors are needed. The acoustic leak detection system can be used on both onshore and offshore pipelines.

The architecture of a typical acoustic leak detection system is similar to the diagram shown in Figure 8 of Chapter 7. Each sensor is housed in a local unit which provides power, manages collected data and communicates with a main computer. The local unit sends signals to the main computer via a communications network and software interface. The main computer processes the signals to determine the existence of a leak.

A clamp-on sensor can detect a shock generated in the fluid by an elastic strain of the pipe wall and transmitted through it and the pipe wall. Therefore, an acoustic system with clamp-on type sensors can detect both leaks and shocks, like those generated by the impact of an outside force such as a mechanical shovel, trenchers, or an anchor on an underwater pipe.

Another variation of an acoustic leak detection system uses an acoustic emission technique. When pipelines are placed under stress, pipe material emits minute pulses of elastic energy (acoustic emission) due to ductile tearing or the plastic strain of defects. This phenomenon occurs at a stress level well below its failure point. These high frequency acoustic waves make it possible to identify the areas within a pipe that contain defects. The acoustic emission technique has been used to analyze the structural integrity of material. The same technique can be applied to pipeline leak detection, particularly for detecting incipient leaks. This leak detection method uses acoustic emission waves generated by tearing noise around an incipient leak at weak spots in the pipeline or by fluid escaping through a leak in a pressurized pipeline.

The advantages of the acoustic leak detection system, if it is installed properly, are as follows:

- It can detect very small leaks in a short time and detection time is almost independent of leak size.
- Leak location is very accurate, in the order of 100 m.
- It can detect outside third party damage by sensing shock waves.
- It is applicable to any type of fluid and pipeline configuration.
- It operates continuously with minimal interruption.

On the other hand, it has the following disadvantages:

- It tends to generate frequent false alarms particularly for small leak detection in the presence of large background noises in the pipeline.
- It can be very expensive for a long transmission line, because it requires an acoustic sensor every 15 km or at most 30 km.
- It may not be able to determine the size of a leak.

3.2.2 Optical Fiber Sensor Cable System

An emerging technology uses an optical fiber sensor to detect leaks (5). It requires the installation of an optical fiber cable along the entire length of the pipeline. It operates through one of the following three ways:

- detection of optical properties
- detection of temperature change
- detection of micro bends

Commercially produced fiber optic cable is coated to keep all wavelengths of light contained in the cable. For leak detection purposes, the fiber optic cable is treated to permit a certain amount of light to be lost when it comes in contact with hydrocarbons. The presence of hydrocarbons causes a change in the refractive index or fluorescent property of the cable. The devices typically measure changes in parts per million.

Optical fiber sensor technology can measure temperatures along the pipeline. In the event of a leak, particularly a gas leak from a gas pipeline, the temperature around the hole drops, resulting in a temperature change in the cable. This change can be detected and measured because of changes in the gas concentration.

A pulsed laser is sent through the optical fiber. The laser light interacts with molecules of the fiber material and is scattered as the pulse travels through the fiber due to differences in density and composition of the fiber. Some of the light is bounced backwards, and its spectrum is analyzed. The analysis and measured propagation time is the information the device needs to detect a leak and its location.

As shown in the accident statistics, the number of damages by third party interference is significant. Preventing third party damages is thus an important subject. Activities outside the pipeline generate pressure waves which can be picked up by micro-strain sensors.

The optical fiber sensor cable consists of micro-strain sensors, optical fiber cable and lasers. A micro-strain sensor can detect micro-strains and monitor vibrations. The vibration signals are sampled at a high frequency and analyzed by the attached computer. The amplitude and frequency characteristics of the signals are analyzed to determine the events that generate such characteristics. If the signal characteristics match the signal patterns for third party interference, this technique generates an appropriate alarm. Alarms can be made available to the SCADA system through the interface between the SCADA and the optical fiber sensor cable system. The optical fiber sensors can sense even weak vibrations in real-time in the soil above and around the sensor cable. If the optical fiber cable is attached to or imbedded near a pipe, it acts as an interference detector for the pipe and its environment.

The basic system operates over spans of up to 50 km between the start sensor, with a computer at one end and the end sensor at the other. The only system component between the ends is the fiber optic cable. For a long pipeline, additional sensing controllers can be placed every 100km, usually coinciding with compressor or pump stations along the pipeline.

In theory, this system requires low maintenance and can be used in various soil types. The system is very sensitive and can detect a person striking the ground with an axe from a distance of three meters. The sensitivity is influenced by the background signals that the system will detect in normal operations. It can accurately locate the event to within 100 meters with good repeatability. The location accuracy depends on the cable's depth and configuration.

The advantages of this leak detection system, if it is installed properly, are as follows:

- It can detect outside third party damage by sensing shock waves in a short time.
- Event location is very accurate.
- It operates continuously with minimal interruption.
- It can detect fluid theft quickly.

On the other hand, it has the following disadvantages:

- The installation cost is reasonable for a new pipeline system, but the cost of installing a fiber optic cable on an existing pipeline is very high.

- Its performance is not fully proven in the field yet. The technology has yet to prove not only that it detects leaks quickly and accurately but also that the fiber optic cable is relatively maintenance free and has a long lifetime in harsh environments.

3.2.3 Vapor Monitoring System

A vapor monitoring leak detection system (6) detects leaks by placing a sensor tube next to the pipeline. In the event of a leak, the hydrocarbon vapors will diffuse into the sensor tube. The system consists of a suction pump, gas detector and a plastic cable or tube that is installed adjacent to the pipeline. Refer to Figure 3.

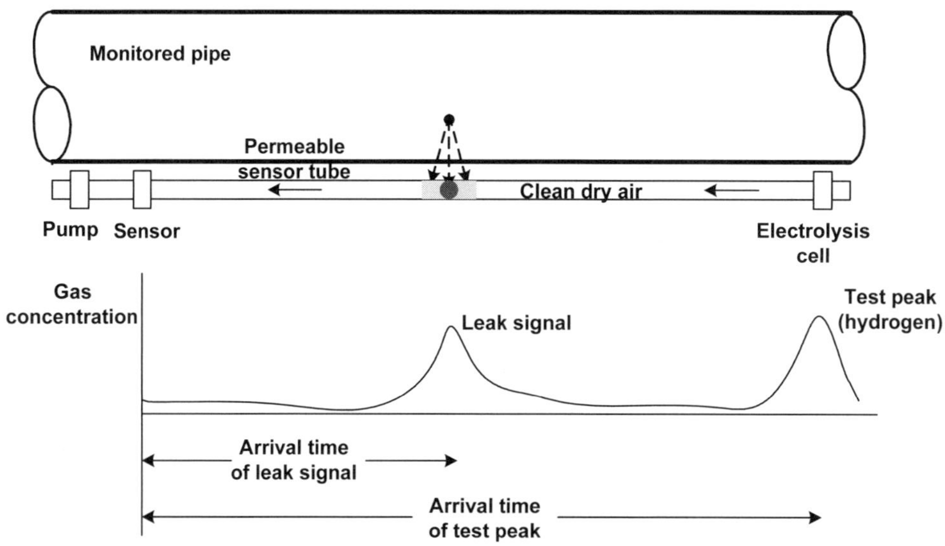

Figure 3 Vapor Sensing Device

The tube is made of an ethylene-vinyl-acetate membrane that is impermeable to liquid but permeable to hydrocarbon molecules that diffuse inside the tube. It contains air at atmospheric pressure and is pressure tight when installed. The detection unit at the end of the tube is equipped with a sensitive semiconductor gas sensor that can detect small amounts of hydrocarbons. The material the sensor is made of varies depending on the type of hydrocarbons needing to be detected. For example, a sensor that is highly sensitive to ethylene is not necessarily sensitive to methane.

The air in the tube is analyzed periodically. A pump at one end pulls the air at a constant speed through the tube into a detection unit. Prior to each pumping action, an electrolysis cell at the other end of the pumping unit injects test gas. This test gas is pulled through the tube with the air. When the detection unit detects the test gas, it marks the complete removal of the air that was contained in the tube and serves as a control marker to indicate that the entire air column has passed through the detection unit. The leak location is determined by measuring the ratio of the travel time of the leak to that of the control marker.

When a leak occurs, some hydrocarbon molecules diffuse into the tube as a result of the hydrocarbon concentration difference between the inside and outside of the tube around the leaking section. In due course, the affected area of the tube will have a higher hydrocarbon concentration than the rest of the tube as shown in the figure above. The speed of diffusion depends on concentration differences and the availability of gas molecules at the outer wall of the tube membrane. When the pump pulls the air, the affected air is also pulled toward the detection unit, which analyzes the hydrocarbon concentration. Because the air is pulled at a constant speed, the system can determine the leak location. Leak size can be estimated from the concentration of hydrocarbons.

This method of leak detection and location can detect a very small leak and locate it accurately. It can be used for both onshore and offshore pipelines as well as multiphase leak detection. In addition, this methodology can be used to detect many different substances. A system based on this technology has been used in an Arctic pipeline (Northstar Development). However, this method may be too slow to react to large leaks, and the installation and operation costs can be very high. This system should be used in conjunction with other leak detection systems in environmentally sensitive areas.

References

(1) Mohitpour, M., Golshan, H., Murray, A., "Pipeline Design and Construction", 2000, ASME Press, New York, N.Y.

(2) Burkle, W.S., "Ultrasonic Testing Enhances Pipe Corrosion Monitoring", OGJ, Sep. 15, 1983

(3) Holm, W.K., "Magnetic Instrumentation Pig Helps NGPL Inspect Pipelines for Potential Leaks", OGJ, June 1, 1981

(4) "Effectiveness of Leak Detection Methods", Project Number 2527, Canadian Petroleum Association, 1991

(5) Jeffrey, D., et al, "An effective and Proven Technique for Continuous Detection and Location of Third Party Interference Along Pipelines", Proceedings of IPC, ASME, 2002

(6) "Northstar Development Project Buried Leak Detection System", Intec Project No. H-0660.03, 1999

4. Measurement Standards

For custody transfer, pipeline companies need to use accepted standards. The standards summarized below are widely accepted in the pipeline industry.

Subject	North American Standards	International Standards
Orifice meter for gas	AGA-3, API 2530	ISO 5167 (1)
Orifice meter for liquid	API MPMS 14.3 (2)	ISO 5167
Turbine meter for gas	AGA-7, ASME MFC-4M	ISO 9951 (3)
Turbine meter for liquid	API 2534 API MPMS 5.3	ISO 2715
Positive displacement meter for gas	AGA-2.1	ISO 2714
Positive displacement meter for liquid	API MPMS 5.2, API 1101	ISO 2714
Ultrasonic flow meter for gas	AGA-9	ISO 12765
Ultrasonic flow meter for liquid	ASME MFC-5M	ISO 12765 (4)
Natural gas supercompressibility	AGA-8 NX-19	ISO 12213-2 (5) ISO 12213-3
Heating value	AGA-5	ISO 6976 (6)
Coriolis mass meters for gas	AGA-11	ISO 10790
Coriolis mass meters for liquid	API MPMS 5.6 & 14.7 ASME MFC-11M	ISO 10790
Natural gas supercompressibility	AGA-8 NX-19	ISO 12213-2 ISO 12213-3
Petroleum liquid volume correction	API MPMS 11 (7)	ISO 9770
Petroleum liquid & LPG measurement standard	API 2540	ISO 5024
Petroleum measurement tables – Reference temperatures of 15°C and 60 °F	API 2540	ISO 91-1
Petroleum liquid volume measurement in tank storage	API 2550	ISO 4269 & ISO 4512
Petroleum & liquid petroleum products calibration of vertical cylindrical tanks	API MPMS 12	ISO 7507
Meter proving	API MPMS 4	ISO 7278
Density and relative density determination	API MPMS 9	ISO 649 & ISO 650

Subject	North American Standards	International Standards
Glossary of terms on measurement	ASME MFC-1M	ISO 4006
Measurement uncertainty for fluid flow in closed conduits	ASME MFC-2M	ISO 5168 & ISO 5725
Weighing method	ASME MFC-9M	ISO 4185

Notes:

(1) ISO has published several guidelines on the use of ISO 5167. For example, ISO/TR 9464 describes "Guidelines for the use of ISO 5167-1", ISO/TR 12767 "Guidelines to the effect of departure from the specifications and operating conditions given in ISO 5167-1", and ISO/TR 15377 "Guidelines for the specification of nozzles and orifice plates beyond the scope of ISO 5167-1". The widely accepted standards for orifice meters include AGA Report No. 3 (AGA-3) Orifice Metering of Natural Gas and Other Related Hydrocarbon Gases or ISO 5167 Measurement of fluid flow by means of pressure differential devices inserted in circular cross-section conduits running full - Part 2. These standards are widely accepted for operating orifice meters by natural gas industry. The standards provide detailed technical specifications such as meter run, plate size, material and piping required for orifice meter design and installation. In addition, they provide methods of calibrating orifice meters and correcting their readings, by taking into account various factors such as type of orifice meter, Reynolds number, static and differential pressure, temperature, etc. The latest version of the AGA-3 was published in 1985 and its modified version in 1992 (with some errata documents issued subsequently). The ISO published several other standards such as ISO 12767 and ISO 15377 with more specifications for orifice and venturi meters.

(2) API MPMS stands for American Petroleum Institute's Manual of Petroleum Measurement Standards.

(3) The applicable standards for turbine meters include AGA Report No. 7 (AGA-7) Measurement of Gas by Turbine Meters or ISO 9951 Measurement of fluid flow in closed conduits – Turbine meters. These standards address the turbine metering and meter run specifications in a similar way to the standards for orifice meters. The volume correction method includes factors such as pressure, temperature and super-compressibility. The current version of AGA-7 was published in 1996, and that of ISO-9951 in 1993.

(4) The applicable standard for ultrasonic flow meters is AGA Report No. 9 (AGA-9) Measurement of Gas by Multipath Ultrasonic Meters or ISO/TR 12765 Measurement of fluid flow in closed conduits. Methods using transit time ultrasonic flow meter. They are available since 1998.

(5) Supercompressibility is required to correct gas volume to a base condition or

a specified pressure and temperature for a given gas composition. AGA Report No. 8 or ISO-12213-Part 2 applies when calculating super-compressibility based on gas composition. The super-compressibility factor is expressed as:

$$S = \sqrt{\frac{Z_b}{Z_f}}$$

Where

S = Supercompressibility factor

Z_b = Compressibility at a base condition

Z_f = Compressibility at flowing conditions

There are two versions of AGA Report No. 8 Compressibility Factors of Natural Gas and Other Related Hydrocarbon Gases: one published in 1985 and the other in 1992. Both AGA-8 1985 and AGA-8 1992 versions use the same equation as above, but the derivation of the compressibility factors are different from each other.

In addition, NX-19 1962 Manual for Determination of Supercompressibility Factors for Natural Gas or ISO-12213-Part 3 provides a method of calculating supercompressibility based on physical properties. Particularly, the NX-19 equation uses four different methods to obtain the adjusted pressure and temperature for the equation of state:

- Specific gravity method,
- Gas analysis method for gas with a high specific gravity,
- Methane method requiring the methane mole fraction and
- Heating value method. Uses the mole fractions of nitrogen and carbon dioxide and the measured heating value of the gas.

Both the flowing and base compressibility factors are determined using the adjusted pressure and temperature calculated using one of the above methods. The NX-19 equation produces slightly less accurate results than the AGA-8 does, but it is adequate for most custody transfer applications. Its applicable pressure and temperature ranges are narrower than those of AGA-8.

(6) Quite often, gas transactions take place based on heating value or energy content. The AGA Report No. 5 (AGA-5) or ISO 6976 Standard can be used for calculating heating value. The AGA-5 method calculates the heating value using the results of the supercompressibility calculation and the volumetric meter calculation derived from metering output (AGA-3, AGA-7 or AGA-9).

(7) The following API correction factors (API 11.2.1) can be used for volume correction:

- **Volume correction factor for temperature**

$$C_T = Exp\left[-\alpha_b \Delta T_f \left(1 + 0.8\alpha_b \Delta T_f\right)\right]$$

where $\alpha_b = \left(K_0 / \rho_b^2\right) + \left(K_1 / \rho_b\right)$

K_o = Constant depending on product and density
K_1 = Constant depending on product and density
ρ_b = base density
ΔT_f = difference of the flowing temperature from the base temperature

- **Volume correction factor for pressure**

$$C_P = 1/\left(1 - 10^{-6} C_f P_f\right)$$

where $C_f = Exp\left[A + BT_f + \left(C / \rho_b^2\right) + \left(DT_f / \rho_b^2\right)\right]$

= compressibility factor
A = -1.6208
B = 0.00021592
C = 0.87096
D = 0.0042092
T_f = flowing temperature
P_f = flowing pressure

- **Net Volume Calculation**
The gross volume is the un-calibrated gross volume multiplied by the meter factor:

$$V_g = VM_f$$

where V_g = gross volume
V = uncalibrated gross volume
M_f = meter factor

The gross volume is corrected to base conditions to obtain the net volume, multiplying the gross volume by the volume correction factors for temperature and pressure, as shown below:

$$V_n = V_g C_T C_P$$

Where V_n = net volume

If there is more than one flow meter at the same meter station, the gross volumes can be different from each other but the volume correction factors should be the same assuming that the meters measure the same product volume.

API 11.2.1 is applicable to products with 0 – 90° API. This includes all grades of crude, diesel, turbo fuel, gasoline, benzene, alkylate, toluene, raffinate, and other heavier hydrocarbon liquids. Also, ASTM D1250 tables can be used for net volume calculation. Listed below are the API standards for the volume correction of lighter hydrocarbon liquids greater than 90 ° API and of ethylene and propylene:

- Liquid Petroleum Gas (LPG) and Natural Gasoline:

  LPG includes propane, normal butane and isobutene. API 11.2.2 or GPA Standard TP15-16 is used to calculate the net volume for this group of light petroleum liquids. API 11.2.2 is used to calculate the compressibility for products with gravity between 0.350 and 0.637.

- Natural Gas Liquid (NGL):

  NGL is composed of several light hydrocarbon liquids. API 11.2.2 or GPA Standard 2145-94 is used to calculate the net volume for NGL.

- Ethylene:

  The custody transfer of ethylene is done in mass. Ethylene can be measured in mass or volume. If it is measured in volume, the volume flow is calculated using API 14.3 for orifice meter, the density using API 2565. The mass of ethylene is obtained by multiplying the volume by the density.

- Propylene:

  The custody transfer of propylene is done in mass. Propylene can also be measured in mass or volume. If it is measured in volume, the volume is calculated using API 14.3, the density using API 11.3.3. The mass of propylene is obtained by multiplying the volume by the density.

5. Glossary

A/D conversion	Analogue to digital conversion. Typically this is used in the context of the conversion of process signal to an electrical signal, which is then converted to a digital representation for use within a computer control system.
Aerial imagery	Digital images acquired from an airplane or a helicopter used for mapping applications
Asynchronous	An asynchronous transmission is a method of transmission in which an event is started by the completion of another event with no fixed time per cycle.
Attributes	Characteristics about a spatial feature
AGA calculation	Gas volume correction of raw gas to base conditions using a set of American Gas Association (AGA) equations such as AGA-8 or NX-19
ALAM	Automatic Look-Ahead Model (ALAM). An ALAM is an automatic pipeline transient flow model executed at regular intervals, for predicting future pipeline state.
Alarm	A warning given by a control system of a limit violation, abnormal change of state, or a failure.
Analog data	Data in a continuous form such as pressure and flow
API	American Petroleum Institute
API gravity	Specific gravity scale for petroleum liquids at 60°F developed by API with reference to the specific gravity of water being equivalent to 10°API. The relationship between the API gravity and specific gravity is given below: API gravity = (141.5/SG at 60°F) - 131.5
ASTM	American Society for Testing and Materials
Audit trail	Log that documents changes that were made or the occurrence of an event in computer records or databases. The log should include the change, the date and time, the person that made each change, and the reason of the change.
Backdoor	An undocumented way of gaining access to a program, online service or an entire computer system

Balancing agreement	A contractual agreement between two or more business entities to account for the differences between measured quantities and the total confirmed nominations
Base conditions	Pressure and temperature reference conditions used in determining a fluid quantity for custody transfer
Batch	A batch refers to a contiguous product entity that remains whole throughout its journey through the pipeline system. A batch has the attributes of product type, volume, identification or name, and lifting and delivery locations and times with flow path.
Batching cycle	Batching cycle is a specific period during which a predefined set of products is transported. Multiple cycles are repeated during the nomination period, usually a month.
Batch interface	Refer to transmix
Bayesian inference	A statistical technique for determining the probability of observing an event conditional on the previous probability of observing other events
Bias error	The difference between the average and true values or measurements. It is directional, but it is difficult to determine a true bias error in practice.
Bleed valves	Valves located on various stages of the axial compressor, used to reduce the developed head during start-up. Usually, there are one or two valves on an axial compressor.
Blending	Mixing of two or more products by injecting one product stream into another
Blow-down valve	Valve used to exhaust gas from a section of pipe when necessitated by repairs, emergency or other conditions
Breakout point	An intermediate location on a pipeline system that joins two or more pipeline sections, where batches can be simultaneously injected into and delivered out of the pipeline or a batch can be tight-lined
BS&W contents	Basic Sediment and Water content of a fluid
Buffer	A temporary product injected between two batches to reduce mixing of the two batches
Bulletin Board	It is an electronic means to share information, capacity and capacity releases, and other key data.

Bundled service	Transportation and sales services packaged together in a single transaction
Bus	A data link between different components of a system
Bypass valve	Valve allowing flow around a metering system or equipment
Calibration	Adjustment of a measuring instrument against a known quantity to improve its performance or to conform to an applicable standard
Capacity release	Release of the right of a shipper with unused firm transportation rights on a pipeline to assign its transportation rights to another party
Cartography	The discipline of designing and making maps
Chart integration	Measurements of volume, pressure and temperature that are collected in the field in a chart form and are validated, corrected to base conditions, and integrated by means of a chart integrator in order to obtain the total volume for a specified period
Choke condition	It occurs when the flow velocity through a compressor is so high that it reaches the sonic speed.
Chromatograph	Gas or liquid analysis instrument
Common carrier	Pipeline that provides transportation service to all parties equally
Compressor Wheel map	Plot of the design speed, pressure and flow conditions of a compressor wheel, usually plotted in adiabatic head versus flow
Confirmed nomination	An agreement a pipeline company has to receive and deliver a specific quantity of fluid under a transportation agreement. The confirmed nomination is in response to a shipper's nomination.
Connectivity	Describing how arcs are connected by recording the from and to nodes for each arc. Arcs that share a common node are connected.
Contract	An agreement between the pipeline company and a shipper which specifies the type of service and minimum/maximum volumes
Covered task	Those tasks which can affect the safety or integrity of a pipeline as discussed in ASME B31Q

CPM	Computational Pipeline Monitoring techniques, which are specified in API 1130
Critical speed	Speed corresponding to an equipment's (e.g. a turbine) natural resonance frequency. Severe vibration can occur at critical speeds.
Curtailment	A reduction in service that is imposed, when the available supply is below the contracted amount
Custody transfer	Custody transfer is the change of ownership of petroleum products at a given transfer point, most likely at a meter station.
Customer	Entity such as a local distribution company or marketer that generates a net outflow of gas or liquid from the pipeline company
Daemons	A computer program that runs continuously in the background without a visible user interface and which performs a system-related service until it is deactivated by a particular event. Daemons are usually spawned automatically by the system and may either live forever or be regenerated at intervals. This may also be called a "system service" or "system agent" in Windows.
Data model	A design process for a database which identifies and organizes data in tables
Datum	A mathematical surface on which a mapping and coordinate system is based
DEM	Digital elevation models are grids where each cell value represents the elevation on the earth's surface.
DCS	Distributed Control System. This is a type of automated control system that is used to monitor and control a process facility.
Delivery	The transfer of a quantity of fluid out of a pipeline system, typically into a tank, either at the end or at an intermediate location. This is the point of custody transfer for fluid moving out of the pipeline company's system.
Density	Mass or weight per unit volume
Digital data	In the context of a SCADA system, it is the on/off or open/close status of devices such as valves.
DRA	Drag Reducing Agent. This is a fluid injected into a pipeline to reduce friction along the pipeline and thus increase the throughput.

EDI	Electronic Data Interchange. This is the data exchange of nominations and other business data through a standard data exchange format.
Effective date	Effective date for a contract is the first day that the contract is legally in effect, and the effective date for a nomination is the date that the nomination is fulfilled.
EOD volume	End of day volume. The EOD volume is the volume estimated at the end of a gas day and used to meet the nomination volume requirement. The volume projection may be based on the daily flow profile.
ESD	Emergency Shutdown Detection. This is a controller that is independent of the station control system that detects conditions requiring an immediate shut down of the pump/compressor station.
ESRI	Environmental Systems Research Institute is an organization that develops widely used GIS software.
ETA	Estimated time of arrival. This is the ETA of a tracked object such as a batch front or scraper at a specified location or facility.
Expiry date	Expiry date of a contract is the last day that the contract is legally in effect.
Eye (of an impeller)	Internal pressure point at the inlet of the first impeller of the compressor. A "suction-to-eye" differential pressure is typically representative of flow.
EPROM:	Erasable Programmable Read Only Memory. This is a computer memory device that is programmed electronically and the program can be erased only by ultraviolet light exposure.
FERC Order 636	The Federal Energy Regulatory Commission (FERC) ordered in 1992 that transmission companies unbundle their transportation, sales and storage services to provide open access to all shippers. It forced pipeline companies to convert from being sellers of gas to being primarily shippers of gas bought and sold by other parties.
Firm service	Transportation service that is guaranteed for the shipper for the contract period except during events of major forces such as pipeline ruptures or earthquakes
Flow computer (FC)	Field device for collecting measurement data in real-time, performing certain calculations such as AGA,

	storing historical measurement and calculation data, and uploading to/downloading from the host SCADA
Flow profile	Normally, an hourly flow pattern for the gas day that the gas pipeline operator tries to deliver
Flow projection	Based on current readings and known flow profile, the flow is projected to estimate the gas volume to be delivered for the gas day which meets the nomination.
Flow totalization	Flows and volumes are totaled for operation and volume accounting.
Fungible batch	A batch that can be combined with other batches of the same product.
Gas day	A 24 hour period measured from a top-of-day to the next top-of-day during which time a daily nomination is implemented fully under an effective contract.
Gas marketer	An entity that sells gas, transportation service, and/or storage service
Gas quality	Specification of natural gas in terms of gas composition or specific gravity, dew point, water vapor content, H_2S and CO_2 contents, O_2 and N_2 contents, and several other compounds
Gas storage	Facility to store natural gas supplies for peak shaving and other purposes. It is usually close to major delivery locations.
Gearbox	Gearing used to change speeds between shafts in mechanical drives. The starter is normally connected to the turbine via a gearbox.
Geodatabase	A database used to hold a collection of GIS datasets
Geodetic datum	Defines the size and shape of the earth and the origin and orientation of the coordinate systems used to map the earth.
Geometric networks	A network of geospatial features
Geoprocessing	Computer programs or models which perform operations on geospatial data
Geospatial	Referring to an objects location or position on the earth
Geotiff	An image in TIF file format with geo-referencing information in the header
GIS	Geographic Information Systems, a computerized information system for storing, manipulating and analyzing spatially referenced information

GISB	Gas Industry Standard Board. It was established in 1994 to address gas industry standards issues for electronic data interchange. In 2002, GISB became the North American Energy Standards Board (NAESB).
Gross volume	Raw volume of fluid at the measured conditions of pressure and temperature before it is corrected and before water and sediment (for liquid measurement) are accounted for
GSM	Global System for Mobile communication. It is the communication standard used for digital cellular telephones.
GPRS	General Packet Radio Service. It is a communication standard that takes GSM messages and transmits them as packets. This allows for higher bandwidth enabling such uses as mobile internet browsing.
Hillshade	An image derived from an illuminated DEM providing artificial shadows to give a 3D effect
HMI	Human Machine Interface. This refers to the interface between a user and a computer system.
Host	The centrally located collection of hardware and software of a SCADA system
Hot end	The exhaust end of a turbine. Hot end drive refers to the power turbine shaft that extends past the exhaust to enable coupling to the compressor.
Hydrate	In natural gas, a hydrate is a solid compound formed when free water vapour combines with light hydrocarbons such as methane.
Hydrocarbon bubble point	The pressure and temperature point of a hydrocarbon liquid, at which vaporization is about to occur
Hydrocarbon dew point	The pressure and temperature point of a hydrocarbon gas, at which condensation is about to occur
HVP products	High vapor pressure products. These are light hydrocarbons whose vapor pressure is higher than 1,400 kPa at 37.8°C.
Imbalance	The volume difference between the actual measured volume and the nominated volume at the custody transfer meter
Imbalance penalty	Penalty imposed by the pipeline company on the shipper for a volume imbalance that exceeds the tolerance level

Incipient leak	A leak that is just about to occur
Increment strapping table	A strapping table of level increments vs. tank volumes, which is used to convert level increments into gross volumes
Injection	The process whereby a fluid is moved from a tank into a pipeline at the head or intermediate location within the pipeline company's system
I/O	Input/output. It is the circuitry that interfaces an electronic controller (PLC, RTU or computer) to the field. Inputs and outputs can be digital/discrete or analogue.
Interface	A common boundary between two or more components of a system that is required to enable communication between them
Interruptible service	A service that can be interrupted if the pipeline capacity is not sufficient to serve a higher priority transportation service. Interruptible service is less expensive than firm service.
Interruption	An event that stops a given computer activity in such a way as to permit resumption at a later time
Intrinsically safe	A designation assigned when equipment can be installed in areas that are designated as being a potential explosion hazard. It limits both electrical and thermal energy to a level below that required to ignite a specific hazardous atmospheric mixture.
LAN	Local Area Network
Landing	The process where fluid is moved out of a pipeline into a tank but still remains within the pipeline company's system
Lateral	A component of a lateral is a pipeline section that connects a mainline junction to an intermediate delivery point.
Latitude	A spherical reference system used to measure locations on the earth's surface. Latitude measures angles in a north-south direction.
LDC	Local Distribution Company. It is a pipeline company that distributes and sells gas to residential and industrial users.

LiDAR	Light detection and ranging is a technique that transmits laser pulses to the earth's surface and collects the return signals to map the terrain elevation.
Lifting	Injection of a batch at the head of a pipeline system
Line fill	The volume of petroleum products within a pipeline or a pipeline segment during transportation
Line pack	The volume of fluid in a pipeline segment or entire pipeline system. Line pack can increase or decrease depending on whether the fluid volumes received are larger than the volumes delivered or vice versa.
Line packing/unpacking	Increasing/decreasing process of line pack
Linear referencing	A method for associating attributes or events to locations or segments of a linear feature
Load	Net outflow gas from a pipeline system expressed in terms of volume or energy
Load forecasting	Forecasting estimated gas loads required at delivery locations on a short-term basis (daily and weekly basis). The loads include both industrial and residential loads which are primarily dependant on weather conditions.
Load sharing	Sharing of load amongst compressor units at a multiple unit station to achieve increased efficiency
Longitude	A spherical reference system used to measure locations on the earth's surface. Longitude measures angles in a east-west direction.
MAOP	Maximum allowable operating pressure
Mainline	A mainline consists of one or more pipeline sections that directly connect an origin point to a final delivery or breakout point. All mainline sections are hydraulically coupled to one another.
Manometer	A U-shaped tube containing a liquid (usually mercury or water) to measure the fluid pressure. The liquid level on one side of the tube changes with respect to the level on the other side with changes in pressure.
Mass flow meter	Mass flow meter is a flow meter measuring the flow rate directly in mass.
Master	Another term used for a SCADA Host

MDQ	Maximum daily quantity is the maximum quantity of gas a shipper can request under a contract on a given gas day.
Metadata	Legacy information about a particular dataset. Typically includes at least the data source, vintage, quality, accuracy, and purpose.
Meter chart	Circular chart that shows the differential pressure and static pressure in an orifice metering system. It is recorded by a flow recorder and is used to measure volumes in an off-line environment.
Meter factor	Correction factor applied to a meter's reading to obtain a corrected reading, typically on a custody flow meter
Meter run	Flow measurement unit consisting of the primary and secondary metering elements in the metering manifold
Meter prover	A meter prover determines the meter factor of a turbine or positive displacement meter i.e. the relationship between the number of counts or revolutions of the meter and the volume flowing through the meter. The number of counts on the meter being 'proved' is related to the volume passing the detectors on the 'prover'.
Metering manifold	A collection of pipe in which a flow meter is mounted
Modem	A device that converts computer generated data streams into analog form so that they can be transmitted over a transmission line
Must take gas	Quantity of gas that a gas purchaser must take under a purchase contract
Net volume	Measured volume corrected to base pressure and temperature, used in accordance with an accepted standard such as API 11.2.2, ASTM tables or ISO 9770
NIST	National Institute of Science and Technology, a U.S. government organization that develops standards and applies technology and measurements
Nomination	Nomination is a request for transportation service including the quantity of petroleum fluid that a shipper requests a pipeline company to transport for the nomination period.
Nomination allocation	A process by which capacity available in a pipeline is distributed to parties in the event that nominations are in excess of the available supply or pipeline capacity.

	Typically the allocation is based on service type, contract type and a company's tariff provisions.
NAESB	North American Energy Standards Board
Operator	Person controlling a pipeline from a central control room using a SCADA system. Some pipeline companies also call this person a "dispatcher" or "controller."
Orders	List of actions that are scheduled to occur in the pipeline. This list specifies such things as date/time, location, product, volume, and flow rate.
Ortho-photography	Digital images created from air photos from which camera and topographic distortions have been removed
Peak shaving	Peak shaving is a mechanism to manage the gas load when the required load is greater than the pipeline capacity. Normally, gas storage or LNG facilities are used for peak shaving purposes.
Photogrammetry	The technique of determining the position and shape of objects from stereo aerial photography or imagery
Pipeline capacity	Pipeline capacity refers to the maximum flow rate that can be transported through the pipeline system in a given period of time under the conditions that prevail in the available facilities.
Pipeline integrity	State of a pipeline that demonstrates the ability to withstand the stresses imposed during operations
Pixel	A cell of a raster
PLC	Programmable Logic Controller is a field device that performs real-time data gathering, calculating, storing and controlling functions including close loop control based on current operating conditions. It can upload to and download data from the host SCADA. A distinguishing function of a PLC, compared to a flow computer or an RTU, is its ability to control valves, regulators and even pump/compressor stations.
Polling	The regular scanning of each RTU by the host SCADA.
Polygon	A vector file that consist of at least 3 points which contain an area
PPA	Pressure Point Analysis, a leak monitoring method
Primary device	Primary device is the device directly or indirectly in contact with the fluid and generating a signal according to a known physical principle when applied to the

	fluid. For example, the primary devices for an orifice metering system include the orifice plate, meter tube, fitting and pressure taps.
Process	An operation that uses equipment to gather, refine, store, or transports a product or group of products
Projections	A mathematical calculation transforming the three dimensional surface of the earth to a two dimensional plane
Prorata	A method of allocating capacity, production, or services in the same ratio as requested
Pumping orders	A list of batches scheduled for injection and delivery at all inlet and outlet batch meters on the pipeline
Protocol	Standard procedure and format that two data communication devices must understand, accept and use to be able to exchange data with each other
Query	Selecting features in a GIS or database by asking a question or a logical expression
Rail road chart	A batch graph that displays product movement in a distance vs. time relationship simultaneously for multiple pipeline routes and batch flow vs. time for specific locations within each route
Random error	A random error, also called precision error, is determined by calculating standard deviations of measured values.
Rangeability	The range of linear flow rate over which the meter can retain its accuracy. The ratio of the maximum to the minimum linear flow rate is called turndown ratio.
Raster display	One using rows and columns of pixels to display objects and text on a screen
Real time	Real time is the actual time that a physical process is taking place.
Receipt	A receipt occurs when fluid is moved from a shipper or feeder into the pipeline system either at the end or at an intermediate location in the pipeline. This location is the point of custody transfer into the pipeline company's system.
Repeatability	The variation in measurements of an item taken by an instrument under the same conditions
Resolution	The amount of detail found in one pixel of the image. For example, an image with one meter resolution

	means that each pixel in the image represents one square meter on the ground.
ROC	Rate of change
Roll-over	The return to zero of an accumulator when the accumulator value reaches the maximum value of the accumulator
Route	The path taken by a batch as it moves through the pipeline, including the lifting and delivery locations of the batch
RTM system	A collection of real time model applications linked to the host SCADA system
RTTM	Real Time Transient Model
RTU	Remote Terminal Unit is a field device for collecting real-time data, calculating Process variables, storing historical data, and performing uploading to and downloading from the host SCADA.
SCADA	Supervisory Control And Data Acquisition
Scan	Process of obtaining and updating real time data by an RTU
Scan rate	The time required to update all real time data and derived values. The required rate depends on the fluid and required response time of the pipeline system. For example, liquid pipelines require a faster scan rate than gas pipelines because of the fast transient times in liquid pipelines.
Secondary device	The device that responds to the signal generated by the primary device and converts it to an output signal that can be displayed. For example, the secondary devices for an orifice metering system include the electronic system, flow recorder or computer, and timer.
Segregated batch	Batch that must be lifted and delivered without having other batches added to, blended with, or taken away from it
Send-out	Send-out is the portion of a load that is not under contract, typically a residential load whose transit may be affected by weather and behavioural conditions.
Serial communication	A method of transferring information amongst computers using a communications cable that transmits data in a serial stream one bit at a time

Set point control	The supervisory control system that sets the target value that a process controller should achieve
Shape file	Editable spatial file format generated in ESRI's software
Shipper	A legal entity that contracts with a pipeline company to transport petroleum fluid over its pipeline system
Side stream injection	Injecting volume of a batch at an intermediate injection location into the main pipeline
Skid	A self-contained collection of instrumentation, equipment and controls for a specific purpose, e.g. a station compressed air skid
Slack flow	A slack flow condition occurs when the pipeline pressure drops below the fluid vapour pressure. In practice, it can arise where a large elevation drop occurs with low back pressure at the downstream of the high elevation point.
Spatial Data Management	Organizing spatial data to ensure data security, integrity, and effective change management
Specific gravity	The ratio of the density of a liquid to the density of water at a given temperature for liquid, or the ratio of the molecular weight of a gas to the molecular weight of air at a given temperature for gas
SPRT	Sequential Probability Ratio Test
Station valves	Valves at a station such as suction, discharge, blow-down, by-pass, or block valve
Strapping table	A table to convert liquid level in a storage tank to gross volume
Strip (side stream) delivery	The delivery of one or more batches at an intermediate delivery location out of the main pipeline
Supervisory control	Method in which information about a process is sent to a remote control location but the controlling action is taken by an operator
Supplier	A petroleum fluid producer or another pipeline company that supplies fluid into the pipeline system
Surge line	Points along the compressor wheel map, which define the transition from stable to reverse flow patterns within the compressor
Synchronous	Method of data transmission in which the events are controlled by a clock

Tariff	The published terms and conditions including transportation rate charged to shippers under which shippers use the transportation services offered by the pipeline company. It details the terms, conditions and rate information applicable to various types of transportation services.
Terminal	A delivery point, usually the final delivery point as opposed to an intermediate terminal
Ticket	A record of metered batch receipt/delivery volume according to the daily batch schedule. For single product operation, a ticket of metered volume is issued daily or at a specified interval.
Ticket allocation	The ticket allocated to a shipper based on actual delivered volume after the volume has been verified and gain/loss calculated
Ticket cutting	Act of issuing a ticket
TIFF world file	A text file that provides geo-referencing information for an image in TIF format
Tight line receipt	Fluid moved directly from a shipper or feeder into the pipeline without going through an intermediate tank
Tight line delivery	Fluid moving directly from a pipeline to a delivery facility outside the pipeline company's system, without going through an intermediate tank
Time stamp	The process of keeping track of the measurement and modification time of real-time data and events. In a DCS or SCADA system this is the time the data was received by the RTU or the host system.
Transfer batch	Batch that transfers from one pipeline to another through a transfer point
Transmix	Commingled product whose volume is from two interfacing products. It is also referred to as a batch interface or contamination batch.
Transport equation	An expression of the motion of diffusion. Diffusion results from molecular interaction between two homogeneous media such as at a batch boundary.
Transportation agreement	An agreement between a shipper and a pipeline company which defines the terms and conditions of the transportation services. A transportation agreement is required to move or store petroleum fluids.

Unbundled service	Service that separates pipeline transmission, sales and storage services to guarantee open access to pipeline capacity for all shippers
Underground storage	Sub-surface facility for storing gas or liquid, which has already been transferred from its original location
Unmanned station	A station which is totally controlled by a central control center without station personnel's intervention
Utility	A general support computer program
Vector display	A geographic object represented by a point, line, or polygon
Visualization	Techniques used to simulate 3D or other natural phenomena by combining different data layers
Volume correction factor for pressure	Factor used to convert gross volume to net volume by taking into account the pressure difference between the operating pressure and base pressure
Volume correction factor for temperature	Factor used to convert gross volume to net volume by taking into account the temperature difference between the operating temperature and base temperature
Web GIS	GIS applications deployed to operate and access over the internet or an intranet

INDEX

B

C

D

E

T

U

V